IEE CONTROL ENGINEERING SERIES 14

SERIES EDITORS: PROF. B. H. SWANICK
PROF. H. NICHOLSON

Optimal Relay and Saturating Control System Synthesis

Previous volumes in this series:

Volume 1 Multivariable control theory
J. M. Layton

Volume 2 Lift traffic analysis, design and control
G. C. Barney and S. M. dos Santos

Volume 3 Transducers in digital systems
G. A. Woolvet

Volume 4 Supervisory remote control systems
R. E. Young

Volume 5 Structure of interconnected systems
H. Nicholson

Volume 6 Power system control
M. J. H. Sterling

Volume 7 Feedback and multivariable systems
D. H. Owens

Volume 8 A history of control engineering, 1800–1930
S. Bennett

Volume 9 Modern approaches to control system design
N. Munro (Editor)

Volume 10 Control of time delay systems
J. E. Marshall

Volume 11 Biological systems, modelling and control
D. A. Linkens (Editor)

Volume 12 Modelling of dynamical systems—1
H. Nicholson (Editor)

Volume 13 Modelling of dynamical systems—2
H. Nicholson (Editor)

Optimal Relay and Saturating Control System Synthesis

E. P. Ryan
Lecturer, School of Mathematics
University of Bath
England

PETER PEREGRINUS LTD.
on behalf of the
Institution of Electrical Engineers

Published by: The Institution of Electrical Engineers, London
and New York
Peter Peregrinus Ltd., Stevenage, UK, and New York

British Library Cataloguing in Publication Data

Ryan, E. P.
Optimal relay and saturating control system synthesis.—(IEE control engineering series; 14)
1. Control theory
2. Mathematical optimisation
I. Title II. Series
629.8′312 QA402.3

ISBN 0-906048-56-7

Typeset by Santype International Limited, Salisbury
Printed in England by A. Wheaton & Co., Ltd., Exeter

Contents

Preface

With the recognition that a well developed linear control theory now exists, more research is being directed towards nonlinear aspects of general control systems. This volume is concerned with one specific class of nonlinear systems, namely, systems with control signal saturation. Within this class, relay control systems, with their attendant practical advantages in implementation, form an important and clearly identifiable category.

Because of their frequent occurrence in practice, saturating control systems have received special attention throughout the period of emergence of control theory as a recognised discipline; this is exemplified by the publication in 1953 of Flügge-Lotz's text on 'Discontinuous automatic control'. The emphasis of many early investigations was on improving system stability properties rather than optimising system performance. With the major advances in optimal control of the late 1950s and early 1960s, the theory and practice of optimal relay and saturating control progressed rapidly and was extensively studied in the important texts (among others) of Pontryagin *et al.* (1962), Athans and Falb (1966), Pavlov (1966), Lee and Markus (1967) and in the later work of Flügge-Lotz (1968).

Since the publication of the above texts over a decade ago, numerous significant contributions have been made to the general area of saturating control systems and, in particular, to the study of feedback control of such systems. However, with the exception of a collection of key papers (Fuller 1970), recent advances have not been brought together to form an up-to-date body of results. This is the main purpose of the present volume, in which optimal synthesis or feedback control of saturating systems is emphasised.

Deterministic control systems, only, are considered. The first four Chapters provide the motivation and mathematical framework within which the optimal control synthesis problem is solved for a wide range of systems in later chapters. In establishing this framework, results from the general theory of optimal control are interpreted in the more specific context of saturating control systems. In order to present, as far as possible, a self-contained account of the underlying properties and structure of such systems, some of the fundamental theorems are rigorously proved. However, for the reader whose interests lie more in the application of these results to the characterisation of feedback controls, it is remarked that a familiarity with the statements of the theorems of Chapters 2 to 4, and not necessarily with the associated proofs, is sufficient for the purposes of subsequent chapters. The contents of Chapters 5 to 9 constitute a catalogue of explicit feedback control solutions to a wide variety of optimal relay and saturating control problems; 'quasioptimal' or 'suboptimal' feedback techniques, appropriate to problems for which an exact feedback control solution is not feasible, are discussed in Chapter 10.

In summary, this volume purports to comprise a reasonably self-contained exposition of deterministic optimal relay and saturating control system synthesis, presented at a level of rigour suited to a broad spectrum of engineers and applied mathematicians.

It is a pleasure to acknowledge my indebtedness to Dr. A. T. Fuller of the University of Cambridge for his invaluable assistance and encouragement in the preparation of this volume. I should also like to thank my colleagues at the University of Bath, in particular, Dr. J. Marshall, C. Dorling, B. Ireland and Dr. S. Salehi, for their helpful suggestions and comments.

E. P. RYAN
Bath
December 1980

Chapter 1

Introduction

1.1 Preliminary remarks

A comprehensive linear theory of control now exists for both finite dimensional (lumped parameter) (Brockett 1970, Layton 1976, Owens 1978, Postlethwaite and MacFarlane 1979, Rosenbrock 1970, Wolovich 1974, Wonham 1974) and infinite dimensional (distributed parameter) dynamical systems (Curtain and Pritchard 1978). It is well known that control of such systems can be optimised via the minimisation of quadratic cost functionals (performance indices) and synthesised by linear feedback. Linear control theory suffers from the fundamental criticism that, in reality, dynamical systems are frequently subject to several complicating factors which may invalidate or at least severely limit its applicability; of these factors, inherent nonlinear effects form a clearly recognisable category. Such nonlinearities are typified by control signal saturation: the outputs of all physical devices are limited to some degree, when a limiting value is reached saturation is said to occur. For example, the output torque of a servomotor is subject to a maximum value; other examples of saturation nonlinearities occur in spacecraft propulsion systems, electronic amplifiers, regulating valves, etc. Attempts to incorporate these nonlinear features in a linear-feedback, quadratic-cost framework via 'heavy-penalty' weighting of the constrained variables often lead to ill conditioned linear feedback gains or a poor design in which the system may, for the most part, operate far from its full capability (Fuller 1967b, Frankena and Sivan 1979). Saturation constraints are more appropriately allied to a formulation of the control problem based on other cost functionals which, in general, give rise to optimal nonlinear feedback control. Moreover, in many situations a control actuator may operate only at discrete levels (with the advantage of reduction in complexity over a continuous-valued actuator) as is the case with relay or 'on-off' control systems, an obvious example being temperature control via thermostatic switching. Clearly such systems cannot be treated within the framework of the optimal linear feedback theory, but are well adapted to optimisation with respect to rapidity of response (time-optimal control), fuel expenditure (fuel-optimal control) and many other measures of system performance which frequently yield a nonlinear discontinuous feedback synthesis characterised in terms of a switching hypersurface in state space governing the discontinuities in control input. On this basis, the objectives of subsequent chapters can loosely be summarised as (i) interpretation of the general theory of optimal control in the more specific context of optimal saturating control systems, (ii) elucidation of the fundamental properties and underlying structure of these systems with main emphasis on optimal feedback synthesis, (iii) explicit characterisation of the optimal synthesis in a variety of cases, and (iv) investigation of 'quasioptimal' or 'suboptimal' feedback techniques appropriate to problems for which the optimal feedback solution is either unavailable or of an impractical level of complexity.

1.2 Brief historical review

Theoretical investigations into relay and saturating control systems may be said to have their origins in a paper by Hazen (1934), although phase plane studies of second-order relay control systems are reported in Léauté (1885, 1891). For references to early investigations in this area, see Fuller (1962, 1967a). The theory and application of saturating control advanced rapidly during the war years, largely to meet the demand for robust, lightweight, aircraft control systems. Much of this work is reported in MacColl (1945), Weiss (1946) and Kochenburger (1950), where the emphasis is on improving system stability properties rather than optimising performance. Early investigations into the latter aspect of optimisation of relay and saturating control systems include MacDonald (1950), Hopkin (1951), Uttley and Hammond (1952), Neiswander and MacNeal (1953), Flügge-Lotz (1953), Bogner and Kazda (1954), Coales and Noton (1956). These studies, based largely on engineering intuition, hinted at a general underlying theory. For example, it was heuristically assumed that, in order to generate a minimum-time system response, the inputs should take their extreme (saturated) values, only. This assumption was based on the intuitive notion that, if the input capabilities were not fully utilised, then the system could be 'driven more rapidly' by judicious use of the spare capacity; this notion anticipated to a considerable extent the bang-bang principle of LaSalle (1953, 1960a, b) and others (e.g. Bellman, Glicksberg and Gross 1956) and the maximum principle of Pontryagin (Pontryagin *et al.*, 1962). Furthermore, Rose (1953) suggested that, for a time-optimal response, the number of control reversals or switches in a relay generated input should be minimised; Bogner and Kazda (1954) obtained results which suggested that this minimum number of switches never exceeds $(n-1)$ when the controlled plant is linear, nonoscillatory and of order n. Again, these results anticipated the general theory.

At this time, the problem of optimal control was also generating considerable research activity among mathematicians. In 1953, Bushaw (see also Bushaw, 1958) gave a mathematical solution to a relatively simple optimal control problem; restricting attention to bang-bang controls (i.e. operating at saturation levels only), he established the existence of a time-optimal control and obtained a feedback synthesis by identifying the switching curve or locus in the phase plane which determines the time-optimal control input discontinuities. Rose (1953), LaSalle (1953, 1960a, b), Bellman, Glicksberg and Gross (1956) showed that, if the class of controls is broadened to allow input values not only at saturation but also at all intermediate levels, the time-optimal solution remains unchanged with the control still taking its extreme values only—a result often referred to as the bang-bang principle of time-optimal control. Also at this time, Krasovskii (1958, 1959) investigated optimal control problems with a range of restrictions on the inputs. The above and other related results were soon subsumed in a general theory of optimal control which developed initially along the two distinct but closely related paths of dynamic programming (Bellman, 1957) and the maximum principle (Boltyanskii, Gamkrelidze and Pontryagin, 1956; Boltyanskii, 1958; Rozonoer, 1959a, b, c; Boltyanskii *et al.*, 1960; Pontryagin *et al.*, 1962). In the context of relay and saturating control systems, Fuller (1960a, 1962, 1963b) has given an expository account of these theoretical developments and their historical background.

The advent of a general theory of optimal control accelerated research into relay and saturating control systems. Important contributions to the area made during the decade 1960–70 are admirably reported in the texts of Pontryagin *et al.* (1962), Athans and Falb (1966), Pavlov (1966), Lee and Markus (1967), Flügge-Lotz (1968), Hermes and LaSalle (1969), Boltyanskii (1971) and in the collections of key papers by Oldenburger (1966b) and Fuller (1970b). Since the publication of these texts over a decade ago, many further advances have been made in the theory and application of saturating control. The purpose of this volume is to present the main features of the underlying theory and to bring together the earlier and more recent advances in an up-to-date body of results. It may be said that the more successful applications of the theory of relay and saturating control have occurred in the field of aerospace systems. Two examples drawn from this field are presented below to illustrate and motivate the class of optimal control problems to be studied in later chapters.

1.3 Examples

Simplified versions of the altitude and attitude control problems will be investigated for a lunar soft landing craft in its descent phase. The primary objective of the altitude control system is to govern the descent engine in such a way that the craft achieves a soft landing, i.e. zero altitude and velocity simultaneously achieved in finite time. During its descent, various external disturbances and engine thrust vector misalignments may produce torques which tend to rotate

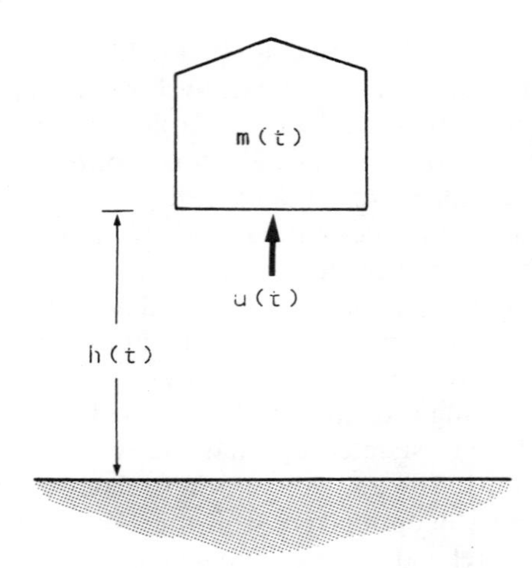

Fig. 1.1

or tumble the craft. The primary objective of the attitude control system is to maintain the desired attitude or spatial orientation by eliminating or minimising these extraneous effects. The altitude control system will be considered initially.

1.3.1 Altitude control system

There may be many possible strategies which achieve the zero-altitude, zero-velocity objective; of these strategies, that which minimises the fuel expended during the descent is clearly of major interest. A characterisation of this minimum-fuel strategy has been obtained by Miele (1962) and subsequently discussed by Meditch (1964b). The configuration is as shown in Fig. 1.1. Let $h(t)$ denote the vertical height above the surface at time t and let $m(t)$ denote the instantaneous mass of the craft so that $\dot{m}(t) \leq 0$ (the mass flow rate) corresponds to the rate of fuel consumption. The thrust $u \geq 0$, developed by the propulsion engine, is assumed to be proportional to the mass flow rate, thereby defining the control for the problem as

$$u = -k\dot{m}; \qquad k > 0 \tag{1.1}$$

which is subject to a maximum value α, i.e. at each time t the control $u(t)$ can take values only in the control restraint set $\Omega = [0, \alpha]$. The fuel F consumed on the descent trajectory, starting at time $t = 0$ and terminating at the (free) terminal time t_f, is given by the mass difference $m(0) - m(t_f)$. Hence the cost functional

$$F(u) = \frac{1}{k}\int_0^{t_f} u(t)\,dt = -\int_0^{t_f} \dot{m}(t)\,dt = m(0) - m(t_f) \tag{1.2}$$

provides the measure of performance for the problem. If g (assumed constant) is the lunar gravitational acceleration, then the vertical motion of the craft is governed by the equation

$$\ddot{h} = -g - \frac{k\dot{m}}{m}$$

$$= -g + \frac{u}{m}; \quad u(t) \in \Omega = [0, \alpha] \tag{1.3}$$

The control problem may now be stated as follows: determine a control function $u^*(t)$, $0 \leq t \leq t_f$, with $u^*(t) \in \Omega$, such that system (1.3) is transferred from an initial state (at $t = 0$) of altitude $h(0) = h^0 > 0$, vertical velocity $\dot{h}(0) = v^0$ and mass $m(0) = m^0 > 0$, to the terminal state $h(t_f) = 0 = \dot{h}(t_f)$ at the free terminal time t_f while minimising the cost functional (1.2).

First, it will be established that the above minimum-fuel problem is equivalent to the minimum-time problem, i.e. a control which achieves the zero-altitude, zero-velocity state in minimum time t_f also does so with minimum fuel expenditure. Now, (1.3) may be rewritten as

$$\ddot{h} = -g - k\frac{d}{dt}(\ln m)$$

which, on integration, yields

$$\dot{h}(t) = -gt - k \ln m(t) + \dot{h}(0) + k \ln m(0)$$

$$= -gt - k \ln \left(\frac{m(t)}{m^0}\right) + v^0$$

Moreover, at the trajectory endpoint,

$$\dot{h}(t_f) = -gt_f - k \ln \left(\frac{m(t_f)}{m^0}\right) + v^0 = 0$$

so that

$$m(t_f) = m^0 \exp\left[k^{-1}(v^0 - gt_f)\right]$$

Hence, the cost of the trajectory may be calculated as

$$F(u) = m^0 - m(t_f) = m^0\{1 - \exp\left[k^{-1}(v^0 - gt_f)\right]\}$$

which is clearly minimised if and only if the terminal time t_f is minimised, i.e. the fuel-optimal and time-optimal control problems are equivalent for this system. The solution to these equivalent problems can be determined through the maximum principle (to be discussed in detail in Chapter 2) which, for the case at hand, is briefly summarised below.

Defining the state vector† $x = (x_1, x_2, x_3)'$, where $x_1 = h$ (altitude), $x_2 = \dot{h}$ (vertical velocity) and $x_3 = m$ (mass), then eqns. 1.1–1.3 may be rewritten as

$$\dot{x}_1 = x_2; \quad \dot{x}_2 = -g + x_3^{-1}u; \quad \dot{x}_3 = -k^{-1}u; \quad u(t) \in \Omega$$

or, more concisely,

$$\dot{x} = f(x, u) = \begin{bmatrix} f_1(x, u) \\ f_2(x, u) \\ f_3(x, u) \end{bmatrix} = \begin{bmatrix} x_2 \\ -g + x_3^{-1}u \\ -k^{-1}u \end{bmatrix} \tag{1.4a}$$

with

$$x(0) = \begin{bmatrix} h^0 \\ v^0 \\ m^0 \end{bmatrix}; \qquad x(t_f) = \begin{bmatrix} 0 \\ 0 \\ m(t_f) \end{bmatrix} \tag{1.4b}$$

The cost functional J for the equivalent time-optimal problem may be expressed in the form

$$J(u) = \int_0^{t_f} dt \qquad (= t_f) \tag{1.5}$$

Defining the Hamiltonian function H (so called because of certain analogies with classical mechanics) as¶

$$H(x, u, \eta) = \langle \eta, f(x, u) \rangle - 1 \tag{1.6a}$$

with the associated Hamiltonian system

$$\dot{x} = \frac{\partial H}{\partial \eta} = f(x, u) \qquad \text{(state equation)} \tag{1.6b}$$

$$\dot{\eta} = -\frac{\partial H}{\partial x} = -[\nabla_x f]^T \eta \qquad \text{(adjoint equation)§} \tag{1.6c}$$

where

$$[\nabla_x f] = \left[\frac{\partial f_i}{\partial x_j}\right] = \begin{bmatrix} \frac{\partial f_1}{\partial x_1} & \frac{\partial f_1}{\partial x_2} & \frac{\partial f_1}{\partial x_3} \\ \frac{\partial f_2}{\partial x_1} & \frac{\partial f_2}{\partial x_2} & \frac{\partial f_2}{\partial x_3} \\ \frac{\partial f_3}{\partial x_1} & \frac{\partial f_3}{\partial x_2} & \frac{\partial f_3}{\partial x_3} \end{bmatrix} \qquad \text{(Jacobian matrix)}$$

† Prime (′) denotes vector transposition, so that $x = (x_1, x_2, x_3)'$ is a column vector.
¶ $\langle \cdot, \cdot \rangle$ denotes the usual inner (scalar) product in $\mathbb{R}^n$, i.e. $\langle \eta, f \rangle = \eta' f = \sum_{i=1}^{n} \eta_i f_i$.
§ T denotes matrix transposition.

then the maximum principle states that, if $u^*(t)$, $0 \leq t \leq t_f$, (generating trajectory $x(t)$, $0 \leq t \leq t_f$) minimises (1.5) subject to (1.4), then there exists a solution $\eta(t) = (\eta_1(t), \eta_2(t), \eta_3(t))'$, $0 \leq t \leq t_f$, of the adjoint equation (1.6c) such that

$$H(x(t), u^*(t), \eta(t)) = \max_{u \in \Omega} H(x(t), u, \eta(t)), \qquad 0 \leq t \leq t_f$$

or equivalently

$$\langle \eta(t), f(x(t), u^*(t)) \rangle = \max_{u \in \Omega} \langle \eta(t), f(x(t), u) \rangle, \qquad 0 \leq t \leq t_f \tag{1.7}$$

Derivation and discussion of this result are postponed until the next Chapter. Suffice it to say here that the maximum principle provides a necessary (and sometimes sufficient) condition for optimality. For the case at hand, (1.7) can be interpreted as saying that the optimal control has the property that it maximises the state flow in the direction of the vector $\eta(t)$ at each point of the trajectory. With the appropriate choice of initial condition $\eta(0) = \eta^0$, the adjoint solution $\eta(\cdot)$ defines a preferred direction $\eta(t)$ at each point of a state trajectory, the optimal choice of control being that which maximises the state flow in the preferred direction.

The control $u^*(t)$ which maximises the inner product

$$\langle \eta(t), f(x(t), u) \rangle = \eta_1(t)x_2(t) - \eta_2(t)g + \eta_2(t)x_3(t)^{-1}u - k^{-1}\eta_3(t)u$$

over the control restraint set $\Omega = [0, \alpha]$ is given by

$$u^*(t) = \begin{cases} \alpha; & \text{if} \quad \xi(t) = \eta_2(t)x_3(t)^{-1} - k^{-1}\eta_3(t) > 0 \\ 0; & \text{if} \quad \xi(t) = \eta_2(t)x_3(t)^{-1} - k^{-1}\eta_3(t) < 0 \end{cases} \tag{1.8}$$

for some solution $\eta(\cdot)$ of the adjoint equation (1.6c) which, in view of (1.4), becomes

$$\begin{bmatrix} \dot{\eta}_1 \\ \dot{\eta}_2 \\ \dot{\eta}_3 \end{bmatrix} = \begin{bmatrix} 0 & 0 & 0 \\ -1 & 0 & 0 \\ 0 & x_3^{-2}u & 0 \end{bmatrix} \begin{bmatrix} \eta_1 \\ \eta_2 \\ \eta_3 \end{bmatrix} \tag{1.9}$$

Note that $u^*(t)$ is indeterminate whenever $\xi(t) = \eta_2(t)x_3(t)^{-1} - k^{-1}\eta_3(t) = 0$. However, if this condition occurs only at a countable number of isolated instants (a set of measure zero) corresponding to control switches, then the actual values assigned to the control at these instants play no essential role. More serious is the situation in which $\xi(t)$ vanishes identically on some finite interval (a set of positive measure), thereby giving rise to the singular condition (discussed further in Chapter 9). For the example under consideration, it can be shown, using special arguments not pursued here (see Meditch 1964b), that the singular condition cannot arise. Consequently, an optimal control u^* is piecewise constant taking the extreme values 0, α only and, for some adjoint solution $\eta(\cdot)$, is determined by (1.8) for almost all $t \in [0, t_f]$ (i.e. for all t other than the isolated zeros of $\xi(t)$, $0 \leq t \leq t_f$). It will now be shown that an optimal control function $u^*(\cdot)$ can contain at most one discontinuity on the interval $(0, t_f)$ or, equivalently, the function $\xi = \eta_2 x_3^{-1} - k^{-1}\eta_3$ can have at most one zero on $(0, t_f)$. Consider the derivative $\dot{\xi}$ of the function ξ, i.e.

$$\dot{\xi} = \dot{\eta}_2 x_3^{-1} - \eta_2 x_3^{-2}\dot{x}_3 - k^{-1}\dot{\eta}_3$$

Now, from (1.9),

$$\dot{\eta}_1 = 0 \Rightarrow \eta_1(t) = \eta_1^0 \text{ (constant)}, \qquad 0 \le t \le t_f$$

$$\dot{\eta}_2 = -\eta_1$$

$$\dot{\eta}_3 = \eta_2 x_3^{-2} u$$

and hence

$$\dot{\xi} = -\eta_1^0 x_3^{-1}$$

As x_3 (the mass of the craft) is a positive quantity, the function ξ is either constant or strictly increasing (decreasing) according as η_1^0 is zero or negative (positive) and consequently has at most one zero on $(0, t_f)$. This identifies the four candidate optimal control strategies, in each of which the control either remains constant ($u \equiv 0$ or $u \equiv \alpha$) on the interval $[0, t_f]$ or the control contains one switch ($0 \to \alpha$ or $\alpha \to 0$ changeover) at some intermediate switching time $t_s \in (0, t_f)$ viz.:

$$\begin{aligned}
&\text{(i)} \quad u(t) = 0; \qquad 0 \le t \le t_f \\
&\text{(ii)} \quad u(t) = \alpha; \qquad 0 \le t \le t_f \\
&\text{(iii)} \quad u(t) = \begin{cases} 0; & 0 \le t < t_s \\ \alpha; & t_s \le t \le t_f \end{cases} \\
&\text{(iv)} \quad u(t) = \begin{cases} \alpha; & 0 \le t < t_s \\ 0; & t_s \le t \le t_f \end{cases}
\end{aligned} \tag{1.10}$$

Of these four candidates (i) and (iv) can easily be eliminated involving, as they do, a final 'free-fall' trajectory ($u \equiv 0$) to the surface which cannot satisfy the required zero-velocity end condition $\dot{x}(t_f) = 0$. Optimal descent trajectories can now be constructed which correspond to the remaining control candidates (ii) and (iii). If $u(t) = \alpha$ on the interval $[0, t_f]$, then the solution of (1.4) is

$$\begin{aligned}
x_3(t) &= x_3(0) - k^{-1}\alpha t \\
&= m^0 - k^{-1}\alpha t \\
x_2(t) &= x_2(0) + \int_0^t [-g + \alpha x_3(s)^{-1}]\, ds \\
&= v^0 - gt - k \ln [1 - (km^0)^{-1}\alpha t] \\
x_1(t) &= x_1(0) + \int_0^t x_2(s)\, ds \\
&= h^0 + v^0 t - \tfrac{1}{2}gt^2 \\
&\quad + k(km^0\alpha^{-1} - t) \ln [1 - (km^0)^{-1}\alpha t] + kt
\end{aligned} \tag{1.11}$$

This solution must satisfy the prescribed end conditions:

$$\begin{aligned} x_2(t_f) &= 0 = v^0 - gt_f - k \ln [1 - (km^0)^{-1}\alpha t_f] \\ x_1(t_f) &= 0 = h^0 + v^0 t_f - \tfrac{1}{2}gt_f^2 \\ &\qquad + k(km^0\alpha^{-1} - t_f) \ln [1 - (km^0)^{-1}\alpha t_f] + kt_f \end{aligned} \tag{1.12}$$

giving relations which determine the initial altitude-velocity pair (h^0, v^0) from which a soft landing can be achieved in minimum time t_f by means of a constant control $u(t) = \alpha, 0 \leq t \leq t_f$, viz.

$$\begin{aligned} h^0 &= -(v^0 + k)t_f + \tfrac{1}{2}gt_f^2 - k \\ &\qquad \times (km^0\alpha^{-1} - t_f) \ln [1 - (km^0)^{-1}\alpha t_f] \\ v^0 &= gt_f + k \ln [1 - (km^0)^{-1}\alpha t_f] \end{aligned} \tag{1.13}$$

Here it is implicitly assumed that there is sufficient fuel available to maintain the maximum thrust α over the interval $[0, t_f]$ which in turn ensures $1 - (km^0)^{-1}\alpha t_f > 0$ so that the logarithmic functions of (1.13) are well defined. As t_f increases from 0, the pair (h^0, v^0) traces a curve Γ in the (x_1, x_2)-plane, as

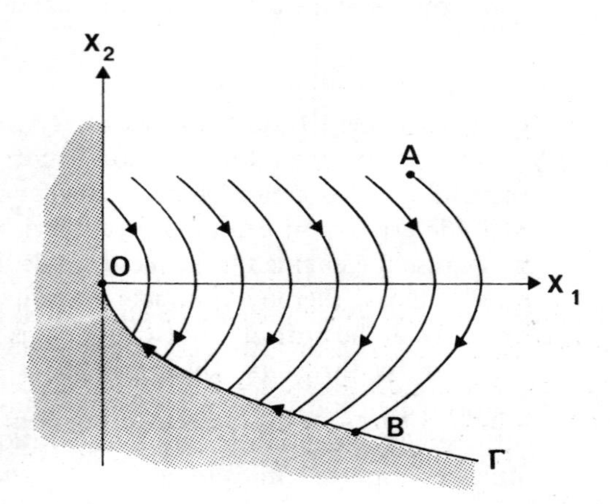

Fig. 1.2 *Switching curve Γ and time/ fuel- optimal trajectories*

depicted in Fig. 1.2, any point of which can be transferred time-optimally to the origin by means of a constant control $u^* = \alpha$.

For a starting pair (h^0, v^0) lying to the right of the curve Γ, it can now be seen that strategy (iii) above is optimal. Referring to Fig. 1.2, for a starting pair $(x_1(0), x_2(0)) = (h^0, v^0)$ at point A the free fall path AB (with $u(t) = 0, 0 \leq t < t_s$) is initially followed until point B on the curve Γ is reached at time $t = t_s$ at which maximum engine thrust α is initiated and the path BO in Γ is followed directly to the zero-altitude, zero-velocity state. For obvious reasons, the curve Γ is usually referred to as the optimal switching curve. The free-fall path AB is a parabolic arc as is easily shown by integrating (1.4) with zero control $u \equiv 0$, viz.

$$x_1(t) = h^0 + v^0 t - \tfrac{1}{2}gt^2$$

$$x_2(t) = v^0 - gt$$

which, on eliminating the parameter t, gives the parabolic relation between x_1 and x_2 as

$$x_1 + \frac{1}{2g} x_2^2 = h^0 + \frac{1}{2g} (v^0)^2$$

The complete family of such time-optimal/fuel-optimal trajectories can be constructed in a similar manner as shown in Fig. 1.2. Note that, for starting points in the shaded region to the left of the curve Γ, time-optimal paths do not exist; in such cases the available engine thrust is insufficient to overcome the gravitational acceleration to ensure a soft touchdown.

Although the above derivation is lacking somewhat in detail and rigour, it does serve to illustrate many of the problems and characteristic features of optimal saturating control which will be formalized in later chapters, as does the following study of attitude control of the craft during its final descent.

1.3.2 Attitude control system

In the previous study of the descent trajectory control system, it was tacitly assumed that the craft's attitude or spatial orientation is correctly maintained throughout the descent phase. In practice, various disturbances act on the system which induce attitude errors; for example, structural bending and propellant 'slosh' effects may cause the engine thrust vector to deviate from passing through the craft centre of mass, thereby producing torques which tend to tumble the craft. The objective of the attitude control system is to minimise such effects. This is accomplished via thrusters which provide attitude correcting torques about the three principal body axes. Clearly, minimisation of fuel expenditure of these thrusters is an essential design consideration. In the case of the Apollo lunar module, as reported in Windall (1970), such a design was devised which uses the descent propulsion engine (when firing) to enable attitude control about the two 'horizontal' body axes without the assistance of the thrusters (attitude control about the third 'vertical' body axis can only be achieved through use of the thrusters). In this design the descent engine is mounted in a two-axis gimbal system with actuators which govern the angle β of the thrust vector relative to the centre of mass. For each of the two gimbal axes the drive system is such that the angle β can be commanded to change at a fixed rate $\pm R$ ($\pm 0{\cdot}2°$/s) so that the gimbal drive signal u (the control for the problem) can take the values $\pm 1, 0$ only, i.e.

$$\dot{\beta} = Ru; \qquad u \in \{-1, 0, +1\} \tag{1.14}$$

Considering a simplified single-axis problem only (i.e. attitude control about

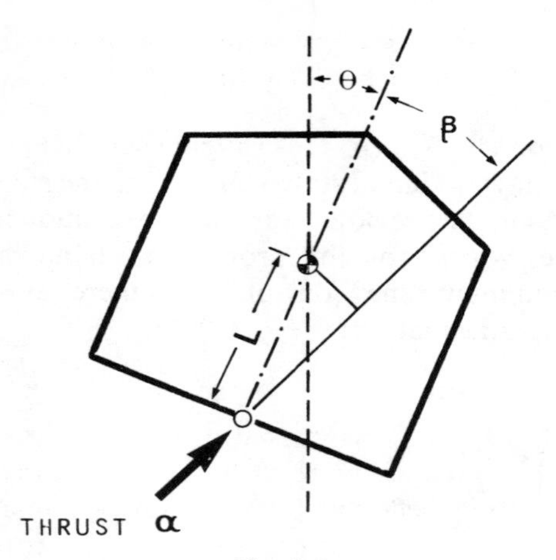

Fig. 1.3

one of the two 'horizontal' body axes with gyroscopic cross-axis coupling assumed to be negligible), let I be the moment of inertia for that axis and suppose θ is the attitude error, then, referring to Fig. 1.3,

$$\begin{aligned} I\ddot{\theta} &= \text{torque} \\ &= \alpha L \sin \beta \end{aligned} \tag{1.15}$$

where α is the thrust magnitude and L is the distance from engine hinge point to the centre of mass. Assuming that the angle β is small so that $\sin \beta \simeq \beta$, differentiating (1.15) yields a third-order linear differential equation relating the gimbal drive control signal u to the attitude error θ

$$\begin{aligned} \dddot{\theta} &= \frac{\alpha L}{I} \dot{\beta} \\ &= (\alpha L R I^{-1})u; \qquad u \in \{-1, 0, +1\} \end{aligned} \tag{1.16}$$

Note that the control can take the values ± 1, 0 only. However, in order to apply the theory, it is assumed that the control can also take arbitrary values between the limits ± 1, i.e. the saturation constraint $|u| \leq 1$ is assumed, although the permitted values ± 1, 0 only will play a role in the final solution. The control objective is to reduce the attitude error θ together with its derivatives $\dot{\theta}$ and $\ddot{\theta}$ to zero as rapidly as possible.

Introducing the state vector $x = (x_1, x_2, x_3)'$ where $x_1 = \theta$, $x_2 = \dot{\theta}$, $x_3 = \ddot{\theta}$ and writing $a = I/\alpha LR$, the equations of motion may be written as

$$\begin{bmatrix} \dot{x}_1 \\ \dot{x}_2 \\ \dot{x}_3 \end{bmatrix} = \begin{bmatrix} 0 & 1 & 0 \\ 0 & 0 & 1 \\ 0 & 0 & 0 \end{bmatrix} \begin{bmatrix} x_1 \\ x_2 \\ x_3 \end{bmatrix} + \begin{bmatrix} 0 \\ 0 \\ \dfrac{1}{a} \end{bmatrix} u; \qquad |u| \leq 1 \tag{1.17a}$$

or

$$\dot{x} = Ax + bu \tag{1.17b}$$

and a control function $u(t)$, $0 \le t \le t_f$ is sought such that system (1.17) is transferred from some (non-zero) initial state $x(0) = x^0$ to the state origin $x(t_f) = 0$ in minimum time t_f. As in the previous example, the solution is determined by the maximum principle, which, for the problem at hand, states that if $u^*(t)$, $0 \le t \le t_f$, is a minimum-time control, then there exists a solution $\eta(t)$, $0 \le t \le t_f$, of the adjoint equation:

$$\dot{\eta} = -A^T\eta \Leftrightarrow \begin{aligned} \dot{\eta}_1 &= 0 \\ \dot{\eta}_2 &= -\eta_1 \\ \dot{\eta}_3 &= -\eta_2 \end{aligned} \tag{1.18}$$

such that

$$\langle \eta(t), bu^*(t) \rangle = \max_{|u| \le 1} \langle \eta(t), bu \rangle \tag{1.19}$$

This condition simplifies to

$$\eta_3(t)u^*(t) = \max_{|u| \le 1} \eta_3(t)u$$

implying that

$$u^*(t) = \begin{cases} +1; & \eta_3(t) > 0 \\ -1; & \eta_3(t) < 0 \end{cases} \tag{1.20}$$

Again, note that the control is indeterminate if $\eta_3(t) = 0$. However, it is easily shown that this condition can occur at two isolated instants at most (for which the actual control value assigned plays no essential role). From (1.18),

$$\eta_1(t) = \eta_1^0$$
$$\eta_2(t) = \eta_2^0 - \eta_1^0 t$$
$$\eta_3(t) = \eta_3^0 - \eta_2^0 t + \tfrac{1}{2}\eta_1^0 t^2$$

where η_1^0, η_2^0, η_3^0 are constants. Note that η_3 is a quadratic function of t and consequently can have at most two zeros on $[0, t_f]$ corresponding to at most two control switches. Hence, the optimal control is piecewise constant, taking its extreme values ± 1 only, with at most three intervals of control constancy (i.e. two discontinuities) on $[0, t_f]$. On reaching the desired zero state, the control is set to zero. The family of minimum-time paths to the origin in state space consists of all those trajectories terminating at the origin and generated by a control of the above form. Consider now the set of optimal trajectories which go to the origin and which are generated by piecewise constant controls with at most one switch (i.e. one switch less than the possible maximum). This set

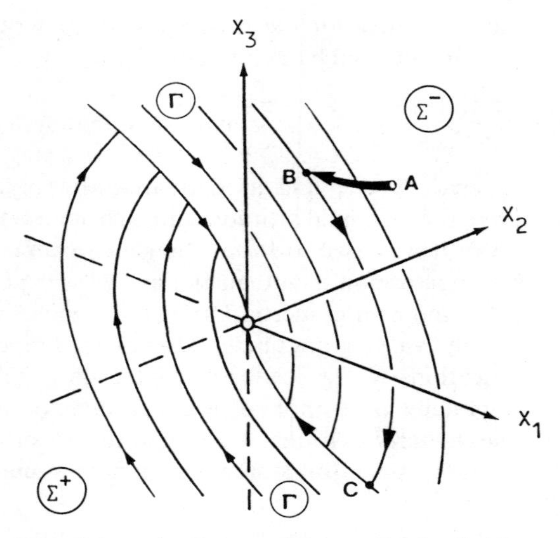

Fig. 1.4 *Schematic diagram of time-optimal switching surface* Γ

constitutes a two-dimensional manifold or surface Γ in the state space which plays a role analogous to the switching curve Γ of the previous example. In particular, Γ partitions the state space into two mutually exclusive regions, Σ^+ and Σ^-, in which the optimal control takes the values +1 and −1, respectively. The general structure is depicted schematically in Fig. 1.4. For an initial state $x(0) = x^A$ at point A in Σ^-, for example, the n-path AB, generated by the constant control $u^* = -1$, is followed until the surface Γ is reached at point B; a control switch then occurs and the p-path BC in Γ, generated by the constant control $u^* = +1$, is followed until the n-path leading to the origin is met at point C where the second control occurs and the state subsequently follows the n-path CO directly to the origin. It will be shown in Chapter 6 that, for this particular example, the time-optimal switching surface Γ and regions Σ^+, Σ^- may be expressed as

$$\Gamma = \{x\colon \xi(x) = 0\}; \quad \Sigma^+ = \{x\colon \xi(x) < 0\}; \quad \Sigma^- = \{x\colon \xi(x) > 0\} \tag{1.21a}$$

where

$$\xi(x) = x_1 + ax_2 x_3 \operatorname{sgn}(\Xi^s) + \frac{a^2 x_3^3}{3} + a^{-1}[ax_2 \operatorname{sgn}(\Xi^s) + \tfrac{1}{2}(ax_3)^2]^{3/2} \operatorname{sgn}(\Xi^s) \tag{1.21b}$$

and

$$\Xi^s = \Xi^s(x_2, x_3) = \begin{cases} x_2 + \frac{1}{2}ax_3|x_3|; & x_2 \neq -\frac{1}{2}ax_3|x_3| \\ x_3 \qquad ; & x_2 = -\frac{1}{2}ax_3|x_3| \end{cases} \tag{1.21c}$$

Hence, the time-optimal control for the problem may be written in feedback form as (this result was first derived by Feldbaum (1955))

$$u^*(x) = \begin{cases} -\operatorname{sgn}(\xi(x)) & ; \quad x \notin \Gamma \\ -\operatorname{sgn}(\Xi^s(x_2, x_3)); & \quad x \in \Gamma \end{cases} \tag{1.22}$$

Thus, if the attitude error $\theta = x_1$ and its derivatives $\dot{\theta} = x_2$, $\ddot{\theta} = x_3$ can be continuously measured, time-optimal attitude control can be achieved via (1.22).

In the above analysis it was assumed that the gain parameter $a = I/\alpha LR$ is constant. Clearly, this assumption is unrealistic; as fuel is expended the spacecraft mass, and hence its moment of inertia I and also L will change so that a is, in practice, a (slowly) time-varying parameter. However, as reported in Windall (1970), certain modifications to the feedback control law can be introduced which render the nominally time-optimal feedback control system relatively insensitive to such parameter variations. A study of the sensitivity of the nominally time-optimal triple integrator system to gain parameter variations is contained in Section 6.6.

1.3.3 Discussion

The previous two examples typify the class of control problems which form the main theme of the text; moreover, the examples illustrate many of the characteristic features of the optimal solutions. In each example, the control was permitted to take values only in some set Ω, representing the physical constraints on the system. Again, in each case, the optimal control took its values only on the boundary of the set Ω. Finally, in each example, the optimal feedback synthesis was obtained by characterising a switching curve or surface in state space which governs the control function discontinuities at which the control jumps from one boundary point of Ω to another (e.g. relay changeover). These features recur in more general settings throughout subsequent chapters, the contents of which are described below.

1.4 Outline of chapter contents

Having discussed and illustrated by example, in the present chapter, the class of optimal control problems to be studied, Chapter 2 provides the mathematical framework for their solution. A mathematical formulation of the general constrained-control problem is presented and conditions for optimality are investigated. The Bellman–Hamilton–Jacobi equation for the value function is derived and applied in the solution of the time-optimal control problem for norm-invariant systems. A restricted form of the Pontryagin maximum principle is then derived from the Bellman–Hamilton–Jacobi equation. This restricted form lends plausibility to a more general statement of the maximum principle which is rigorously proved only in relation to a more specific optimal control problem, namely, the time-optimal control problem for linear, autonomous systems.

In Chapter 3, the underlying theory of the previous chapter is applied to establish the fundamental structure and properties of optimal saturating control systems. Switching hypersurfaces and dynamic behaviour of systems under discontinuous feedback are discussed; conditions for regular switching and sliding (chattering) are obtained. Invariance and equivalence properties of certain systems are established; in subsequent chapters, these properties prove useful in both the analysis and synthesis of optimal solutions and also provide a basis for some suboptimal control techniques in Chapter 10.

The next four chapters are devoted to time-optimal control. Chapter 4 contains a detailed analysis of the linear, autonomous, time-optimal control problem. Theorems on the number of switchings for time-optimal bang-bang control are presented. Properties of isochronal surfaces and the domain of null controllability are derived. An idealised predictive control strategy is discussed, its time-optimality is proved and its application to optimal control synthesis is described. In Chapters 5, 6 and 7, the time-optimal synthesis problem is studied and solved for a variety of systems of up to fourth order. Sensitivity aspects of these nominally time-optimal feedback systems are also considered. These three chapters are intended to provide a reasonably complete catalogue of currently available closed-form feedback solutions to linear time-optimal control problems.

Feedback controls which minimise other cost functionals are derived in Chapters 8 and 9. The fuel-optimal control problem and some of its variants are studied in Chapter 8. The quadratic-cost problem (with its characteristic feature of singular solutions) is considered initially in Chapter 9; certain nonquadratic cost problems are subsequently investigated, e.g. integral costs with integrands inhomogeneous in system states and minimax control systems.

Chapter 10 considers some of the difficulties of precise optimal control implementation and describes some of the numerous techniques of suboptimal or quasioptimal control which must be resorted to in situations where the exact optimal solution is either unavailable or of impracticable complexity. Because of the great diversity of approaches to quasioptimal control, no attempt is made to cover the field. Instead, certain techniques which complement the optimal control studies of earlier chapters are described. Specifically, open-loop computational techniques for time-optimal control are briefly discussed. Predictive control strategies, introduced in a different context in Chapter 4, receive more detailed study. Adaptive control methods, which exploit the insensitivity of sliding trajectories to system parameter variation, are described. Finally, the concluding chapter outlines related areas which fall outside the scope of the book.

Chapter 2

Formulation of the control problem and conditions for optimality

2.1 Formulation of the optimal control problem

Let the state of a dynamical system at time t be represented by a real n-dimensional vector $x(t) = (x_1(t), x_2(t), \ldots, x_n(t))'$ in the state space $\mathbb{R}^n$ (endowed with the usual Euclidean norm $\|x\| = [x_1^2 + x_2^2 + \cdots + x_n^2]^{1/2}$ and inner product $\langle x, y \rangle = x_1 y_1 + x_2 y_2 + \cdots + x_n y_n$).

The state evolution is governed by a differential equation of motion, defined on some time interval, of the form

$$\dot{x} = f(t, x, u(t)) \tag{2.1a}$$

with initial condition

$$x(t_0) = x^0 \tag{2.1b}$$

Here, $u(t) = (u_1(t), u_2(t), \ldots, u_m(t))' \in \mathbb{R}^m$ is an m-dimensional vector of control inputs at time t. The class of functions U from which the input vector function $u(\cdot)$ is chosen depends on the particular problem under consideration. For the problems treated in later chapters, the class of piecewise continuous functions usually proves to be sufficiently broad to contain the appropriate controls, although this class is occasionally extended to that of bounded measurable functions, e.g. to establish the existence of time-optimal controls in Section 2.5. The constraint fundamental to the subsequent development is that, for all t, the control vector $u(t)$ is constrained to take values only in a closed and bounded (compact) set $\Omega \subset \mathbb{R}^m$. For the most part, Ω will be taken as the m-dimensional unit cube $|u_i| \leq 1$, $i = 1, 2, \ldots, m$, representing (normalised) saturation constraints on the system inputs. Single-input systems will receive detailed treatment, in which case $u(t) \in \Omega = [-1, 1]$, with single-input relay systems forming an important subclass. In summary, the set U of admissible control functions will usually be taken as the set of piecewise continuous functions which take values in the *control restraint set* Ω, i.e. an admissible control is a function $u(\cdot)$, defined on some interval $[t_0, t_f]$ (which may differ for different elements of U), which takes values $u(t)$ in Ω and which is continuous for all values of t in the interval, excluding only a finite number of isolated instants where the function can have discontinuities of the first kind (i.e. the left and right limits exist at a point of discontinuity). From a practical viewpoint, the value of a piecewise continuous control at an instant of discontinuity t_s plays no essential role; however, for definiteness, the value of control at t_s will be defined as the right limit, i.e. $u(t_s) = \lim_{t \downarrow t_s} u(t)$. Occasionally, the analysis will necessitate broadening the set U to the more general class of bounded measurable functions with values in Ω; this class contains not only the earlier set of piecewise continuous functions, but also all functions obtained by composition and countable limit operations on these functions.

In the region of interest, the function $f = (f_1, f_2, \ldots, f_n)'$ is assumed to be continuous with respect to all its arguments, continuously differentiable with respect to x and sufficiently well behaved to ensure that, for an admissible piecewise continuous control $u \in U$ defined on $[t_0, t_f]$, there exists a *unique* solution, or *state trajectory*, of (2.1), written as

$$x(t) = \phi(t, t_0, x^0, u) \tag{2.2}$$

which is differentiable at all points of (t_0, t_f) other than the finite number of isolated points of control discontinuity (a set of measure zero), i.e. (2.2) satisfies (2.1) for almost all (a.a.) t. In the case of bounded measurable controls, (2.2) is interpreted as the *unique absolutely continuous* solution satisfying (2.1) almost everywhere (a.e.) on $[t_0, t_f]$. For example, the condition

$$\langle x, f(t, x, u)\rangle \leq k[1 + \|x\|^2] \tag{2.3}$$

for some constant $k > 0$ and for all $t \in [t_0, t_f]$, $x \in \mathbb{R}^n$, $u \in \Omega$ prevents finite escape time ('blowing up') of solutions† (see Filippov 1962) and, together with the above continuity conditions on f, ensures the existence of (2.2).

Much of the treatment in later chapters will be concerned with linear time-invariant (autonomous) systems for which

$$\begin{aligned} \dot{x} &= f(x, u(t)) = Ax + Bu(t) \\ x(0) &= x^0 \end{aligned} \tag{2.4a}$$

where, in view of time-invariance, the starting time $t_0 = 0$ may be assumed without loss of generality, and where A and B are constant matrices of appropriate

† For any solution $x(\cdot)$ of (2.1), (2.3) yields the result

$$\frac{d}{dt}\|x\|^2 = 2\langle x, \dot{x}\rangle \leq 2k[1 + \|x\|^2]$$

or

$$\|x\|^2 \leq [1 + \|x^0\|^2] \exp(2k(t - t_0)) - 1$$

dimension. In this case, the unique solution (2.2) becomes

$$x(t) = \phi(t, x^0, u) = \exp(At)x^0 + \int_0^t \exp(A(t-s))Bu(s)\,ds \tag{2.4b}$$

where, in view of time-invariance, the function ϕ no longer exhibits t_0-dependence.

Note that the above system description is exclusively *deterministic*, i.e. stochastic effects are not investigated.

Free end time problems, only, will be considered, i.e. the terminal time t_f is *not prescribed*. It is assumed that the control objective is to drive the system optimally from the initial state x^0 to some target G in state space, where G is either a prescribed point (i.e. $G = \{x^f\}$) or a continuously differentiable (C^1) manifold in $\mathbb{R}^n$ of dimension $1 \le q \le n$ (i.e. $G = \{x: g(x) = 0\}$ g continuously differentiable). Optimal *regulator* problems are emphasised, in which case G is simply the origin in state space, i.e. the control objective is to drive the system optimally to the zero state.

The state transition from x^0 to the target G is to be accomplished optimally, in the sense that a cost functional (performance index) of the form

$$J(u) = \int_{t_0}^{t_f} L(t, x(t), u(t))\,dt; \qquad t_f \text{ free} \tag{2.5}$$

is to be minimised, where L is assumed to be continuous with respect to all its arguments and continuously differentiable with respect to x. For example,

(i) $L \equiv 1 \Rightarrow J(u) = (t_f - t_0)$ corresponds to the time-optimal control problem, a recurring theme in later chapters, for which the transfer from x^0 to G is to be accomplished in minimum time $(t_f - t_0)$;
(ii) $L = \sum_{i=1}^m |u_i(t)|$ corresponds to the fuel-optimal problem;
(iii) $L = \mu + (1-\mu)\sum_{i=1}^m |u_i(t)|$ yields a time-fuel-optimal problem which reduces to (i) for $\mu = 1$ and to (ii) for $\mu = 0$;
(iv) $L = \langle x(t), Qx(t)\rangle + \langle u(t), Ru(t)\rangle$ corresponds to the quadratic-cost problem.

The optimal control problem to be considered can now be summarised as follows. Determine an admissible control $u^* \in U$ which generates a state trajectory $x(t) = \phi(t, t_0, x^0, u^*)$, $t_0 \le t \le t_f$, emanating from x^0 (at t_0) and terminating at a point of the target G at the free terminal time t_f, i.e. $x(t_f) = \phi(t_f, t_0, x^0, u^*) \in G$, such that cost functional (2.5) is minimised, i.e. $J(u^*) = \min_{u \in U} J(u)$. When the functions f and L do not depend explicitly on t, i.e. if $f = f(x, u)$ and $L = L(x, u)$, then the system and optimal control problem are said to be *autonomous* or *time-invariant*.

Two fundamental approaches to characterising an optimal solution to the above control problem, viz. the Bellman–Hamilton–Jacobi equation and Pontryagin's maximum principle, will now be discussed.

2.2 Bellman–Hamilton–Jacobi equation

Assume that for every initial pair (t_0, x^0) in a region $\mathscr{R}$ of (t, x)-space the above problem has unique solution. Let $V(t_0, x^0)$ denote the value of the cost functional for this solution, i.e. $V(t_0, x^0)$ is the minimum cost of a trajectory starting at state x^0 at time t_0. The function V so defined for all starting pairs $(t, x) \in \mathscr{R}$ is called the *value function* for the problem.

Definition 2.2.1 Value function: If $V(t, x)$ denotes the minimum cost of an admissible trajectory starting at $(t, x) \in \mathscr{R}$, i.e.

$$V(t, x) = \min_{u \in U} \int_t^{t_f} L(s, \phi(s, t, x, u), u(s))\, ds;\ \phi(t_f, t, x, u) \in G \tag{2.6}$$

then the function $V: \mathscr{R} \to \mathbb{R}$, so defined, is the *value function* for the problem. If $x \in G$, i.e. if the initial state x is contained in the target set, then the value function is nonpositive,† i.e.

$$V(t, x) \leq 0; \qquad x \in G \tag{2.7}$$

Under the assumption that V is continuously differentiable on $\mathscr{R}$, a partial differential equation that V must satisfy (the Bellman–Hamilton–Jacobi equation) will now be derived.

From an arbitrary starting pair $(t, x) \in \mathscr{R}$, allow the system to evolve to $(t + \Delta t, x + \Delta x)$ under a possibly nonoptimal continuous admissible control $u(s)$, $t \leq s \leq t + \Delta t$, i.e. the control $u(\cdot)$ transfers the system from state x at time t to state $x + \Delta x$ at time $t + \Delta t$. Now adopt the optimal control u^* from this point onwards to the target set G, so that the transition from $(t + \Delta t, x + \Delta x)$ to $(t_f, x(t_f) \in G)$ is optimal. Then, the cost of the overall trajectory from (t, x) is given by

$$\int_t^{t+\Delta t} L(s, x(s), u(s))\, ds + \int_{t+\Delta t}^{t_f} L(s, x(s), u^*(s))\, ds$$

where

$$x(s) = \begin{cases} \phi(s, t, x, u); & t \leq s \leq t + \Delta t \\ \phi(s, t + \Delta t, x + \Delta x, u^*); & t + \Delta t \leq s \leq t_f \end{cases}$$

As the control u^* is optimal on $[t + \Delta t, t_f]$, the second integral corresponds to the minimum cost of a trajectory starting at state $x + \Delta x$ at time $t + \Delta t$, i.e. $V(t + \Delta t, x + \Delta x)$. Moreover, as the control u on $[t, t + \Delta t]$ is possibly nonoptimal, the cost of the overall trajectory from (t, x) must be greater than, or equal to, the minimum cost $V(t, x)$, hence

$$\int_t^{t+\Delta t} L(s, x(s), u(s))\, ds + V(t + \Delta t, x + \Delta x) \geq V(t, x)$$

† If $V(t, x) > 0$ for some $x \in G$, then a contradiction results (as *zero* cost $V(t, x) = 0$ can be incurred by taking x as the endpoint of the trivial trajectory with $t_f = t_0$). Frequently, the cost integrand is a non-negative function in which case *equality* holds in (2.7).

or

$$V(t + \Delta t, x + \Delta x) - V(t, x) + \int_t^{t+\Delta t} L(s, x(s), u(s))\, ds \geq 0$$

Since V is assumed to be continuously differentiable (C^1), applying Taylor's theorem to the left hand side yields

$$\frac{\partial V}{\partial t}(t, x)\,\Delta t + \left\langle \frac{\partial V}{\partial x}(t, x), \Delta x \right\rangle + \int_t^{t+\Delta t} L(s, x(s), u(s))\, ds + o(\|(\Delta t, \Delta x)\|) \geq 0$$

where

$$\frac{o(\|(\Delta t, \Delta x)\|)}{\|(\Delta t, \Delta x)\|} \to 0 \qquad \text{as} \qquad \|(\Delta t, \Delta x)\| \to 0$$

Now, using the continuity of L and u on $[t, t + \Delta t]$, dividing through by Δt and passing to the limit $\Delta t \to 0$ ($\Delta x \to 0$) yields the inequality

$$\frac{\partial V}{\partial t}(t, x) + \left\langle \frac{\partial V}{\partial x}(t, x), f(t, x, u(t)) \right\rangle + L(t, x, u(t)) \geq 0 \tag{2.8}$$

If the above analysis is repeated, but this time employing the *optimal* control u^* over the *entire* interval $[t, t_f]$, then equality holds throughout and, in particular

$$\frac{\partial V}{\partial t}(t, x) + \left\langle \frac{\partial V}{\partial x}(t, x), f(t, x, u^*(t)) \right\rangle + L(t, x, u^*(t)) = 0 \tag{2.9}$$

Thus, combining (2.8) and (2.9) yields the Bellman–Hamilton–Jacobi equation for the value function:

$$\frac{\partial V}{\partial t}(t, x) + \min_{u(t) \in \Omega} \left[\left\langle \frac{\partial V}{\partial x}(t, x), f(t, x, u(t)) \right\rangle + L(t, x, u(t)) \right] = 0 \tag{2.10}$$

or equivalently,

$$\frac{\partial V}{\partial t}(t, x) - \max_{u(t) \in \Omega} \left[- \left\langle \frac{\partial V}{\partial x}(t, x), f(t, x, u(t)) \right\rangle - L(t, x, u(t)) \right] = 0 \tag{2.11}$$

Finally, from (2.6), at the endpoint of an optimal trajectory $x(s) = \phi(s, t, x, u^*)$, $t \leq s \leq t_f$, the boundary condition

$$V(t_f, x(t_f)) = 0 \tag{2.12}$$

must hold.

Equations (2.9) and (2.12) characterise the relationship between an optimal control u^* and the value function V, i.e. if u^* is optimal, then the value function must satisfy (2.9) with boundary condition (2.12). The question of sufficiency now arises: if V is *any* solution of the Bellman–Hamilton–Jacobi equation (2.10) which also satisfies (2.7) and u^* is an admissible control satisfying (2.9) and (2.12), then is u^* optimal? The following theorem provides the affirmative answer to this question.

Theorem 2.2.1: Let $V(\cdot,\cdot)$ be a continuously differentiable function on $\mathscr{R}$ which satisfies (2.10) and (2.7). If an admissible control $u^*(t), t_0 \leq t \leq t_f$, generates a trajectory $x(t) = \phi(t, t_0, x^0, u^*)$ from $x(t_0) = x^0$ to $x(t_f) \in \mathrm{G}$ such that (2.9) is satisfied for almost all $t \in [t_0, t_f]$ with boundary condition (2.12), then u^* is optimal and $V(t_0, x^0)$ is the minimum cost.

Proof (sketch):
Noting that

$$V(t_f, x(t_f)) - V(t_0, x^0) = \int_{t_0}^{t_f} \left[\frac{d}{dt} V(t, x(t))\right] dt$$

$$= \int_{t_0}^{t_f} \left[\frac{\partial V}{\partial t}(t, x(t)) + \left\langle \frac{\partial V}{\partial x}, f(t, x(t), u^*(t))\right\rangle\right] dt$$

then the cost $J(u^*) = \int_{t_0}^{t_f} L(t, x(t), u^*(t))\, dt$ of the candidate optimal trajectory may be expressed as

$$\begin{aligned} J(u^*) &= V(t_0, x^0) - V(t_f, x(t_f)) \\ &\quad + \int_{t_0}^{t_f} \left[\frac{\partial V}{\partial t}(t, x(t)) + \left\langle \frac{\partial V}{\partial x}, f(t, x(t), u^*(t))\right\rangle \right. \\ &\quad \left. + L(t, x(t), u^*(t))\right] dt \\ &= V(t_0, x^0) \end{aligned}$$

since $u^*(\cdot)$, $x(\cdot)$ satisfy (2.9) a.e. and (2.12).

Consider now any admissible control $\tilde{u}(t)$, $t_0 \leq t \leq \tilde{t}_f$, generating a trajectory $\tilde{x}(t) = \phi(t, t_0, x^0, \tilde{u})$ from $\tilde{x}(t_0) = x^0$ to $\tilde{x}(\tilde{t}_f) \in \mathrm{G}$. In a similar manner to above, the cost $J(\tilde{u})$ of this trajectory may be expressed as

$$\begin{aligned} J(\tilde{u}) &= V(t_0, x^0) - V(\tilde{t}_f, \tilde{x}(\tilde{t}_f)) \\ &\quad + \int_{t_0}^{\tilde{t}_f} \left[\frac{\partial V}{\partial t}(t, \tilde{x}(t)) + \left\langle \frac{\partial V}{\partial x}, f(t, \tilde{x}(t), \tilde{u}(t))\right\rangle + L(t, \tilde{x}(t), \tilde{u}(t))\right] dt \end{aligned}$$

Now, as the function $V(\cdot, \cdot)$ satisfies (2.10) and (2.7), it immediately follows that

$$J(\tilde{u}) \geq V(t_0, x^0) = J(u^*)$$

i.e. u^* is an optimal control and $V(t_0, x^0)$ is the minimum cost.

For refinements and extensions of the above results to more general settings, see e.g. Vinter and Lewis (1978, 1980).

To illustrate the application of the Bellman–Hamilton–Jacobi approach, the time-optimal control problem for norm-invariant systems will be considered in the next section. For solutions to other optimal saturating control problems obtained via the Bellman–Hamilton–Jacobi equation, see e.g. Wonham (1963), Fuller (1966).

2.3 Time-optimal control of norm-invariant systems

By way of motivation, first consider the rotation of a rigid body in free space under the control of thrusters directed along the three principal orthonormal body axes. Let $\omega = (\omega_1, \omega_2, \omega_3)'$ be the vector of angular velocities about these axes and let

$$I = \begin{bmatrix} I_1 & 0 & 0 \\ 0 & I_2 & 0 \\ 0 & 0 & I_3 \end{bmatrix}$$

be the inertia matrix (or inertia tensor) of the body, where I_1, I_2, I_3 are the moments of inertia about the principal body axes. The basic equation is the momentum balance

$$I\omega = Ph$$

where h is the vector of angular momenta with respect to an inertially fixed orthonormal reference frame and P is the attitude of the body with respect to this frame, i.e. P is a 3×3 orthogonal matrix which relates the set of body axes to the inertial frame and which satisfies the matrix differential equation

$$\dot{P} = \begin{bmatrix} 0 & \omega_3 & -\omega_2 \\ -\omega_3 & 0 & \omega_1 \\ \omega_2 & -\omega_1 & 0 \end{bmatrix} P = S(\omega)P$$

Let $u = (u_1, u_2, u_3)'$ denote the controlling torques about the three principal body axes, then, referring u to the inertial frame and applying Newton's second law, yields the relation

$$\dot{h} = P^{-1}u$$

Finally, differentiating the momentum balance equation yields the Euler equations of motion, viz.

$$
\begin{aligned}
I\dot{\omega} &= \dot{P}h + P\dot{h} \\
&= S(\omega)Ph + u \\
&= S(\omega)I\omega + u
\end{aligned}
$$

or

$$
\begin{aligned}
I_1\dot{\omega}_1 &= (I_2 - I_3)\omega_2\omega_3 + u_1 \\
I_2\dot{\omega}_2 &= (I_3 - I_1)\omega_1\omega_3 + u_2 \\
I_3\dot{\omega}_3 &= (I_1 - I_2)\omega_1\omega_2 + u_3
\end{aligned}
$$

Writing $x = (x_1, x_2, x_3)' = I_1\omega_1, I_2\omega_2, I_3\omega_3)'$, these equations may be expressed as

$$
\begin{aligned}
\dot{x}_1 &= (I_3^{-1} - I_2^{-1})x_2x_3 + u_1 \\
\dot{x}_2 &= (I_1^{-1} - I_3^{-1})x_1x_3 + u_2 \\
\dot{x}_3 &= (I_2^{-1} - I_1^{-1})x_1x_2 + u_3
\end{aligned}
\tag{2.13}
$$

For the uncontrolled (homogeneous) system (i.e. with $u \equiv 0$), a straightforward calculation reveals that

$$
\begin{aligned}
\frac{d}{dt}\|x\| &= \frac{d}{dt}\sqrt{x_1^2 + x_2^2 + x_3^2} \\
&= \|x\|^{-1}(x_1\dot{x}_1 + x_2\dot{x}_2 + x_3\dot{x}_3) = \|x\|^{-1}\langle x, \dot{x}\rangle \\
&= 0
\end{aligned}
$$

so that $\|x(t)\| =$ constant for all t; this is the defining property of a *norm-invariant* system.

Consider now a general controlled norm-invariant system of the form

$$
\begin{aligned}
&\dot{x} = \psi(t, x) + u; \qquad x(t), u(t) \in \mathbb{R}^n \\
&x(t_0) = x^0 \neq 0
\end{aligned}
\tag{2.14}
$$

where ψ is such that the system is norm invariant, i.e. the solution $x(t) = \phi(t, t_0, x^0, 0)$ of the *homogeneous* equation (i.e. with $u \equiv 0$) satisfies

$$
\|x(t)\| = \|x^0\| \ \forall\ t \geq t_0 \qquad (u \equiv 0) \tag{2.15a}
$$

so that

$$
\frac{d}{dt}\|x\| = \|x\|^{-1}\langle x, \dot{x}\rangle = 0 \Rightarrow \langle x, \psi\rangle = 0 \tag{2.15b}
$$

For the *inhomogeneous* (controlled) equation, consider the time-optimal problem: If the control input is subject to a 'norm' saturation constraint, viz.

$$\|u\| = \sqrt{(u_1^2 + u_2^2 + \cdots + u_n^2)} \leq 1 \tag{2.16}$$

(i.e. the control restraint set Ω is the closed unit ball in $\mathbb{R}^n$), determine a control u^* which steers the system from $x(t_0) = x^0$ to the state origin $x(t_f^*) = 0$ in minimum time t_f^*, i.e. such that

$$J(u) = \int_{t_0}^{t_f} dt = (t_f - t_0) \tag{2.17}$$

is minimised. For this problem, the Bellman–Hamilton–Jacobi equation (2.11) becomes

$$\frac{\partial V}{\partial t}(t, x) + \left\langle \frac{\partial V}{\partial x}(t, x), \psi(t, x) \right\rangle + 1 - \max_{\|u\| \leq 1} \left[-\left\langle \frac{\partial V}{\partial x}(t, x), u \right\rangle \right] = 0 \tag{2.18}$$

with boundary condition

$$V(t_f, x(t_f)) = 0 \tag{2.19}$$

It will now be shown that (2.18)–(2.19) admits the solution

$$V = V(x) = \|x\| \tag{2.20}$$

Clearly, on the conjectured solution,

$$\frac{\partial V}{\partial t} = 0 \quad \text{and} \quad \frac{\partial V}{\partial x} = \|x\|^{-1} x$$

and, from property (2.15) of norm invariance,

$$\left\langle \frac{\partial V}{\partial x}, \psi \right\rangle = \|x\|^{-1} \langle x, \psi \rangle$$

$$= 0$$

Consequently, the left-hand side of (2.18) reduces to

$$1 - \|x\|^{-1} \max_{\|u\| \leq 1} [-\langle x, u \rangle]$$

By the Schwartz inequality,

$$-\langle x, u \rangle \leq \|x\| \, \|u\|$$

$$\leq \|x\| \qquad (\text{since } \|u\| \leq 1)$$

and *equality* holds if

$$u = u^* = -\|x\|^{-1} x \qquad (x \neq 0) \tag{2.21}$$

i.e.

$$\max_{\|u\| \leq 1} [-\langle x, u\rangle] = -\langle x, u^*\rangle = \|x\|$$

so that (2.20) satisfies the Bellman–Hamilton–Jacobi equation (2.18); moreover, (2.20) clearly satisfies the boundary condition (2.19). The corresponding optimal control u^* is given as an explicit function of the state, viz.

$$u^* = u^*(x) = \begin{cases} -\|x\|^{-1}x; & x \neq 0 \\ 0; & x = 0 \end{cases} \tag{2.22}$$

Note that, for non-zero x, the optimal control takes values on the boundary of the control restraint set Ω. This is a characteristic feature of time-optimal control systems.

For more detailed treatments of norm-invariant systems see e.g. Athans, Falb and Lacoss (1963, 1964), Athans and Falb (1966), Akulenko (1978) and, for related results obtained via the second method of Lyapunov, see Nahi (1964).

2.4 The maximum principle

The Bellman–Hamilton–Jacobi approach to optimal control as outlined in previous sections necessitated the rather restrictive condition that the value function be continuously differentiable. In even the simplest of control problems, the value function can *fail* to satisfy this condition. In such cases, an alternative approach based on the maximum principle of Pontryagin frequently proves more appropriate; this approach is adopted almost exclusively in characterising the optimal solutions to the specific control problems studied in later chapters. In the context of the general control problem posed in Section 2.2, the maximum principle will be stated only; however, a proof will be provided in the case of the linear time-optimal control problem (a recurring theme throughout the book).

Initially, to provide some insight into the general result, a restricted version of the maximum principle is derived from the Bellman–Hamilton–Jacobi equation (a similar approach was adopted in Desoer (1961)); while this derivation requires assumptions which are not generally justifiable, it does lend some plausibility to the subsequent statement of the general result.

Specifically, from (2.9),

$$\frac{\partial V}{\partial t}(\bar{t}, \bar{x}) + \left\langle \frac{\partial V}{\partial x}(\bar{t}, \bar{x}), f(\bar{t}, \bar{x}, u^*(\bar{t}))\right\rangle + L(\bar{t}, \bar{x}, u^*(\bar{t})) = 0 \tag{2.23}$$

at an arbitrary point $\bar{x} = x(\bar{t}) = \phi(\bar{t}, t_0, x^0, u^*)$ on an optimal trajectory. Keeping $\bar{t}$ and $u^*(\bar{t})$ fixed, consider how the left-hand side of (2.23) varies when fixed $\bar{x}$ is replaced by variable x, i.e. define the real-valued function W of the vector x as

$$W(x) = \frac{\partial V}{\partial t}(\bar{t}, x) + \left\langle \frac{\partial V}{\partial x}(\bar{t}, x), f(\bar{t}, x, u^*(\bar{t}))\right\rangle + L(\bar{t}, x, u^*(\bar{t}))$$

As $u^*(\bar{t})$ is optimal at the point $(\bar{t}, \bar{x})$ but not necessarily optimal at the point $(\bar{t}, x)$ with $x \neq \bar{x}$, it follows that $W(x) \geq 0$ for $x \neq \bar{x}$ and the function attains its minimum value (zero) at $x = \bar{x}$, i.e.

$$\min W(x) = W(\bar{x}) = 0$$

Now assume that V is *twice* continuously differentiable (C^2), then W is continuously differentiable (C^1), i.e. $\partial W/\partial x$ exists and is continuous. As W attains its minimum value (zero) at $x = \bar{x}$, its gradient must also be zero at this point, i.e.

$$\begin{aligned}\frac{\partial W}{\partial x_i}(\bar{x}) &= \frac{\partial^2 V}{\partial x_i\, \partial t}(\bar{t}, \bar{x}) + \sum_{j=1}^{n} \frac{\partial^2 V}{\partial x_i\, \partial x_j}(\bar{t}, \bar{x}) f_j(\bar{t}, \bar{x}, u^*(\bar{t})) \\ &\quad + \sum_{j=1}^{n} \frac{\partial V}{\partial x_j}(\bar{t}, \bar{x}) \frac{\partial f_j}{\partial x_i}(\bar{t}, \bar{x}, u^*(\bar{t})) + \frac{\partial L}{\partial x_i}(\bar{t}, \bar{x}, u^*(\bar{t})) \\ &= 0; \qquad i = 1, 2, \ldots, n\end{aligned}$$

and, as $(\bar{t}, \bar{x})$ is an arbitrary point on an optimal trajectory in (t, x)-space, the above equation holds at *all* such points (t, x) of an optimal trajectory, i.e.

$$\begin{aligned}&\frac{\partial^2 V}{\partial x_i\, \partial t}(t, x) + \sum_{j=1}^{n} \left[\frac{\partial^2 V}{\partial x_i\, \partial x_j}(t, x) f_j(t, x, u^*(t)) \right. \\ &\qquad \left. + \frac{\partial V}{\partial x_j}(t, x) \frac{\partial f_j}{\partial x_i}(t, x, u^*(t))\right] \\ &\qquad + \frac{\partial L}{\partial x_i}(t, x, u^*(t)) = 0; \qquad i = 1, 2, \ldots, n \end{aligned} \tag{2.24}$$

Introducing the vector-valued function (of t) $\eta(\cdot) = (\eta_1(\cdot), \eta_2(\cdot), \ldots, \eta_n(\cdot))'$ defined at each time t along an optimal trajectory as

$$\eta(t) = -\frac{\partial V}{\partial x}(t, x) \tag{2.25}$$

then, as V is twice continuously differentiable, η is continuously differentiable so that

$$\begin{aligned}\frac{d\eta_i}{dt} &= -\frac{\partial^2 V}{\partial t\, \partial x_i} - \sum_{j=1}^{n} \frac{\partial^2 V}{\partial x_i\, \partial x_j} \frac{dx_j}{dt} \\ &= -\frac{\partial^2 V}{\partial t\, \partial x_i} - \sum_{j=1}^{n} \frac{\partial^2 V}{\partial x_i\, \partial x_j} f_j(t, x, u^*(t))\end{aligned} \tag{2.26}$$

Combining (2.24) and (2.26) yields the result

$$\frac{d\eta_i}{dt} = -\sum_{j=1}^{n} \eta_j \frac{\partial f_j}{\partial x_i}(t, x, u^*(t)) + \frac{\partial L}{\partial x_i}(t, x, u^*(t)); \quad i = 1, 2, \ldots, n$$

or equivalently,

$$\dot{\eta} = -[\nabla_x f(t, x, u^*(t))]^T \eta + \nabla_x L(t, x, u^*(t)) \tag{2.27}$$

where ∇_x denotes the gradient with respect to x, and $[\nabla_x f]^T$ is an $n \times n$ matrix, with element ij given by $\partial f_j/\partial x_i$, which is evaluated along the optimal solution $x(\cdot)$, $u^*(\cdot)$ (as is the vector $\nabla_x L$) so that (2.27) is a linear, time-varying, inhomogeneous differential equation, referred to as the *adjoint equation.*

Defining the real-valued function H as

$$H = H(t, x, u, \eta) = -L(t, x, u) + \langle \eta, f(t, x, u) \rangle \tag{2.28}$$

referred to as the *Hamiltonian* function, then (2.27) may be written more concisely as

$$\dot{\eta} = -\frac{\partial H}{\partial x}(t, x, u^*(t), \eta) \tag{2.29}$$

Finally, combining (2.11), (2.25) and (2.28) yields the relation

$$\begin{aligned} H(t, x, u^*(t), \eta(t)) &= -L(t, x, u^*(t)) - \left\langle \frac{\partial V}{\partial x}, f(t, x, u^*(t)) \right\rangle \\ &= \max_{u \in \Omega} \left[-L(t, x, u) - \left\langle \frac{\partial V}{\partial x}, f(t, x, u) \right\rangle \right] \\ &= \max_{u \in \Omega} H(t, x, u, \eta(t)) \end{aligned} \tag{2.30}$$

at all points (t, x) of an optimal trajectory. This constitutes a version of the maximum principle and provides a necessary condition for optimality, viz. if $u^*(t)$, $t_0 \le t \le t_f$, is an optimal control which generates an optimal state trajectory with state x at time t given by $x = \phi(t, t_0, x^0, u^*)$, then there exists a solution $\eta(t)$, $t_0 \le t \le t_f$, of the adjoint equation (2.29) such that (2.30) holds along the trajectory.

Note that the state equation (2.1) may be interpreted in terms of the Hamiltonian function as

$$\dot{x} = \frac{\partial H}{\partial \eta}(t, x, u^*(t), \eta) \tag{2.31}$$

The pair of equations (2.29 and 2.31) constitute the *Hamiltonian system* of $2n$ first-order scalar differential equations for which n boundary conditions are provided through the prescribed initial state x^0 in (2.1). Other boundary conditions are obtained by considering conditions at the trajectory endpoint. Specifically, for the free-end-time problem under consideration (t_f free) it can be shown that the following *transversality conditions* hold at the endpoint of an optimal trajectory (Rozonoer, 1959 a, b, c; Pontryagin *et al.*, 1962):

(i) $\langle \eta(t_f), x - x(t_f) \rangle = 0 \qquad \forall\, x \in P(x(t_f))$ (2.32a) where $P(x(t_f))$ denotes the tangent hyperplane to the smooth target manifold $G = \{x: g(x) = 0\}$ at the trajectory endpoint $x(t_f)$, i.e.

$$P(x(t_f)) = \left\{x: \left\langle \frac{\partial g}{\partial x}(x(t_f)), x - x(t_f) \right\rangle = 0\right\} \tag{2.32b}$$

In other words, the adjoint vector η is orthogonal to the smooth target manifold G at an optimal trajectory endpoint;†

(ii) $$H(t_f, x(t_f), u^*(t_f), \eta(t_f)) = 0 \tag{2.33}$$

i.e. the maximised Hamiltonian takes the value zero at an optimal trajectory endpoint.

The validity of (2.33) may be loosely illustrated as follows. Since t_f is free and $V(t_f, x(t_f)) = 0$ for *every* optimal trajectory endpoint, it may be seen that

$$\frac{\partial V}{\partial t}(t, x(t))\bigg|_{t=t_f} = 0$$

Now, from (2.9) and (2.25),

$$\begin{aligned}\frac{\partial V}{\partial t}(t, x(t)) - [\langle \eta(t), f(t, x(t), u^*(t))\rangle - L(t, x(t), u^*(t))]\\ = \frac{\partial V}{\partial t}(t, x(t)) - H(t, x(t), u^*(t), \eta(t))\\ = 0\end{aligned}$$

and, in particular, at $t = t_f$

$$H(t_f, x(t_f), u^*(t_f), \eta(t_f)) = \frac{\partial V}{\partial t}(t, x(t))\bigg|_{t=t_f} = 0$$

The time variation of the Hamiltonian will now be investigated; in particular, it will be shown that, on an optimal trajectory, the *total* time derivative $\dot{H}$ of H is equal to its *partial* time derivative $\partial H/\partial t$. Specifically, for $s, t \in (t_0, t_f)$, the following relation holds

$$H(s, x(s), u^*(s), \eta(s)) \geq H(s, x(s), u^*(t), \eta(s))$$

If $s > t$, then subtracting $H(t, x(t), u^*(t), \eta(t))$ from both sides and dividing by $s - t > 0$ yields

$$\frac{H(s, x(s), u^*(s), \eta(s)) - H(t, x(t), u^*(t), \eta(t))}{s - t} \geq \frac{H(s, x(s), u^*(t), \eta(s)) - H(t, x(t), u^*(t), \eta(t))}{s - t} \tag{2.34}$$

† For the most part, the target set will be taken as a prescribed *point* in the state space, i.e. $G = \{x^f\}$, in which case (2.32) is redundant and the $2n$ boundary conditions for the Hamiltonian system of equations are provided by the prescribed initial *and* final states: x_i^0 *and* x_i^f, $i = 1, 2, \ldots, n$.

Noting that the smoothness condition (i.e. twice continuously differentiable) on the value function V ensures existence of the derivatives appearing below, taking the limit as $s \downarrow t$ in (2.34) gives

$$\dot{H} \geq \frac{\partial H}{\partial t} + \left\langle \frac{\partial H}{\partial x}, \dot{x} \right\rangle + \left\langle \frac{\partial H}{\partial \eta}, \dot{\eta} \right\rangle$$

which, in view of (2.29) and (2.31), reduces to

$$\dot{H} \geq \frac{\partial H}{\partial t} \tag{2.35a}$$

If $s < t$, then, dividing by $s - t < 0$, the inequality is reversed in (2.34) and taking the limit as $s \uparrow t$ yields the relation

$$\dot{H} \leq \frac{\partial H}{\partial t} \tag{2.35b}$$

Combining (2.35a) and (2.35b) gives the required result

$$\dot{H}(t, x(t), u^*(t), \eta(t)) = \frac{\partial H}{\partial t}(t, x(t), u^*(t), \eta(t)) \tag{2.35c}$$

Finally, if the functions f and L (and hence H) do not depend explicitly on t, i.e. if the control problem is *autonomous* so that $H = H(x, u, \eta)$, then $\dot{H} = \partial H/\partial t = 0$ which, when combined with (2.33), yields the useful result

$$H(x(t), u^*(t), \eta(t)) = 0; \quad 0 \leq t \leq t_f \tag{2.36}$$

i.e. the Hamiltonian vanishes along a solution of an *autonomous, free end time* optimal control problem.

The above derivation of the maximum principle necessitated very restrictive smoothness conditions on the value function V and, consequently, is intended only to lend plausibility to the following general statement of the maximum principle which provides a necessary condition for optimality for a considerably larger class of systems than those which exhibit the above smoothness conditions.

First, the set of admissible controls U is extended to the class of bounded measurable functions (which includes the earlier class of piecewise continuous controls) taking values in the compact control restraint set $\Omega \subset \mathbb{R}^m$ (see e.g. Lee and Markus, 1967; Naylor and Sell, 1971; Wheeden and Zygmund, 1977; Curtain and Pritchard, 1977, for introductory accounts of measurable functions). While the introduction of such bounded measurable controls proves important in establishing existence and other results in optimal control theory (where the analysis may require certain sequences of controls to converge), it is stressed that the optimal controls themselves frequently turn out to be simply piecewise continuous or piecewise constant. As before, the functions f and L are assumed to be continuous in t, x, u and continuously differentiable with respect to x and the Hamiltonian function is defined as in (2.28).

Theorem 2.4.1 The maximum principle: Defining the maximised Hamiltonian function H^* as

$$H^*(t, x, \eta) = \max_{u \in \Omega} H(t, x, u, \eta) \tag{2.37}$$

if $u^*(t)$, $t_0 \leq t \leq t_f$, is an optimal control generating an optimal trajectory $x(t) = \phi(t, t_0, x^0, u^*)$ satisfying (2.1) almost everywhere on $[t_0, t_f]$ then, corresponding to $u^*(\cdot)$ and $x(\cdot)$, there exists an (absolutely continuous) function $\eta(t)$, $t_0 \leq t \leq t_f$, such that

(i) *adjoint equation:*

$$\dot{\eta}(t) = -\frac{\partial H}{\partial x}(t, x(t), u^*(t), \eta(t)) \qquad \text{for a.a. } t \in [t_0, t_f] \tag{2.38}$$

(ii) *maximisation of the Hamiltonian:*

$$H(t, x(t), u^*(t), \eta(t)) = H^*(t, x(t), \eta(t)) \qquad \text{for a.a. } t \in [t_0, t_f] \tag{2.39}$$

Moreover, at the trajectory endpoint $x(t_f) = \phi(t_f, t_0, x^0, u^*) \in G$ (at the free terminal time t_f) the following *transversality conditions* hold:

(iii) $$\langle \eta(t_f), x - x(t_f) \rangle = 0 \quad \text{for all} \quad x \in P(x(t_f)) \tag{2.40a}$$

where $P(x(t_f))$ denotes the tangent hyperplane to the smooth target manifold $G = \{x: g(x) = 0\}$ at the trajectory endpoint $x(t_f)$, viz.

$$P(x(t_f)) = \left\{x: \left\langle \frac{\partial g}{\partial x}(x(t_f)), x - x(t_f) \right\rangle = 0\right\} \tag{2.40b}$$

i.e. the adjoint vector is orthogonal to the target manifold at the trajectory endpoint;

(iv) $$H(t_f, x(t_f), u^*(t_f), \eta(t_f)) = H^*(t_f, x(t_f), \eta(t_f)) = 0 \tag{2.41}$$

Corollary 2.4.1: If the optimal control problem is *autonomous*, i.e. if $f = f(x, u)$ and $L = L(x, u)$ do not depend explicitly on t, then H^* vanishes identically along an optimal trajectory, i.e.

$$H^* = H^*(x(t), \eta(t)) = 0; \qquad t_0 \leq t \leq t_f \tag{2.42}$$

Proof of the maximum principle in this and even more general settings can be found in many texts (e.g. Pontryagin *et al.*, 1962; Athans and Falb, 1966; Lee and Markus, 1967; Berkovitz, 1974; Fleming and Rishel, 1975; Gamkrelidze, 1978; see also Fuller (1963b) for a bibliography of early investigations into the maximum principle). However, as the linear time-optimal control problem figures predominantly in later chapters, the maximum principle will be proved in this special case.

2.5 Linear autonomous time-optimal control systems

Attention will now be restricted to linear, autonomous control systems in $\mathbb{R}^n$ and, without loss of generality, the process starting time t_0 is taken as zero ($t_0 = 0$), viz.

$$\dot{x} = Ax + Bu; \qquad x(0) = x^0 \in \mathbb{R}^n \tag{2.43}$$

where the set of admissible control functions U is taken as the class of bounded measurable functions with values $u(t)$ in a *convex compact* control restraint set $\Omega \subset \mathbb{R}^m$. Then, for $u \in U$, the unique absolutely continuous solution (satisfying (2.43) almost everywhere) is given by

$$x(t) = \phi(t, x^0, u) = \exp(At)x^0 + \int_0^t \exp(A(t-s))Bu(s)\,ds \tag{2.44}$$

The problem is to select a control $u^* \in U$ which steers the system from x^0 to some prescribed terminal state x^f, i.e.

$$x(t_f^*) = \phi(t_f^*, x^0, u^*) = x^f$$

in minimum time t_f^*, i.e. $J(u) = \int_0^{t_f} dt$ is to be minimised.

It is assumed throughout that the pair (A, B) is controllable with respect to *unconstrained* controls $u(t) \in \mathbb{R}^m$, i.e. the familiar rank condition

$$\text{Rank}\,[B : AB : A^2B : \cdots : A^{n-1}B] = n \tag{2.45}$$

is assumed to hold, in which case the system is said to be *proper* (Hermes and LaSalle, 1969). For this problem, the Hamiltonian function is given by

$$H(x, u, \eta) = -1 + \langle \eta, Ax \rangle + \langle \eta, Bu \rangle \tag{2.46}$$

and the maximum principle (Theorem 2.4.1) states that if $u^*(t), 0 \leq t \leq t_f^*$, is an optimal control generating a time-optimal trajectory $x(t) = \phi(t, x^0, u^*)$ from $x(0) = x^0$ to $x(t_f^*) = x^f$, then, corresponding to $x(\cdot)$ and $u^*(\cdot)$, there exists a nontrivial solution $\eta(t), 0 \leq t \leq t_f^*$, of the adjoint equation:

(i) $$\dot{\eta} = -A^T\eta \tag{2.47}$$

such that the maximum principle:

(ii) $$\begin{aligned} H(x(t), u^*(t), \eta(t)) &= \max_{u \in \Omega} H(x(t), u, \eta(t)) \\ &= H^*(x(t), \eta(t)) \end{aligned} \tag{2.48}$$

holds for almost all $t \in [0, t_f^*]$.

In view of (2.46), the maximum principle (2.48), is equivalent to the relation

$$\langle \eta(t), Bu^*(t) \rangle = \max_{u \in \Omega} \langle \eta(t), Bu \rangle \qquad \text{for almost all } t \in [0, t_f^*] \tag{2.49}$$

which will now be established. Similar treatments can be found in Kreindler (1963), Lee and Markus (1967), Hermes and LaSalle (1969), Hájek (1971), Markus (1976) and Curtain and Pritchard (1977).

Definition 2.5.1 Attainable set $K_{x^0}(t_f)$: The attainable set $K_{x^0}(t_f)$ is the set of all possible trajectory endpoints $x(t_f) = \phi(t_f, x^0, u)$ i.e. the set of all states to which x^0 can be steered in time t_f by means of an admissible control $u \in U$, i.e.

$$K_{x^0}(t_f) = \left\{x: x = \exp(At_f)x^0 + \int_0^{t_f} \exp(A(t_f - s))Bu(s)\, ds; \quad u \text{ admissible}\right\}$$

Note that $K_{x^0}(t_f)$ is well defined for all real t_f so that negative values of t_f are admissible, in which case the attainable set corresponds to the set of states which can be reached from x^0 in *reverse* or *backwards* time t_f.

Theorem 2.5.1 : The attainable set $K_{x^0}(t_f)$ is (i) convex, (ii) compact and (iii) varies continuously with t_f.

Proof
(i) *Convexity :* To establish convexity, it is required to show that the line segment $(1 - \mu)x^1 + \mu x^2$, $0 \leq \mu \leq 1$, joining two arbitrary points x^1 and x^2 in $K_{x^0}(t_f)$, also lies in $K_{x^0}(t_f)$.

Let $u^1, u^2 \in U$ be controls which steer x^0 to x^1 and x^2, respectively. Then, since Ω is convex, the control u^3 defined as $u^3 = (1 - \mu)u^1 + \mu u^2$ is also an admissible control.

Now

$$\begin{aligned} x^1 &= \phi(t_f, x^0, u^1) \\ &= \exp(At_f)x^0 + \int_0^{t_f} \exp(A(t_f - s))Bu^1(s)\, ds \end{aligned}$$

and

$$\begin{aligned} x^2 &= \phi(t_f, x^0, u^2) \\ &= \exp(At_f)x^0 + \int_0^{t_f} \exp(A(t_f - s))Bu^2(s)\, ds \end{aligned}$$

Hence,

$$\begin{aligned} (1 - \mu)x^1 + \mu x^2 &= \exp(At_f)x^0 + \int_0^{t_f} \exp(A(t_f - s))B \\ &\quad \times [(1 - \mu)u^1(s) + \mu u^2(s)]\, ds \\ &= \exp(At_f)x^0 + \int_0^{t_f} \exp(A(t_f - s))Bu^3(s)\, ds \end{aligned}$$

and since u^3 is an admissible control, it immediately follows that

$$(1 - \mu)x^1 + \mu x^2 \in K_{x^0}(t_f)$$

i.e. $K_{x^0}(t_f)$ is convex.

(*ii*) *Compactness:* To establish compactness (i.e. that the attainable set $K_{x^0}(t_f)$ is closed and bounded in $\mathbb{R}^n$) it is required to show that every sequence of points $\{x^k\} = x^1, x^2, \ldots, x^k, \ldots$ in $K_{x^0}(t_f)$ contains a subsequence $\{x^{k_r}\}$ which converges to some limit point $\bar{x}$ in $K_{x^0}(t_f)$. Let the sequence $\{x^k \in K_{x^0}(t_f)\}$ correspond to the sequence of trajectory endpoints generated by a sequence of admissible (i.e. bounded measurable) controls $\{u^k(s), 0 \leq s \leq t_f\}$, i.e.

$$x^k = \exp(At_f)x^0 + \int_0^{t_f} \exp(A(t_f - s))Bu^k(s)\, ds \tag{2.50}$$

The concept of weak convergence (tailored to the problem at hand) is now introduced.

Weak convergence: A sequence of bounded measurable (L^∞) functions $\{u^k(s), 0 \leq s \leq t_f\}$, with values in $\mathbb{R}^m$, is said to be weekly convergent to a bounded measurable function $\bar{u}(s), 0 \leq s \leq t_f$, if

$$\lim_{k \to \infty} \int_0^{t_f} \langle v(s), u^k(s) \rangle\, ds = \int_0^{t_f} \langle v(s), \bar{u}(s) \rangle\, ds$$

for every integrable (L^1) test function $v(s), 0 \leq s \leq t_f$, with values in $\mathbb{R}^m$. It can be shown (Lee and Markus, 1967; Lemma 1A, Chapter 2) that every sequence of bounded measurable control functions $\{u^k(s), 0 \leq s \leq t_f\}$, taking values $u^k(s) \in \Omega$ in the *compact* control restraint set $\Omega \subset \mathbb{R}^m$ contains a weakly convergent subsequence $\{u^{k_r}(\cdot)\}$ with weak limit $\bar{u}(s) \in \Omega, 0 \leq s \leq t_f$; in other words, the set of bounded measurable controls on $[0, t_f]$ with values in Ω is (sequentially) weakly compact.

As the $n \times m$ matrix-valued function $\exp(A(t_f - s))B$ is continuous on $[0, t_f]$, its constituent row vectors are integrable $\mathbb{R}^m$-valued functions and hence, by the above weak compactness property of the controls, it may be concluded that

$$\lim_{r \to \infty} \int_0^{t_f} \exp(A(t_f - s))Bu^{k_r}(s)\, ds = \int_0^{t_f} \exp(A(t_f - s))B\bar{u}(s)\, ds \tag{2.51}$$

Writing $\bar{x} = \phi(t_f, x^0, \bar{u}) \in K_{x^0}(t_f)$ as the trajectory endpoint corresponding to the admissible control $\bar{u}(s) \in \Omega, 0 \leq s \leq t_f$, it follows from (2.50) and (2.51) that

$$\begin{aligned}
\lim_{r \to \infty} x^{k_r} &= \exp(At_f)x^0 + \lim_{r \to \infty} \int_0^{t_f} \exp(A(t_f - s))Bu^{k_r}(s)\, ds \\
&= \exp(At_f)x^0 + \int_0^{t_f} \exp(A(t_f - s))B\bar{u}(s)\, ds \\
&= \bar{x} \in K_{x^0}(t_f)
\end{aligned}$$

i.e. the sequence of points $\{x^k \in K_{x^0}(t_f)\}$ contains a subsequence $\{x^{k^r}\}$ converging to $\bar{x}$ in $K_{x^0}(t_f)$, so that $K_{x^0}(t_f)$ is compact.

(*iii*) $K_{x^0}(t_f)$ *varies continuously with* t_f *:* To establish continuity it is required to show that given any $\epsilon > 0$ there exists a $\delta > 0$ such that the 'distance' between the sets $K_{x^0}(t_1)$ and $K_{x^0}(t_2)$ is less than ϵ whenever $|t_2 - t_1| < \delta$; this necessitates the definition of distance $\rho(K(t_1), K(t_2))$ between two compact sets in $\mathbb{R}^n$, which is given as follows (the Hausdorff metric for the set of nonempty compact subsets of $\mathbb{R}^n$).

First, the distance $d(x, A)$ of a point x from a compact set A is defined as

$$d(x, A) = \min_{a \in A} \{\|x - a\|\}$$

then the distance ρ between the two compact sets $K_{x^0}(t_1)$ and $K_{x^0}(t_2)$ is defined as

$$\rho(K_{x^0}(t_1), K_{x^0}(t_2)) = \frac{1}{2}\left[\max_{x^1 \in K_{x^0}(t_1)} d(x^1, K_{x^0}(t_2)) + \max_{x^2 \in K_{x^0}(t_2)} d(x^2, K_{x^0}(t_1))\right]$$

It is easily shown that $\rho \geq 0$ satisfies the axioms of a metric, viz.

(i) $\rho(x, y) = \rho(y, x)$; (ii) $\rho(x, z) \leq \rho(x, y) + \rho(y, z)$;

(iii) $\rho(x, y) = 0 \Leftrightarrow x = y$

To prove continuity of the map $t_f \mapsto K_{x^0}(t_f)$ it must be shown that, given $\epsilon > 0$, there exists $\delta > 0$ such that

$$\rho(K_{x^0}(t_2), K_{x^0}(t_1)) < \epsilon \text{ whenever } |t_2 - t_1| < \delta \tag{2.52}$$

In turn, to establish (2.52) it is sufficient to show that

(i) each point $x^1 \in K_{x^0}(t_1)$ is within ϵ of some point $\hat{x}^2 \in K_{x^0}(t_2)$ (i.e. $\|x^1 - \hat{x}^2\| < \epsilon$)

(ii) each point $x^2 \in K_{x^0}(t_2)$ is within ϵ of some point $\tilde{x}^1 \in K_{x^0}(t_1)$ (i.e. $\|x^2 - \tilde{x}^1\| < \epsilon$)

whenever $|t_2 - t_1| < \delta$. To see that (i) and (ii) imply (2.52), first note, from (i), since $\|x^1 - \hat{x}^2\| < \epsilon$ where $\hat{x}^2 \in K_{x^0}(t_2)$ it follows that

$$\min_{x^2 \in K_{x^0}(t_2)} \|x^2 - x^1\| = d(x^1, K_{x^0}(t_2)) < \epsilon$$

and since this is to hold for all $x^1 \in K_{x^0}(t_1)$ it may be concluded that

$$\max_{x^1 \in K_{x^0}(t_1)} d(x^1, K_{x^0}(t_2)) < \epsilon$$

Similarly, from (ii) it follows that

$$\max_{x^2 \in K_{x^0}(t_2)} d(x^2, K_{x^0}(t_1)) < \epsilon$$

and, hence, (i) and (ii) imply the inequality

$$\rho(K_{x^0}(t_1), K_{x^0}(t_2)) < \tfrac{1}{2}(\epsilon + \epsilon) = \epsilon$$

To establish (i) and (ii), let $u(t) \in \Omega$, $0 \leq t \leq t_1 + T$, where T is an arbitrary positive constant, be an admissible control generating a trajectory $x(t) = \phi(t, x^0, u)$, $0 \leq t \leq t_1 + T$. Then, for $t_2 \in (0, t_1 + T)$

$$\begin{aligned} x(t_2) - x(t_1) &= [\exp(At_2) - \exp(At_1)]x^0 + \int_0^{t_2} \exp(A(t_2 - s))Bu(s)\, ds \\ &\quad - \int_0^{t_1} \exp(A(t_1 - s))Bu(s)\, ds \\ &= [\exp(At_2) - \exp(At_1)]\left[x^0 + \int_0^{t_1} \exp(-As)\mathrm{B}u(s)\, ds\right] \\ &\quad + \int_{t_1}^{t_2} \exp(A(t_2 - s))Bu(s)\, ds \end{aligned}$$

and, on taking norms,

$$\begin{aligned} \|x(t_2) - x(t_1)\| &= \Bigg\| [\exp(At_2) - \exp(At_1)] \\ &\quad \times \left[x^0 + \int_0^{t_1} \exp(-As)Bu(s)\, ds\right] \\ &\quad + \int_{t_1}^{t_2} \exp(A(t_2 - s))Bu(s)\, ds \Bigg\| \\ &\leq \|\exp(At_2) - \exp(At_1)\| \\ &\quad \times \left[\|x^0\| + \left\|\int_0^{t_1} \exp(-As)Bu(s)\, ds\right\|\right] \\ &\quad + \left\|\exp(At_2) \int_{t_1}^{t_2} \exp(-As)Bu(s)\, ds\right\| \\ &\leq \|\exp(At_2) - \exp(At_1)\| \\ &\quad \times \left[\|x^0\| + \int_0^{t_1} \|\exp(-As)\|\,\|B\|\,\|u(s)\|\, ds\right] \\ &\quad + \|\exp(At_2)\| \left\|\int_{t_1}^{t_2} \exp(-As)Bu(s)\, ds\right\| \end{aligned} \tag{2.53}$$

Now, there exists a constant $c_1 > 0$ such that

$$\|\exp(As)\| < c_1 \quad \text{and} \quad \|\exp(-As)\| < c_1, \quad 0 \le s \le t_1 + T$$

and, since $\|u(s)\|$ is bounded, there exists a constant $c_2 > 0$ such that

$$\|x^0\| + \int_0^{t_1} \|\exp(-As)\| \|B\| \|u(s)\| \, ds < c_2$$

Also, as an integral is a continuous function (as is $\exp(At)$) it follows that, given $\epsilon > 0$, there exists $\delta > 0$ such that

$$\left\| \int_{t_1}^{t_2} \exp(-As) Bu(s) \, ds \right\| < \frac{\epsilon}{2c_1} \quad \text{and} \quad \|\exp(At_2) - \exp(At_1)\| < \frac{\epsilon}{2c_2}$$

whenever $|t_2 - t_1| < \delta$. Using the above bounds in (2.53), it may be concluded that

$$\|x(t_2) - x(t_1)\| < c_1\left[\frac{\epsilon}{2c_1}\right] + c_2\left[\frac{\epsilon}{2c_2}\right] = \epsilon \tag{2.54}$$

whenever $|t_2 - t_1| < \delta$. With reference to (i) above, let $x^1 = \phi(t_1, x^0, u) \in K_{x^0}(t_1)$ be the endpoint corresponding to an (arbitrary) admissible control $u(t)$, $0 \le t \le t_1$. Extend this function to a control $\hat{u}(t)$, $0 \le t \le t_1 + T$, by defining

$$\hat{u}(t) = \begin{cases} u(t); & 0 \le t \le t_1 \\ u(t_1); & t_1 < t \le t_1 + T \end{cases}$$

and, for $t_2 \in (0, t_1 + T)$, write $\hat{x}^2 = \phi(t_2, x^0, \hat{u})$. Then, from (2.54),

$$\|\hat{x}^2 - x^1\| < \epsilon \quad \text{whenever} \quad |t_2 - t_1| < \delta$$

This establishes (i).

Now, let $x^2 = \phi(t_2, x^0, u) \in K_{x^0}(t_2)$ be the endpoint corresponding to an (arbitrary) admissible control $u(t)$, $0 \le t \le t_2$. Extend this function to a control $\tilde{u}(t)$, $0 \le t \le t_1 + T$, by defining

$$\tilde{u}(t) = \begin{cases} u(t); & 0 \le t \le t_2 \\ u(t_2); & t_2 < t \le t_1 + T \end{cases}$$

and write $\tilde{x}^1 = \phi(t_1, x^0, \tilde{u})$. Then, from (2.54),

$$\|x^2 - \tilde{x}^1\| < \epsilon \quad \text{whenever} \quad |t_2 - t_1| < \delta$$

This establishes (ii). Consequently, given $\epsilon > 0$, the distance ρ between the sets $K_{x^0}(t_1)$ and $K_{x^0}(t_2)$ is less than ϵ whenever $|t_2 - t_1| < \delta$ (where δ depends on ϵ and t_1 and satisfies $0 < \delta(\epsilon, t_1) < T$).

Hence, $K_{x^0}(t_f)$ varies continuously with t_f. The following corollary will be used later.

Corollary 2.5.1: If x is an interior point of $K_{x^0}(t_2)$ for some $t_2 > 0$, then x is an interior point of $K_{x^0}(t_1)$ for some $t_1 \in (0, t_2)$.

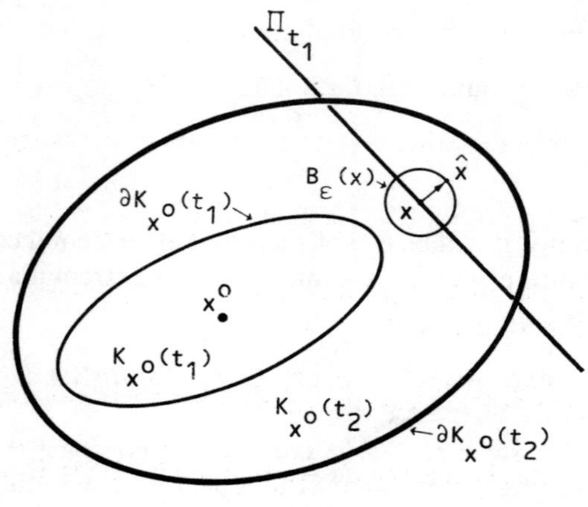

Fig. 2.1

Proof

If x is in the interior of $K_{x^0}(t_2)$ then there exists an ϵ-neighbourhood $B_\epsilon(x)$ of x (of radius $\epsilon > 0$) which also lies in the interior of $K_{x^0}(t_2)$. Now, contrary to the Corollary, suppose that for each $t_1 \in (0, t_2)$ x is not an interior point of $K_{x^0}(t_1)$, then for each $t_1 \in (0, t_2)$ a hyperplane Π_{t_1} through x may be constructed such that $K_{x^0}(t_1)$ lies to one side of Π_{t_1}, as depicted in Fig. 2.1. But, as $B_\epsilon(x) \in K_{x^0}(t_2)$, there must be a point $\hat{x} \in K_{x^0}(t_2)$ the distance of which from $K_{x^0}(t_1)$ is at least ϵ (e.g. $\hat{x}$ of Fig. 2.1), i.e.

$$d(\hat{x}, K_{x^0}(t_1)) \geq \epsilon \ \forall\ t_1 \in (0, t_2)$$

which contradicts the continuity of the map $t \mapsto K_{x^0}(t)$. Hence the above supposition is false and $x \in \text{int } K_{x^0}(t_1)$ for some $t_1 \in (0, t_2)$.

The concept of an *extremal control* will now be introduced; it will subsequently be shown that a control is extremal if, and only if, it satisfies (2.49). Finally, it will be shown that an *optimal* control is *extremal*, thereby establishing the maximum principle for autonomous, linear, time-optimal control problems.

Definition 2.5.2 Extremal control: If an admissible control $u^*(t)$, $0 \leq t \leq t_f$, generates a trajectory $x(t) = \phi(t, x^0, u^*)$, $0 \leq t \leq t_f$, with endpoint $x(t_f)$ in the boundary $\partial K_{x^0}(t_f)$ of the attainable set $K_{x^0}(t_f)$, then $u^*(\cdot)$ is an extremal control, i.e. under an extremal control $u^*(t)$, $0 \leq t \leq t_f$, the system is transferred from x^0 to an extreme point of the attainable set $K_{x^0}(t_f)$.

Theorem 2.5.2: An admissible control $u^*(t)$, $0 \leq t \leq t_f$, is extremal if and only if there exists a nontrivial solution $\eta(t) = \exp(-A^T t)\eta^0$, $0 \leq t \leq t_f$, of the adjoint equation

$$\dot{\eta} = -A^T \eta$$

such that

$$\langle \eta(t), Bu^*(t) \rangle = \max_{u \in \Omega} \langle \eta(t), Bu \rangle$$

for a.a. $t \in [0, t_f]$.

Proof

(*a*) *Necessity :* Initially, it will be established that an extremal control necessarily satisfies the maximum principle. Assume $u^* \in U$ is extremal and so steers x^0 to $x^f \in \partial K_{x^0}(t_f)$ where

$$x^f = x(t_f) = \exp(At_f)x^0 + \int_0^{t_f} \exp(A(t_f - s))Bu^*(s)\, ds \tag{2.55}$$

Referring to Fig. 2.2, since $K_{x^0}(t_f)$ is convex and compact (by theorem 2.5.1), there exists a supporting hyperplane† Π to $K_{x^0}(t_f)$ at the boundary point x^f. Let $\eta^f \neq 0$ denote an *exterior* normal to Π at x^f, then

$$\langle \eta^f, (\bar{x} - x^f) \rangle \leq 0 \;\forall\; \bar{x} \in K_{x^0}(t_f) \tag{2.56}$$

since the vectors η^f and $(\bar{x} - x^f)$ lie on opposite sides of the hyperplane Π. For the adjoint solution $\eta(t) = \exp(-A^T t)\eta^0$, select the vector η^0 such that

$$\eta^0 = \exp(A^T t_f)\eta^f \Leftrightarrow \eta^f = \exp(-A^T t_f)\eta^0$$

then, from (2.55) and (2.56),

$$\begin{aligned}\langle \eta^f, x^f \rangle &= \left\langle \exp(-A^T t_f)\eta^0, \left[\exp(At_f)x^0 + \int_0^{t_f} \exp(A(t_f - s))Bu^*(s)\, ds\right]\right\rangle \\ &= \langle \exp(-A^T t_f)\eta^0, \exp(At_f)x^0 \rangle \\ &\quad + \int_0^{t_f} \langle \exp(-A^T t_f)\eta^0, \exp(A(t_f - s))Bu^*(s) \rangle\, ds \\ &= \langle \eta^0, x^0 \rangle + \int_0^{t_f} \langle \exp(-A^T s)\eta^0, Bu^*(s) \rangle\, ds\end{aligned}$$

i.e.

$$\langle \eta^f, x^f \rangle = \langle \eta^0, x^0 \rangle + \int_0^{t_f} \langle \eta(s), Bu^*(s) \rangle\, ds \tag{2.57}$$

A proof by contradiction is now initiated by the supposition that u^* does *not* satisfy the maximum principle, i.e. it is supposed that

(S1): $\langle \eta(t), Bu^*(t) \rangle < \max_{u \in \Omega} \langle \eta(t), Bu \rangle$

on some finite time interval (2.58)

† See, for example, Eggleston (1958) for proof of this and related properties of convex sets.

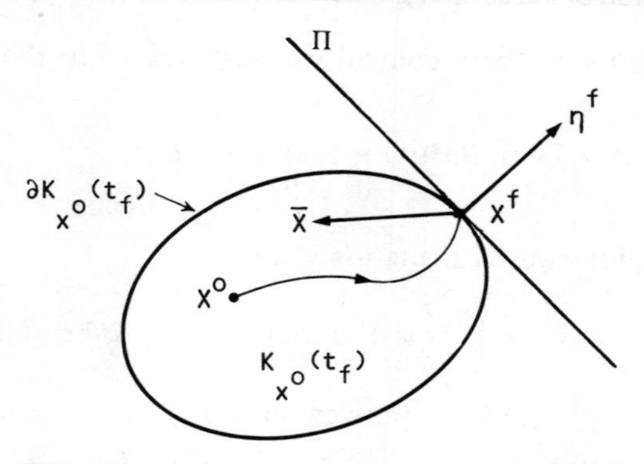

Fig. 2.2 *Attainable set $K_{x^0}(t_f)$ and supporting hyperplane* Π

Define the (admissible) control $\bar{u}$ (non-extremal by the above supposition) as

$$\langle \eta(t), B\bar{u}(t) \rangle = \max_{u \in \Omega} \langle \eta(t), Bu \rangle \qquad \text{for a.a. } t \in [0, t_f] \tag{2.59}$$

which generates a trajectory with endpoint $\bar{x} = \phi(t_f, x^0, \bar{u})$. Proceeding as in the derivation of (2.57) yields the result

$$\langle \eta^f, \bar{x} \rangle = \langle \eta^0, x^0 \rangle + \int_0^{t_f} \langle \eta(s), B\bar{u}(s) \rangle \, ds$$

and, subtracting (2.57), gives

$$\begin{aligned} \langle \eta^f, \bar{x} - x^f \rangle &= \int_0^{t_f} [\langle \eta(s), B\bar{u}(s) \rangle - \langle \eta(s), Bu^*(s) \rangle] \, ds \\ &> 0 \qquad \text{(in view of (2.58), (2.59))} \end{aligned} \tag{2.60}$$

But, by supposition, $\bar{u}$ is non-extremal and, hence, the corresponding trajectory endpoint $\bar{x}$ must lie in the *interior* of $K_{x^0}(t_f)$; this, in turn, implies that $\langle \eta^f, \bar{x} - x^f \rangle \leq 0$ which contradicts (2.60). Consequently, supposition (2.58) is false, so that an extremal control u^* must satisfy the maximum principle.

(*b*) *Sufficiency:* It will now be established that, if an admissible control $u^*(t)$, $0 \leq t \leq t_f$, satisfies the maximum principle for some non-trivial adjoint solution $\eta(t)$, $0 \leq t \leq t_f$, then it generates a trajectory endpoint $x(t_f)$ in the boundary $\partial K_{x^0}(t_f)$ of the attainable set $K_{x^0}(t_f)$ and hence is extremal. Again, a proof by contradiction is initiated by supposing that $x(t_f)$ lies in the *interior* of $K_{x^0}(t_f)$, i.e.

$$\text{(S2):} \qquad x(t_f) = \phi(t_f, x^0, u^*) \in \text{int } K_{x^0}(t_f) \tag{2.61}$$

then there exists a point $\tilde{x} \in K_{x^0}(t_f)$ such that

$$\langle \eta(t_f), x(t_f) - \tilde{x} \rangle < 0 \tag{2.62}$$

Now let $\tilde{u}(t)$, $0 \le t \le t_f$, be a control which drives x^0 to the point $\tilde{x}$, then it follows that

$$\langle \eta(t), B\tilde{u}(t) \rangle \le \langle \eta(t), Bu^*(t) \rangle = \max_{u \in \Omega} \langle \eta(t), Bu \rangle \quad \text{for a.a. } t \in [0, t_f] \tag{2.63}$$

As before, a straightforward calculation yields

$$\langle \eta(t_f), x(t_f) - \tilde{x} \rangle = \int_0^{t_f} [\langle \eta(s), Bu^*(s) \rangle - \langle \eta(s), B\tilde{u}(s) \rangle] \, ds$$

$$\ge 0 \qquad \text{(in view of (2.63))}$$

which contradicts (2.62). Hence, supposition (2.61) is false and $x(t_f)$ must lie on the boundary $\partial K_{x^0}(t_f)$ of the attainable set, implying that u^* is an extremal control.

This completes the proof of Theorem 2.5.2.

Theorem 2.5.3 Existence of a time-optimal control: If there exists an admissible control which transfers system (2.43) from the initial state x^0 to a prescribed terminal state $x^f \neq x^0$ in finite time, then there exists a time-optimal control $u^*(t)$, $0 \le t \le t_f^*$, which transfers the system from x^0 to x^f in *minimum time* t_f^*. Moreover, u^* is an extremal control.

Proof

Define t_f^* as

$$t_f^* = \inf \{t_f : K_{x^0}(t_f) \text{ contains } x^f\}$$

Since the convex, compact set $K_{x^0}(t_f)$ varies continuously with t_f and contains x^f for some finite t_f, t_f^* exists and is first time at which $K_{x^0}(t_f)$ meets x^f. A control $u^* \in U$ defined on $[0, t_f^*]$ which generates the endpoint x^f is clearly time-optimal. That x^f is a boundary point of $K_{x^0}(t_f^*)$ is a direct consequence of corollary 2.5.1 since, if x^f were in the interior of $K_{x^0}(t_f^*)$ then, by corollary 2.5.1, $x^f \in K_{x^0}(t_f)$ for some earlier time $t_f < t_f^*$, thereby contradicting the definition of t_f^*. Finally, since $x^f \in \partial K_{x^0}(t_f^*)$, u^* must be extremal. This completes the proof.

While theorem 2.5.2 establishes the maximum principle as a *necessary* and *sufficient* condition for an extremal control and theorem 2.5.3 shows that a time-optimal control (when one exists) is extremal, the composition of these theorems provides only a necessary condition for time-optimality. For the regulator problem ($x^f = 0$) of the next section, the maximum principle is shown to be not only a *necessary* but also a *sufficient* condition† for *time-optimal regulation* in the sense that if an adjoint solution $\eta(t) = \exp(-A^T)\eta^0$ and control $u^*(t)$, $0 \le t \le t_f^*$, can be found which satisfy the maximum principle (2.49) and such that $u^*(\cdot)$ generates a trajectory $x(\cdot)$ from $x(0) = x^0$ to $x(t_f^*) = 0$, then $u^*(\cdot)$ is optimal and t_f^* is the minimum time of transition from x^0 to the origin.

† For general discussions on sufficient conditions in optimal control theory see, e.g. Mereau and Powers (1976), Peterson and Zalkind (1978).

However, there may be many such controls u^* which achieve this time-optimal state transition, which raises the question of *uniqueness* which is also considered in the next section (see also Harvey and Lee, 1962; Lee and Markus, 1967; Hermes and LaSalle, 1969), viz. when is a time-optimal control *uniquely* determined by the maximum principle?

2.6 Linear autonomous time-optimal regulating systems

Linear autonomous time-optimal regulator problems (or problems reducible to a regulator formulation) are of major concern in later chapters. In this case, the prescribed target state is given as $x^f = 0$, i.e. the *origin* in $\mathbb{R}^n$. The control problem is to drive the system from a non-zero initial state x^0 to the origin in minimum time and subsequently maintain the system at the zero state. Now, if the zero state is to be maintainable, the system equation must admit the zero solution $x \equiv 0$ for which admissibility of the zero control function $u \equiv 0$ is clearly sufficient. Consequently, for the regulator problem the control restraint set Ω is assumed to contain the origin (in $\mathbb{R}^m$) in its interior, i.e. for the regulator problem, Ω is assumed to be a convex, compact neighbourhood of the origin in $\mathbb{R}^m$. Moreover, in order to present some definitions and results which prove useful in solving many of the specific problems posed in later chapters, it will be assumed that Ω is the *unit cube* in $\mathbb{R}^m$, i.e.

$$\Omega = \{u \in \mathbb{R}^m\colon\ |u_i| \leq 1;\ i = 1, 2, \ldots, m\} \tag{2.64}$$

Consider the attainable set $K_0(-t)$, i.e. the set of states to which the origin can be steered in *backwards* or reverse time $-t$ by means of an admissible control. Equivalently, the set $K_0(-t)$ is the set of all initial states from which it is possible to attain the origin in forwards time t by means of an admissible control. From theorems 2.5.2 and 2.5.3 it follows that, if $u^*(t)$, $0 \leq t \leq t_f^*$, is a time-optimal (and hence extremal) control which transfers x^0 to the origin in minimum time t_f^*, then $x^0 \in \partial K_0(-t_f^*)$, i.e. if t_f^* is the minimum time of transition from x^0 to the origin, then x^0 is a boundary point of $K_0(-t_f^*)$.† However, it is not immediately evident that the point x^0 cannot remain on the boundary of the set $K_0(-t)$ over some finite time interval $[t_f^*, t_f]$. In other words, the possibility that $x^0 \in \partial K_0(-t)$, $t_f^* \leq t \leq t_f$, cannot be immediately discounted, in which case there exists an *extremal* control $\tilde{u}(t)$, $0 \leq t \leq t_f$, transferring x^0 to the origin in a time $t_f > t_f^*$ which exceeds the minimum time t_f^*; this illustrates that the maximum principle of theorem 2.5.2, while being necessary and sufficient for an extremal control, may not be sufficient for time-optimal control. It will now be established that, under the controllability assumption (2.45) (i.e. for

† To see this, note that the control $u^*(t)$, $0 \leq t \leq t_f^*$, transfers x^0 to the origin and reversing the process, the control $\tilde{u}(\tau) = u^*(t_f^* - \tau)$, $0 \leq \tau \leq t_f^*$ will transfer the system from the origin to x^0. But, as u^* is extremal, $\tilde{u}$ is also extremal and hence transfers the system from the origin to the boundary $\partial K_0(-t_f^*)$ of the attainable set $K_0(-t_f^*)$.

proper systems), x^0 cannot remain on the boundary of the attainable set $K_0(-t)$ over a finite interval. In particular, it will be shown that $K_0(-t)$ is an *expanding set* in the sense that, for all $t > 0$, the set $K_0(-s)$ is contained in the *interior* of the set $K_0(-t)$ for all $0 \leq s < t$. It immediately follows that, if $K_0(-t)$ is expanding and $x^0 \in \partial K_0(-t_f^*)$, then t_f^* is the minimum time of transition from x^0 to the origin, as will be seen.

Lemma 2.6.1 : For the linear, autonomous, proper system defined by (2.43), (2.45) and (2.64), the set $K_0(-t)$ is an expanding set.

Proof
It will now be shown that for all t_1, t_2 such that $0 \leq t_1 < t_2$, the set $K_0(-t_1)$ is contained in the interior of the set $K_0(-t_2)$. By definition 2.5.1,

$$K_0(-t_1) = \left\{x: x = \int_0^{-t_1} \exp(-A(t_1 + s))Bu(s)\, ds;\right.$$

$$\left. u(s),\ -t_1 \leq s \leq 0, \text{ admissible}\right\}$$

$$= \left\{x: x = -\int_0^{t_1} \exp(-A\sigma)Bu[\sigma]\, d\sigma;\right.$$

$$\left. u[\sigma] = u(\sigma - t_1),\ 0 \leq \sigma \leq t_1 \text{ admissible}\right\}$$

As the control set Ω contains the origin in its interior, every admissible control $u[\sigma]$ on $[0, t_1]$ can be extended to an admissible control $u^e[\sigma]$ on $[0, t_2]$ defined as follows

$$u^e[\sigma] = \begin{cases} u[\sigma]; & 0 \leq \sigma \leq t_1 \\ 0 \quad ; & t_1 < \sigma \leq t_2 \end{cases}$$

so that

$$-\int_0^{t_1} \exp(-A\sigma)Bu[\sigma]\, d\sigma = x = -\int_0^{t_2} \exp(-A\sigma)Bu^e[\sigma]\, d\sigma \in K_0(-t_2)$$

Hence $x \in K_0(-t_1) \Rightarrow x \in K_0(-t_2)$ or $K_0(-t_1) \subseteq K_0(-t_2)$.

Consider now an arbitrary admissible control $u^1[\sigma]$ on $[0, t_1]$ generating a point

$$x^1 \in K_0(-t_1) \Rightarrow x^1 \in K_0(-t_2)$$

To establish that $K_0(-t_1)$ is contained in the *interior* of $K_0(-t_2)$, a proof by contradiction is initiated by the supposition that x^1 is a boundary point of $K_0(-t_2)$. Under this supposition, there exists a supporting hyperplane Π to the convex, compact set $K_0(-t_2)$ at the point x^1. Let $\eta^1 \neq 0$ denote an *exterior* normal to Π at x^1 so that

$$\langle \eta^1, \bar{x} - x^1 \rangle \leq 0 \ \forall \ \bar{x} \in K_0(-t_2) \tag{2.65}$$

Now define the admissible control $u^2[\sigma] = (u_1^2[\sigma], u_2^2[\sigma], \ldots, u_m^2[\sigma])'$ on $[0, t_2]$ as follows

$$u_j^2[\sigma] = \begin{cases} u_j^1[\sigma]; & 0 \leq \sigma \leq t_1 \\ -\operatorname{sgn} \langle \eta^1, \exp(-A\sigma)b^j \rangle; & t_1 < \sigma \leq t_2 \end{cases}$$

$$j = 1, 2, \ldots, m$$

where b^j is the jth column vector of the $n \times m$ matrix B. Denote, by $x^2 \in K_0(-t_2)$, the trajectory endpoint corresponding to the admissible control $u^2[\sigma], 0 \leq \sigma \leq t_2$. Then

$$\begin{aligned} \langle \eta^1, x^2 - x^1 \rangle &= \left\langle \eta^1, -\int_0^{t_2} \exp(-A\sigma)Bu^2[\sigma]\, d\sigma \right. \\ &\qquad \left. + \int_0^{t_1} \exp(-A\sigma)u^1[\sigma]\, d\sigma \right\rangle \\ &= \left\langle \eta^1, -\int_{t_1}^{t_2} \exp(-A\sigma)Bu^2[\sigma]\, d\sigma \right\rangle \\ &= \int_{t_1}^{t_2} -\langle \eta^1, \exp(-A\sigma)Bu^2[\sigma] \rangle\, d\sigma \end{aligned}$$

or

$$\langle \eta^1, x^2 - x^1 \rangle = \int_{t_1}^{t_2} \sum_{j=1}^{m} |\langle \eta^1, \exp(-A\sigma)b^j \rangle|\, d\sigma \geq 0 \tag{2.66}$$

Since $x^2 \in K_0(-t_2)$, combining (2.65) and (2.66) yields the result

$$\int_{t_1}^{t_2} \sum_{j=1}^{m} |\langle \eta^1, \exp(-A\sigma)b^j \rangle|\, d\sigma = 0$$

which, in turn, implies that

$$\langle \exp(-A^T\sigma)\eta^1, b^j \rangle = \langle \eta^1, \exp(-A\sigma)b^j \rangle = 0;$$

$$j = 1, 2, \ldots, m; \qquad t_1 \leq \sigma \leq t_2$$

Hence, all derivatives of $\langle \eta^1, \exp(-A\sigma)b^j \rangle$ must also vanish on the interior of the interval $[t_1, t_2]$ and, in particular

$$\langle \exp(-A^T\sigma)\eta^1, A^k b^j \rangle = 0 \begin{cases} k = 0, 1, 2, \ldots, n-1 \\ j = 1, 2, \ldots, m \end{cases}; \qquad t_1 < \sigma < t_2$$

or equivalently,

$$[B : AB : \cdots : A^{n-1}B]^T \exp(-A^T\sigma)\eta^1 = 0, \qquad t_1 < \sigma < t_2 \tag{2.67}$$

But, $0 \neq \eta^1 \in \mathbb{R}^n$ and hence $\exp(-A^T\sigma)\eta^1 \neq 0$ so that, if (2.67) is to hold, then rank $[B : AB : \cdots : A^{n-1}B] < n$ which contradicts the controllability assumption (2.45). Hence, the supposition that x^1 is a boundary point of $K_0(-t_2)$ is false so that x^1 is an *interior* point of $K_0(-t_2)$. Consequently, $K_0(-t)$ is an expanding set.

The following theorem may now be deduced.

Theorem 2.6.1 Proper systems: Sufficient condition for time-optimality: If there exist adjoint and admissible control functions $\eta(t) = \exp(-A^Tt)\eta^0$ and $u^*(t)$, $0 \leq t \leq t_f^*$, which satisfy the maximum principle (2.49) and such that $u^*(\cdot)$ generates a trajectory $x(\cdot)$ from $x(0) = x^0$ to $x(t_f^*) = 0$, then $u^*(\cdot)$ is optimal and t_f^* is the minimum time of transition from x^0 to the origin for the linear, autonomous, proper system (2.43), (2.45) and (2.64).

Proof
Clearly, by theorem 2.5.2, $u^*(t)$, $0 \leq t \leq t_f^*$ is extremal and $x^0 \in \partial K_0(-t_f^*)$. Suppose u^* is not time-optimal, then by theorem 2.5.3 there exists an optimal control $\tilde{u}(t)$, $0 \leq t \leq \tilde{t}_f$ which transfers x^0 to the origin in minimum time $\tilde{t}_f < t_f^*$. But $\tilde{u}$ is an extremal control (by theorem 2.5.3) so that $x^0 \in \partial K_0(-\tilde{t}_f)$ for $\tilde{t}_f < t_f^*$ which contradicts the fact that $K_0(-t)$ is an expanding set (by lemma 2.6.1). Hence the original supposition that u^* is not optimal is false, and t_f^* is the minimum time of transition from x^0 to the origin. ■

The question of uniqueness of optimal control will now be considered. To this end, a further restriction on the system is introduced. Specifically, it is assumed that

$$\text{Rank } [b^j : Ab^j : A^2b^j : \cdots : A^{n-1}b^j] = n \tag{2.68}$$

for all *column vectors* b^j, $j = 1, 2, \ldots, m$ of the $n \times m$ matrix B. The latter condition is usually referred to as the *normality condition* or *condition for generality of position* which implies that each component of control is always effective and which ensures uniqueness of extremal (and hence optimal) controls as shown below. It is remarked that, in the case of a scalar input ($m = 1$), the normality condition (2.68) is simply the controllability condition (2.45); however, in the case of multi-input systems ($m > 1$), controllability is a weaker condition than normality (i.e. (2.68) $\Rightarrow$ (2.45) or all normal systems are proper but not vice versa, unless $m = 1$).

Theorem 2.6.2: A time-optimal regulating control $u^*(t)$, $0 \leq t \leq t_f^*$, (when it exists) for the normal, linear, autonomous system (2.43), (2.64), (2.68) is piecewise constant with $|u_j(t)| = 1$, $j = 1, 2, \ldots, m$, and is *uniquely* determined by the maximum principle.

Proof
Suppose $u^*(t)$, $0 \leq t \leq t_f^*$, transfers the system from $x(0) = x^0$ to the origin, then, from theorems 2.5.2, 2.5.3 and 2.6.1, u^* is time-optimal if and only if there exists

an adjoint solution $\eta(t) = \exp(-A^T t)\eta^0, 0 \leq t \leq t_f^*, (\eta^0 \neq 0)$ such that

$$\begin{aligned}\langle \eta(t), Bu^*(t) \rangle &= \max_{u \in \Omega} \langle \eta(t), Bu \rangle \\ &= \max_{u \in \Omega} \langle \exp(-A^T t)\eta^0, Bu \rangle \\ &= \sum_{j=1}^{m} \max_{|u_j| \leq 1} \langle \eta^0, \exp(-At)b^j \rangle u_j \\ &\quad \text{for a.a. } t \in [0, t_f^*] \end{aligned} \tag{2.69}$$

where b^j denotes the jth column vector of the matrix B and u_j denotes the jth component of the control vector u. Now each $\langle \eta^0, \exp(-At)b^j \rangle$ is a real analytic function and hence (a) is identically zero, or (b) has a finite number of zeros on the interval $[0, t_f^*]$. Consequently, the right-hand side of (2.69) is uniquely maximised (almost everywhere) by the controls

$$u_j^*(t) = \operatorname{sgn} \langle \eta(t), b^j \rangle = \operatorname{sgn} \langle \eta^0, \exp(-At)b^j \rangle; j = 1, 2, \ldots, m \tag{2.70}$$

provided that $\langle \eta^0, \exp(-At)b^j \rangle$ does not vanish identically. It will now be shown that the latter cannot occur as otherwise the normality condition (2.68) is contradicted. Specifically, suppose

$$(S): \qquad \langle \eta^0, \exp(-At)b^j \rangle \equiv 0 \tag{2.71}$$

then all its derivatives must also vanish, i.e.

$$\langle \eta^0, A^k \exp(-At)b^j \rangle \equiv 0; \qquad k = 1, 2, \ldots$$

and, on setting $t = 0$,

$$\langle \eta^0, b^j \rangle = \langle \eta^0, Ab^j \rangle = \cdots = \langle \eta^0, A^{n-1}b^j \rangle = 0$$

or (2.72)

$$[b^j : Ab^j : \cdots : A^{n-1}b^j]^T \eta^0 = 0; \qquad \eta^0 \neq 0$$

For (2.72) to hold for $\eta^0 \neq 0$, the vectors $b^j, Ab^j, \ldots, A^{n-1}b^j$ must be linearly dependent (i.e. rank $[b^j : Ab^j : \cdots : A^{n-1}b^j] < n$) which contradicts the normality condition (2.68). Hence, supposition (2.71) is false and the optimal control is uniquely (almost everywhere) determined by (2.70).

Example 2.6.1 : For the linear autonomous system (2.43) with

$$A = \begin{bmatrix} 0 & 1 \\ 0 & 0 \end{bmatrix}; \qquad B = \begin{bmatrix} 0 \\ 1 \end{bmatrix}; \qquad \Omega = [-1, 1]$$

i.e. the double integrator system with scalar saturable control:

$$\dot{x}_1 = x_2; \qquad \dot{x}_2 = u; \qquad |u| \leq 1 \tag{2.73}$$

consider the minimum-time regulator problem of transferring (2.73) from the initial state $x(0) = (x_1(0), x_2(0))' = (x_1^0, x_2^0)' = x^0$ to the origin $x(t_f^*) = 0$ in minimum time t_f^*. This system clearly satisfies the normality condition and hence the optimal control $u^*(t)$, $0 \le t \le t_f^*$ (if it exists) is uniquely determined by (2.70), i.e.

$$u^*(t) = \operatorname{sgn} \langle \eta(t), B \rangle = \operatorname{sgn} (\eta_2(t)) \tag{2.74}$$

where $(\eta_1(t), \eta_2(t))' = \eta(t) = \exp(-A^T t)\eta^0$ is a nontrivial solution of the adjoint equation

$$\dot{\eta} = -A^T \eta \Leftrightarrow \dot{\eta}_1 = 0; \quad \dot{\eta}_2 = -\eta_1$$

Specifically

$$\eta_1(t) = \eta_1^0; \qquad \eta_2(t) = \eta_2^0 - \eta_1^0 t$$

whence

$$u^*(t) = \operatorname{sgn} (\eta_2^0 - \eta_1^0 t) \tag{2.75}$$

which clearly implies that $|u^*(t)| = 1$ a.e. on $[0, t_f^*]$ with at most *one* discontinuity (corresponding to a zero of the function $\eta_2(t)$) on $(0, t_f^*)$. On integrating equations (2.73), it is found that on an interval on which $u^*(t) = 1$:

$$x_1(t) = \tfrac{1}{2}(x_2(t))^2 + c_1 \tag{2.76}$$

where c_1 is a constant of integration. Similarly, on an interval on which $u(t) = -1$:

$$x_1(t) = -\tfrac{1}{2}(x_2(t))^2 + c_2 \tag{2.77}$$

where c_2 is a constant of integration. Thus every optimal trajectory consists of an appropriate combination of the parabolic arcs (2.76) and (2.77). The family of (i) the parabolae (2.76), the positive control ($u \equiv +1$) paths or p-paths, and (ii) the parabolae (2.77), the negative control ($u \equiv -1$) paths or n-paths, are shown in Fig. 2.3, where the arrows denote the direction of motion for increasing t. It

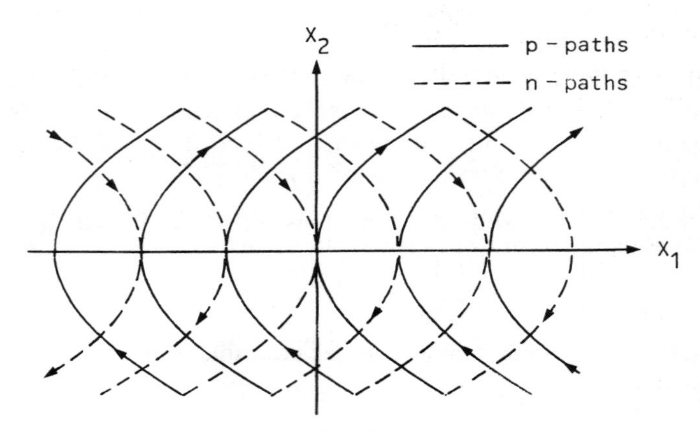

Fig. 2.3

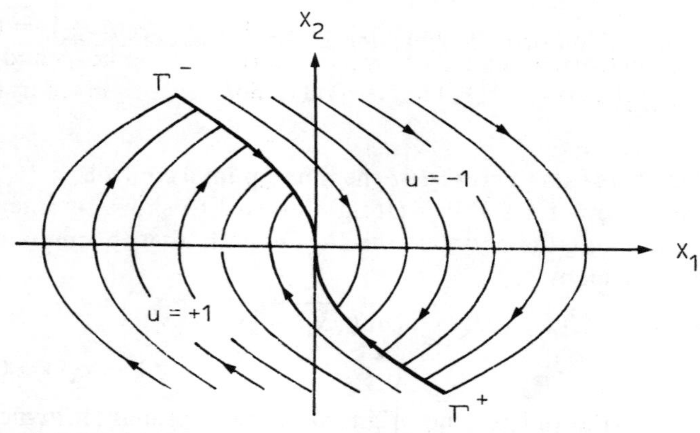

Fig. 2.4 *Example 2.6.1: time-optimal state portrait of double integrator system*

can easily be verified that the only (optimal) path to the origin from a given non-zero starting point x^0 and generated by a control satisfying $|u^*| = 1$ almost everywhere with at most one control switch ($+1 \to -1$ or $-1 \to +1$) is that member of the family of trajectories of Fig. 2.4 which passes through the given point x^0. The family of optimal trajectories of Fig. 2.4 constitutes the *optimal state portrait* or *optimal vector field*; moreover, this family fills the entire state space, implying that a (unique) minimum-time trajectory to the origin exists for *all* starting points $x^0 \in \mathbb{R}^2$. This latter property of time-optimal controllability to the origin is treated more generally in Chapter 4. The curve Γ of Fig. 2.4, consisting of the origin and the parabolic semi-arcs Γ^+ and Γ^- (respectively, the p-path and n-path leading to the origin) has the characterisation

$$\Gamma = \{x\colon x_1 + \tfrac{1}{2}x_2|x_2| = 0\} \tag{2.78}$$

On an optimal trajectory (emanating from $x^0 \notin \Gamma$) the control must switch on reaching Γ, the control switch being a $+1$ to -1 (pn-switch) or -1 to $+1$ (np-switch) depending on whether x^0 lies in the region Σ^+ below Γ or in the region Σ^- above Γ, where

$$\Sigma^+ = \{x\colon x_1 + \tfrac{1}{2}x_2|x_2| < 0\}; \qquad \Sigma^- = \{x\colon x_1 + \tfrac{1}{2}x_2|x_2| > 0\} \tag{2.79}$$

Γ is usually referred to as the *switching curve, switching boundary* or *switching locus*. The time-optimal control can be written in feedback form as

$$u^* = u^*(x) = \begin{cases} +1; & x \in \Sigma^+ \cup \Gamma^+ \\ -1; & x \in \Sigma^- \cup \Gamma^- \end{cases} \tag{2.80}$$

or, equivalently,

$$u^* = u^*(x) = -\operatorname{sgn}\,[\Xi(x)] \tag{2.81a}$$

where the *time-optimal switching function* $\Xi\colon \mathbb{R}^2 \to \mathbb{R}$ is defined as

$$\Xi(x) = \begin{cases} \xi(x) = x_1 + \tfrac{1}{2}x_2|x_2|; & \xi(x) \neq 0 \\ x_2; & \xi(x) = 0 \end{cases} \tag{2.81b}$$

(2.80) and (2.81) are said to *synthesise* the time-optimal control.

For a starting point $x \notin \Gamma$ it is straightforward to show, by integrating the state equations along the optimal (one-switch) path, that the minimum time t_f^* to the origin is given by

$$t_f^* = x_2 \operatorname{sgn}(\xi(x)) + 2\sqrt{x_1 \operatorname{sgn}(\xi(x)) + \tfrac{1}{2}x_2^2};$$
$$x \notin \Gamma \Leftrightarrow \xi(x) \neq 0 \tag{2.82}$$

while, for $x \in \Gamma$, the time to the origin along the optimal (switchless) path is given by

$$t_f^* = |x_2|; \qquad x \in \Gamma \tag{2.83}$$

Combining (2.81b), (2.82) and (2.83) gives the value function (i.e. minimum cost (time t_f^*) from x) in the form

$$V(x) = x_2 \operatorname{sgn}(\Xi(x)) + 2\sqrt{x_1 \operatorname{sgn}(\Xi(x)) + \tfrac{1}{2}x_2^2} = t_f^* \tag{2.84}$$

Fig. 2.5 depicts some contours of constant cost or optimal *isochrones*† in the state plane which exhibit 'corners' on the switching curve Γ. The value function (2.84) is clearly not differentiable on Γ (but is differentiable elsewhere). This illustrates the restricted nature of the maximum principle derived from the Bellman–Hamilton–Jacobi equation in Section 2.4 which presupposed that the value function V is twice continuously differentiable and hence is not valid for the simple problem at hand. However, the more general statement of the maximum principle (theorem 2.4.1) and its proof for the time-optimal problem of Section 2.5 does not contain the value function V or its derivatives in the formulation and can easily be applied to determine the (unique) optimal solution as shown above. This example is considered in more detail, together with other second-order examples, in Chapter 5.

Example 2.6.2: Consider the equations of motion (2.13) of a controlled body in inertial space, derived in Section 2.3, i.e.

$$\dot{x}_1 = (I_3^{-1} - I_2^{-1})x_2x_3 + u_1$$
$$\dot{x}_2 = (I_1^{-1} - I_3^{-1})x_1x_3 + u_2$$
$$\dot{x}_3 = (I_2^{-1} - I_1^{-1})x_1x_2 + u_3$$

Suppose the body is rotationally symmetric about axis 3, so that $I_1 = I_2 = I$, and consequently $\dot{x}_3 = u_3$. In addition, suppose the craft is spin-stabilised about this axis of symmetry, i.e. rotating with a fixed angular velocity so that $x_3 = c$ (a

† Each point of an isochrone can be steered to the origin in the *same* minimum time, hence the terminology. Isochrones are discussed more generally in Section 4.2.

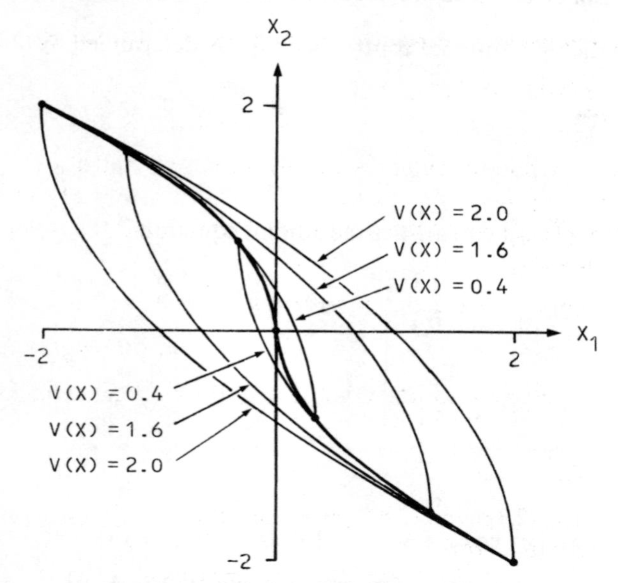

Fig. 2.5 *Example 2.6.1: optimal isochrones*

constant) and $u_3 = 0$, then the linear autonomous equations of motion governing the rotations about axes 1 and 2 are

$$\begin{aligned} \dot{x}_1 &= c(I_3^{-1} - I^{-1})x_2 + u_1 \\ \dot{x}_2 &= c(I^{-1} - I_3^{-1})x_1 + u_2 \end{aligned} \tag{2.85}$$

For simplicity suppose the equations have been normalised so that

$$c(I_3^{-1} - I^{-1}) = 1$$

then the system equations become

$$\dot{x} = \begin{bmatrix} 0 & 1 \\ -1 & 0 \end{bmatrix} x + \begin{bmatrix} 1 & 0 \\ 0 & 1 \end{bmatrix} u \tag{2.86}$$

Now suppose that the controlling torques u_1 and u_2 are subject to the (normalised) saturation constraints

$$|u_j| \leq 1; \qquad j = 1, 2$$

The control problem is to reduce the rotations about axes 1 and 2 from an initial value $x(0) = x^0$ to zero $x(t_f^*) = 0$ in minimum time t_f^*. It is straightforward to verify that this problem is normal, viz.

$$\text{rank } [b^1 : Ab^1] = \text{rank} \begin{bmatrix} 1 & 0 \\ 0 & -1 \end{bmatrix} = 2$$

$$\text{rank } [b^2 : Ab^2] = \text{rank} \begin{bmatrix} 0 & 1 \\ 1 & 0 \end{bmatrix} = 2$$

consequently the time-optimal control is uniquely determined by (2.70) for some $\eta^0 \neq 0$, i.e.

$$\begin{aligned} u_j^*(t) &= \operatorname{sgn} \langle \eta(t), b^j \rangle \\ &= \operatorname{sgn} (\eta_j(t)) \qquad j = 1, 2 \end{aligned}$$

where $\eta(\cdot) = (\eta_1(\cdot), \eta_2(\cdot))'$ satisfies the adjoint equation

$$\dot{\eta} = -A^T\eta = \begin{bmatrix} 0 & 1 \\ -1 & 0 \end{bmatrix}\eta; \qquad \eta(0) = \eta^0$$

In particular

$$\eta(t) = \begin{bmatrix} \cos t & \sin t \\ -\sin t & \cos t \end{bmatrix}\eta^0 \tag{2.87}$$

Hence

$$\begin{aligned} u_1^*(t) &= \operatorname{sgn} [\eta_1^0 \cos t + \eta_2^0 \sin t] = \operatorname{sgn} [\sin (t + \delta)] \\ u_2^*(t) &= \operatorname{sgn} [-\eta_1^0 \sin t + \eta_2^0 \cos t] = \operatorname{sgn} [\cos (t + \delta)] \end{aligned} \tag{2.88}$$

for some choice of initial adjoint vector $\eta^0 = (\eta_1^0, \eta_2^0)'$ with $\delta = \tan^{-1} [\eta_1^0/\eta_2^0]$. From (2.88), the following deductions can be made. Viewed individually, each control component is bang-bang with $|u_i| = 1$ almost everywhere and with switches (changes in sign) separated π units in time; while, viewed collectively, the control switches are separated $\pi/2$ units in time (i.e. a u_1 switch at time τ, followed by u_2 switch at time $\tau + \pi/2$, followed by u_1 switch at time $\tau + \pi$, etc.). Hence, on an optimal trajectory, the control vector is constant on an interval of length $\pi/2$ at most, while an individual component can remain constant on an interval of length π at most. Equipped with this information, the family of time-optimal trajectories (the optimal state portrait) can be constructed in state space (the phase plane) as follows.

Denote, by the prefixes pp, pn, np, nn respectively, quantities related to the four possible values $(+1, +1)'$, $(+1, -1)'$, $(-1, +1)'$, $(-1, -1)'$ that the control vector $u^*(t)$ takes almost everywhere. Referring to Fig. 2.6, the pp, pn, np, and nn-paths in the state plane constitute four families of concentric circles centred on points with co-ordinates $(+1, -1)$, $(-1, -1)$, $(+1, +1)$, $(-1, +1)$, respectively; these paths are traversed in a clockwise sense with increasing time. On any such path, the time taken on one complete cycle is 2π units, and the time taken to traverse an arc of any circular path corresponds to the angle subtended by that arc at the corresponding centre. In particular, the state traverses a *quadrant* of a circle in a time $\pi/2$. Now, from (2.88), one or other of the control components u_i must change sign when

$$t + \delta = \frac{k\pi}{2} \qquad (k \text{ an integer})$$

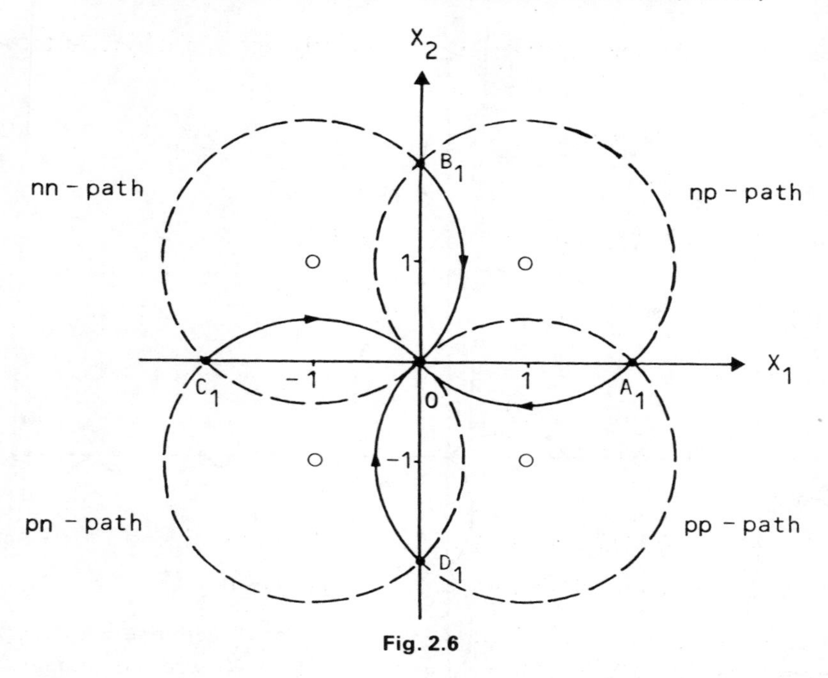

Fig. 2.6

and any sequence of such switches has the ordering

$$\rightarrow \begin{bmatrix} 1 \\ 1 \end{bmatrix} \rightarrow \begin{bmatrix} 1 \\ -1 \end{bmatrix} \rightarrow \begin{bmatrix} -1 \\ -1 \end{bmatrix} \rightarrow \begin{bmatrix} -1 \\ 1 \end{bmatrix} \quad \text{or} \quad \text{pn} \;{}^{\swarrow\,\text{pp}\,\nwarrow}_{\searrow\,\text{nn}\,\nearrow}\; \text{np}$$

each switch being separated $\pi/2$ units in time.

Since the origin is to be reached in finite time, every optimal trajectory must terminate with an arc of the pp, pn, np, or nn-path which leads to the origin; moreover, the duration of such an arc cannot exceed $\pi/2$ time units. Hence, every optimal trajectory must terminate with one of the quarter arcs D_1O, C_1O, A_1O, B_1O, or some subarc thereof, leading to the origin, as shown in Fig. 2.7. Suppose, for definiteness, that a particular optimal trajectory switches at point P_1 onto the np-path (arc A_1O) to the origin. Prior to this switch the trajectory must have followed an nn-path of duration $\pi/2$, so that the optimal switch prior to P_1 must have occurred at the point P_2, determined such that the angle subtended at the nn-centre $(-1, +1)$ by the nn-arc P_2P_1 is $\pi/2$. Repeating this construction for all possible switch points P_1 on arc A_1O, it is easily seen that every optimal switch point on the quarter circular arc A_1O maps uniquely to an earlier optimal switch point on the quarter circular arc B_1B_2 of Fig. 2.7, i.e. B_1B_2 is obtained by a $+\pi/2$ rotation of OA_1 about the nn-centre $(-1, 1)$. If the above optimal trajectory (P_2P_1O) is traced further back then, prior to the optimal switch at P_2, a pn-path of duration $\pi/2$ was followed, so that the

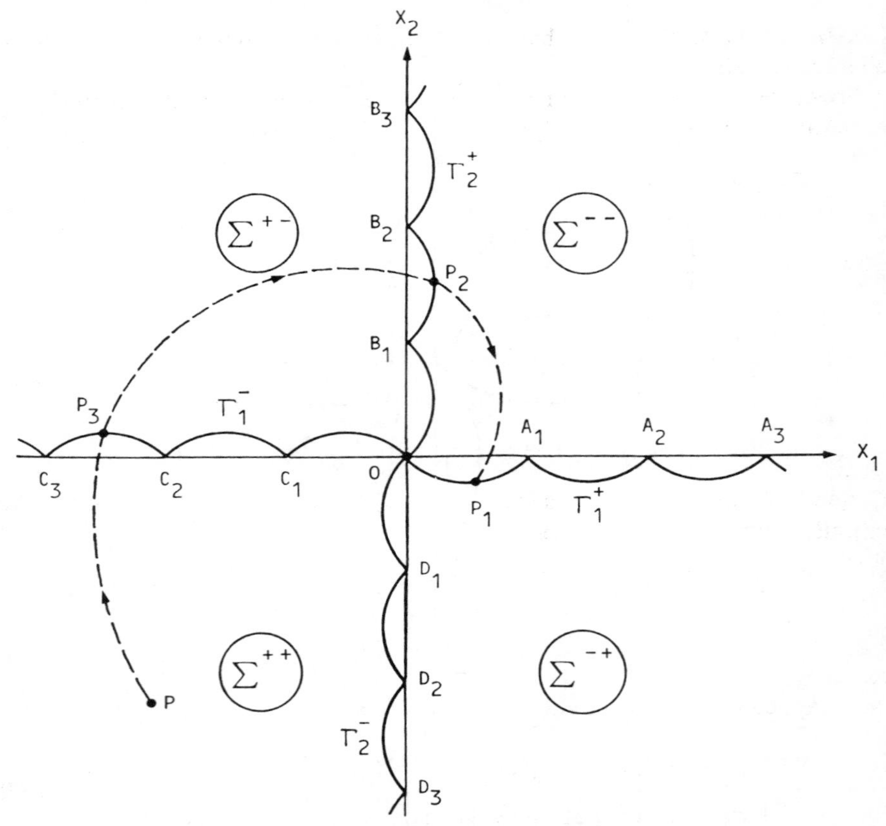

Fig. 2.7 *Example 2.6.2: time-optimal switching curves* Γ_1, Γ_2

optimal switch prior to P_2 is located at P_3 on the quarter circular arc C_2C_3 of Fig. 2.7 which is obtained by a $+\pi/2$ rotation of B_1B_2 about the pn-centre $(-1, -1)$. Continuing this construction (and repeating the procedure for optimal trajectories which terminate with one of the quarter circular arcs OB_1, OC_1, OD_1 or some subarc thereof), the set of optimal switch points is seen to consist of the arcs OA_1, OB_1, OC_1, OD_1 together with the sequences $\{A_i A_{i+1}\}$, $\{B_i B_{i+1}\}$, $\{C_i C_{i+1}\}$, $\{D_i D_{i+1}\}$ of arcs obtained from these by $+\pi/2$ rotations about the appropriate centres. The 'end-to-end' composition of these quarter circular arcs along the x_1 and x_2 state axes form the time-optimal switching curves Γ_1 and Γ_2, respectively, consisting of the origin, $\Gamma_1^+ = \bigcup_i A_i A_{i+1}$ extending along the positive x_1 semi-axis, $\Gamma_1^- = \bigcup_i C_i C_{i+1}$ extending along the negative x_1 semi-axis, $\Gamma_2^+ = \bigcup_i B_i B_{i+1}$ extending along the positive x_2 semi-axis, and $\Gamma_2^- = \bigcup_i D_i D_{i+1}$ extending along the negative x_2 semi-axis. For example, again referring to Fig. 2.7, for an initial state $x(0) = x^P$ at the point P, $PP_3P_2P_1O$ is the minimum-time path to the state origin. As in Example 2.6.1, it may be seen that the family of optimal trajectories (the optimal state portrait) fills the state space, implying that a minimum-time path to the origin exists from all starting points $x \in \mathbb{R}^2$.

From the geometry of their construction, it may be deduced that the time-optimal switching curves Γ_1, Γ_2 have the characterisations

$$\begin{aligned} \Gamma_1 &= \{x\colon \xi_1(x) = 0\} \\ \Gamma_2 &= \{x\colon \xi_2(x) = 0\} \end{aligned} \tag{2.89a}$$

where

$$\begin{aligned} &\xi_1(x) = x_2 + \{[2 - (|x_1| - 2n - 1)^2]^{1/2} - 1\} \operatorname{sgn}(x_1); \\ &\qquad 2n < |x_1| \leq 2n + 2; \qquad n = 0, 1, 2, \ldots \\ &\xi_2(x) = x_1 - \{[2 - (|x_2| - 2n - 1)^2]^{1/2} - 1\} \operatorname{sgn}(x_2); \\ &\qquad 2n < |x_2| \leq 2n + 2; \qquad n = 0, 1, 2, \ldots \end{aligned} \tag{2.89b}$$

Γ_1 and Γ_2 form the boundaries of the regions $\Sigma^{++}, \Sigma^{+-}, \Sigma^{-+}, \Sigma^{--}$ wherein the optimal control takes the values

$$\begin{bmatrix} +1 \\ +1 \end{bmatrix}, \quad \begin{bmatrix} +1 \\ -1 \end{bmatrix}, \quad \begin{bmatrix} -1 \\ +1 \end{bmatrix}, \quad \begin{bmatrix} -1 \\ -1 \end{bmatrix},$$

respectively with

$$\left.\begin{aligned} \Sigma^{++} &= \{x\colon \xi_1(x) < 0;\ \xi_2(x) < 0\} \\ \Sigma^{+-} &= \{x\colon \xi_1(x) > 0;\ \xi_2(x) < 0\} \\ \Sigma^{-+} &= \{x\colon \xi_1(x) < 0;\ \xi_2(x) > 0\} \\ \Sigma^{--} &= \{x\colon \xi_1(x) > 0;\ \xi_2(x) > 0\} \end{aligned}\right\} \tag{2.90}$$

Consequently, the optimal feedback control synthesis may be expressed as

$$u^*(x) = \begin{bmatrix} u_1^*(x) \\ u_2^*(x) \end{bmatrix} = \begin{cases} \begin{bmatrix} +1 \\ +1 \end{bmatrix}; & x \in \Sigma^{++} \cup \Gamma_2^- \\ \begin{bmatrix} +1 \\ -1 \end{bmatrix}; & x \in \Sigma^{+-} \cup \Gamma_1^- \\ \begin{bmatrix} -1 \\ +1 \end{bmatrix}; & x \in \Sigma^{-+} \cup \Gamma_1^+ \\ \begin{bmatrix} -1 \\ -1 \end{bmatrix}; & x \in \Sigma^{--} \cup \Gamma_2^+ \end{cases} \tag{2.91}$$

or, equivalently,

$$u_1^*(x) = -\operatorname{sgn}(\Xi_2(x)); \qquad u_2^*(x) = -\operatorname{sgn}(\Xi_1(x)) \tag{2.92a}$$

where the time-optimal *switching functions* Ξ_i: $\mathbb{R}^2 \to \mathbb{R}$ are defined as follows:

$$\Xi_1(x) = \begin{cases} \xi_1(x); & \xi_1(x) \neq 0 \\ -x_1; & \xi_1(x) = 0 \end{cases} \tag{2.92b}$$

$$\Xi_2(x) = \begin{cases} \xi_2(x); & \xi_2(x) \neq 0 \\ x_2; & \xi_2(x) = 0 \end{cases} \tag{2.92c}$$

2.7 Discussion

The maximum principle, as stated in theorem 2.4.1, sets up a framework for solving the optimal control problems posed in later chapters. It provides a *necessary* condition for optimality and facilitates identification of the candidate solutions which *may* be optimal; however, the question as to whether one (or more) of these candidate solutions is, in fact, optimal may not be immediately answerable. In the case of the linear, time-optimal regulator problem of Section 2.6, the latter question was resolved and the maximum principle was established as a *necessary and sufficient* condition for optimality using rather specialized arguments (theorems 2.5.1, 2.5.2, 2.5.3 and 2.6.1); moreover, for the *normal* regulator problem, the optimal solution was shown to be *unique* (theorem 2.6.2).

In some other specific cases, it can be shown (as in theorem 2.6.2) that the maximum principle *uniquely* determines a candidate solution which may be optimal; if, in addition, the existence† of an optimal control can be established in such cases then the unique candidate solution must be optimal. Attempts to extend this approach to more general problems gives rise to major difficulties in establishing the requisite existence and uniqueness results.

Boltyanskii's concept of a *regular synthesis* (Boltyanskii, 1966) to some extent avoids the difficulties alluded to above and provides the result that, if an admissible control synthesis derived on the basis of the maximum principle is regular, then it is optimal; in other words, the maximum principle is a necessary and sufficient condition for optimality if it admits a regular synthesis. The technical conditions which define a regular synthesis (essentially a finite cellular decomposition of the domain of controllability such that a solution of the feedback system can be defined on each cell of the decomposition and extended from cell to cell until the target set G is attained in finite time) are discussed in detail by Boltyanskii (1966, 1971), Mirica (1969, 1976), Fleming and Rishel (1975). Suffice it to remark here that all time-optimal and time-fuel-optimal syntheses derived in Chapters 5, 6, 7, 8 are regular, as are those of Examples 2.6.1 and 2.6.2.

† For example, Filippov (1962) (see also Hermes and LaSalle, 1969) has established the following existence result for the *nonlinear* time-optimal control problem. Denoting, by $K(t, x)$, the set generated by $f(t, x, u)$ as u ranges over the compact control restraint set Ω, (i) if there exists at least one admissible control which drives the nonlinear system (2.1) from x^0 to x^f in finite time, (ii) if the function f satisfies (2.3), and (iii) if the set $K(t, x)$ is convex for every t and x, then there exists a time-optimal control which drives (2.1) from x^0 to x^f in minimum time. See e.g. Markus and Lee (1962), Lee and Markus (1967) for other existence results.

Chapter 3

Structure and properties of optimal saturating control systems

3.1 Discontinuous feedback control: switching surfaces

Examples 2.6.1 and 2.6.2 of the previous chapter illustrated that, in many cases, time-optimal control can be synthesised by a discontinuous feedback† control $u^*(x) = -\text{sgn}(\Xi(x))$ defined via a characterisation of surfaces of discontinuity, or switching surfaces, $\Gamma_i = \{x\colon \xi_i(x) = 0\}$ which partition the state space into regions of specified control values. Feedback synthesis via switching surfaces arises, not only in the context of time-optimal control, but also in many other saturating control problems such as quadratic-cost and minimum-fuel problems, as will be seen in Chapters 8 and 9. Moreover, many suboptimal approaches to feedback control are based *a priori* on a switching surface structure to be parametrically optimised via some set of design parameters.

Discontinuous feedback systems give rise to many problems pertaining to the interpretation of solutions of the state equation. The major difficulty is that a control defined via discontinuous feedback converts the original inhomogeneous state equation into a homogeneous differential equation with discontinuous right-hand side for which neither existence nor uniqueness of solutions follows from the standard theory of differential equations. For such discontinuous differential equations, it is widely recognised that an appropriate concept of solution (which admits an intuitive and practical interpretation) is that of Filippov (1960a, b) who extended the work of André and Seibert (1956, 1960).

3.2 Filippov solutions: regular switching and sliding motion

By way of introduction, the relay controlled double integrator plant of Fig. 3.1, under linear feedback, will be studied.

† For results relating to general properties of time-optimal feedback systems see, e.g., Brunovský (1974, 1976), Hájek (1973), Meeker (1980), Moroz (1970) and Yeung (1977).

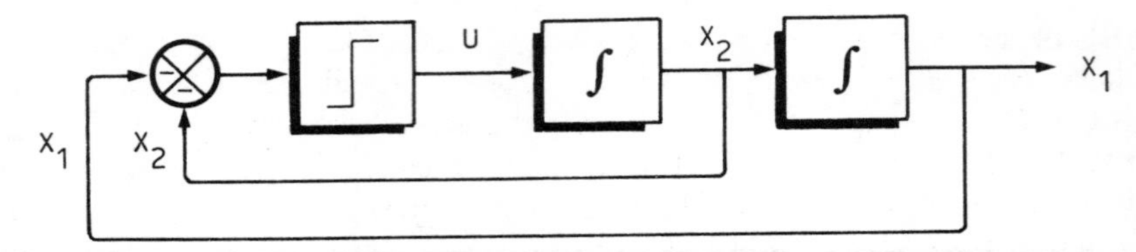

Fig. 3.1 *Relay-controlled double integrator plant with linear feedback*

Example 3.2.1 : The state equations (with discontinuous right-hand side) are

$$\dot{x}_1 = x_2; \qquad \dot{x}_2 = u(x) = -\text{sgn}\,(x_1 + x_2) \tag{3.1}$$

so that $u = +1$ for $x \in \Sigma^+ = \{x: x_1 + x_2 < 0\}$ and $u = -1$ for $x \in \Sigma^- = \{x: x_1 + x_2 > 0\}$ with $\Gamma = \{x: x_1 + x_2 = 0\}$ defining a switching line in the phase plane, as shown in Fig. 3.2. The question arises as to how solutions of (3.1) are to be defined on Γ. For example, referring to Fig. 3.2, consider a trajectory emanating from the point A. Initially, the parabolic path AB is followed, corresponding to an arc of the solution curve (n-path) through A of the linear equations

$$\dot{x}_1 = x_2; \qquad \dot{x}_2 = -1 \tag{3.2}$$

From B, the parabolic arc BC is followed, corresponding to an arc of the solution curve (p-path) through B of the linear equations

$$\dot{x}_1 = x_2; \qquad \dot{x}_2 = +1 \tag{3.3}$$

Hence, arc ABC of the overall solution curve for the original system (3.1), with discontinuous right-hand side, is defined simply as the composition of arcs AB

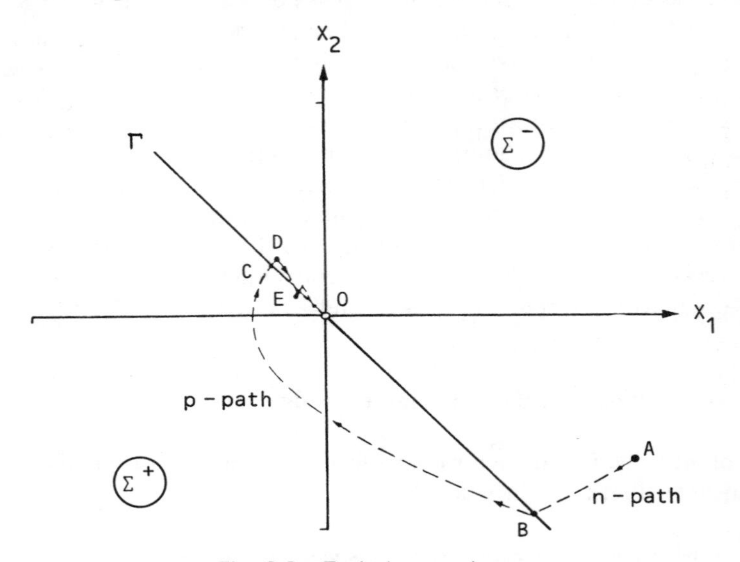

Fig. 3.2 *Typical state trajectory*

and BC of the solution curves for the linear systems (3.2) and (3.3), and no real difficulty arises at the point $B \in \Gamma$. On the other hand, the behaviour at the point $C \in \Gamma$ poses questions as to the interpretation of the subsequent solution of (3.1). The difficulty arises from the fact that, in contrast to the behaviour at B, the state flow at $C \in \Gamma$ is directed towards the switching line Γ from both sides; any attempt to continue the solution into the region Σ^- fails as an n-path into Σ^- from C does not exist. In practice, owing to unavoidable component inertia, small delays are always present in switching operations so that the switch point C will not lie precisely on Γ but instead the state will penetrate a finite distance into the region Σ^-, on a p-path of (3.3), before the switch is effected at D. On switching, the system will then recross on an n-path of (3.2) and repenetrate a finite distance into the region Σ^+ before the next switch is effected at E; this behaviour is repeated, crossing and recrossing Γ, giving rise to a 'chattering' control trajectory as shown in Fig. 3.2. The concept of a Filippov solution may be regarded as an idealisation of this chattering behaviour for which the delay in switching is infinitesimal, in which case from $C \in \Gamma$ the state point *slides* along Γ towards the origin under an effective sliding control input $u_s = -x_2$. Hence, path ABCO corresponds to the overall Filippov solution of (3.1) emanating from A. To construct such a solution, the velocity vector $\dot{x}$ at a point C on Γ is defined as follows.

Denote, by $\dot{x}_{+1}$, the velocity vector at C for system (3.3) (with full positive control) and, by $\dot{x}_{-1}$, the velocity vector at C for system (3.2) (with full negative control). Consider now the velocity vector $\dot{x}_u$ for intermediate input values $-1 \leq u \leq +1$; it is readily seen that, as u increases from -1 to $+1$, the velocity vector $\dot{x}_u$ generates the straight line segment joining $\dot{x}_{-1}$ to $\dot{x}_{+1}$. For a Filippov solution, the actual velocity vector $\dot{x}$ at $C \in \Gamma$ is defined to be the vector $\dot{x}_{u_s}$, tangential to Γ, corresponding to some intermediate sliding control value $u_s \in (-1, +1)$ as shown in Fig. 3.3. In this way the sliding path CO is followed, at each point of which the control takes that value† which ensures that the corresponding velocity vector remains tangential to Γ. These basic notions will now be generalised.

Consider now a general autonomous system with discontinuous feedback control $u = u(x)$, viz.

$$\dot{x} = f(x, u(x)) \triangleq F(x) \tag{3.4}$$

where $F(x)$ is a discontinuous function with discontinuities occurring on some switching surfaces $\Gamma_i = \{x\colon \xi_i(x) = 0\}$ in state space. How is the dynamic behaviour of such a system to be defined on Γ_i? For the purposes of later chapters the

† For the example under consideration, the sliding control input at each point of CO is given by $u_s = -x_2$ and the origin is approached exponentially along Γ with

$$\| x(t) \| = \| x^c \| e^{-t}; \qquad t \geq 0; \qquad x(0) = x^c \in \Gamma$$

Such sliding trajectories occur only on the bounded segment $\Gamma_s \subset \Gamma$ of the switching line Γ, viz.

$$\Gamma_s = \{x\colon x_1 + x_2 = 0; \qquad |x_2| \leq 1\}$$

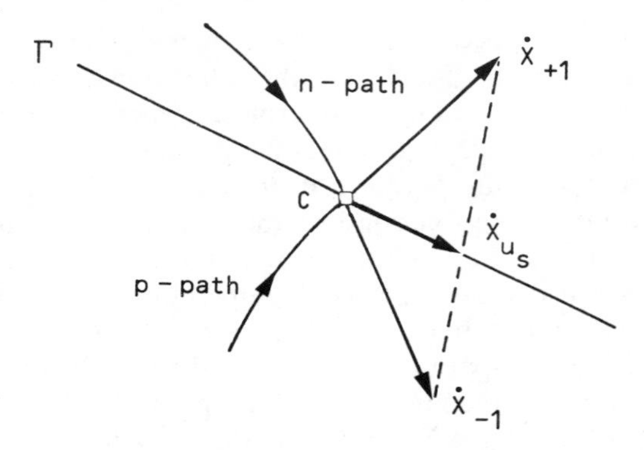

Fig. 3.3

concept of a Filippov solution proves entirely satisfactory. However, as a general treatment of Filippov solutions is beyond the scope of the present text, only the salient features are briefly discussed here. The reader is referred to André and Seibert (1956, 1960), Filippov (1960a, b; 1967), Siljak (1969), Brunovský (1974, 1976), Utkin (1978) and Hájek (1979a, b) for details.

The underlying idea is to replace the original discontinuous differential equation (3.4) by the differential inclusion

$$\dot{x} \in \mathscr{H}(x) \tag{3.5a}$$

where

$$\mathscr{H}(x) = \bigcap_{\delta > 0} \bigcap_{\text{meas}(\mathscr{N})=0} \overline{\text{cnvx}}\ F(B_\delta(x) \backslash \mathscr{N});$$

$$B_\delta(x) = \{y \colon \| y - x \| < \delta\} \tag{3.5b}$$

An (absolutely continuous) function $x(\cdot)$ on an interval I is then said to be a Filippov solution of (3.4) if it satisfies (3.5) a.e. on I. Loosely speaking, on a Filippov solution, the velocity vector $\dot{x}$ which determines the state flow belongs to the closed convex hull (denoted by $\overline{\text{cnvx}}$) containing all values that the right-hand side $F(x)$ of (3.4) generates as x ranges over the entire δ-neighbourhood $B_\delta(x)$ $(\delta \to 0^+)$, with the possible exception of sets $\mathscr{N}$ of zero measure (meas $(\mathscr{N}) = 0$). The latter possibility of overlooking sets of measure zero enables points of the switching surfaces Γ_i (where the velocity vector is not defined) in the δ-neighbourhood to be disregarded so that solutions on Γ_i can be determined. In terms of the introductory example 3.2.1, at the point $C \in \Gamma$, $\overline{\text{cnvx}}$ $F(B_\delta(x) \backslash \mathscr{N})$ can be identified as the straight line segment of Fig. 3.3, joining $\dot{x}_{-1}$ to $\dot{x}_{+1}$. As all discontinuities of (3.4) are assumed to occur on the switching surfaces Γ_i, then, for $x \notin \cup_i \Gamma_i$ (i.e. off the switching surfaces), the set $\mathscr{H}(x)$ contains the single element $F(x)$ and the Filippov solution through x coincides locally with the usual concept of a solution of the continuous equation $\dot{x} = F(x)$ through x, as on the parabolic arcs AB $(\mathscr{H}(x) = \{(x_2, +1)'\})$ and BC

($\mathscr{H}(x) = \{(x_2, -1)'\}$) of Fig. 3.2 excluding points B, C $\in \Gamma$. Hence, in the present context, Filippov solutions play an essential role only on the switching surfaces Γ_i. For notational convenience assume that there is only one switching surface $\Gamma = \{x: \xi(x) = 0\}$ and consider the Filippov solutions in a neighbourhood B_δ of a point $x \in \Gamma$. Assume furthermore that the $(n-1)$-dimensional manifold Γ is continuously differentiable in the neighbourhood with $\nabla\xi = \partial\xi/\partial x \neq 0$. Writing

$$B_\delta = \Sigma_B^+ \cup \Gamma_B \cup \Sigma_B^- \tag{3.6a}$$

where

$$\begin{aligned} &\Sigma_B^+ = \{x \in B_\delta: \xi(x) < 0\}; \qquad \Gamma_B = \{x \in B_\delta: \xi(x) = 0\}; \\ &\Sigma_B^- = \{x \in B_\delta: \xi(x) > 0\} \end{aligned} \tag{3.6b}$$

with

$$F(x) = \begin{cases} F^+(x); & x \in \Sigma_B^+ \\ F^-(x); & x \in \Sigma_B^- \end{cases} \tag{3.6c}$$

where F^+ and F^- are continuous on $\Sigma_B^+ \cup \Gamma_B$ and $\Sigma_B^- \cup \Gamma_B$, respectively. (Referring to the introductory example, $F^+(x) = (x_2, +1)'$ and $F^-(x) = (x_2, -1)'$). Now, if the limit $(\delta \to 0)$ direction vectors F^+ and F^- point towards the surface Γ_B on one side and away from the surface on the other as in Fig. 3.4, then there is no difficulty in interpreting the Filippov solution through $x \in \Gamma_B$ as simply standard solutions of the continuous systems $\dot{x} = F^+(x)$ and $\dot{x} = F^-(x)$ pieced together at x. Such dynamic behaviour at a surface of discontinuity will be referred to as *regular switching motion.* Note that, for such motion, the definition of the vector F on the surface of discontinuity Γ is irrelevant as, on a regular switching path, the isolated points on Γ constitute a set of measure zero.

A second and more interesting possibility is depicted in Fig. 3.5 where the limit $(\delta \to 0)$ direction vectors F^+ and F^- are both directed towards the surface Γ_B (when viewed from Σ_B^+ and Σ_B^-, respectively), i.e. the state vector field is directed towards Γ_B from both sides of the surface of discontinuity, viz.

$$\xi(x) \uparrow 0 \text{ as } x \text{ approaches } \Gamma_B \text{ from } \Sigma_B^+ \Rightarrow \dot{\xi}(x)^+ = \langle \nabla\xi(x), F^+(x) \rangle > 0 \tag{3.7a}$$

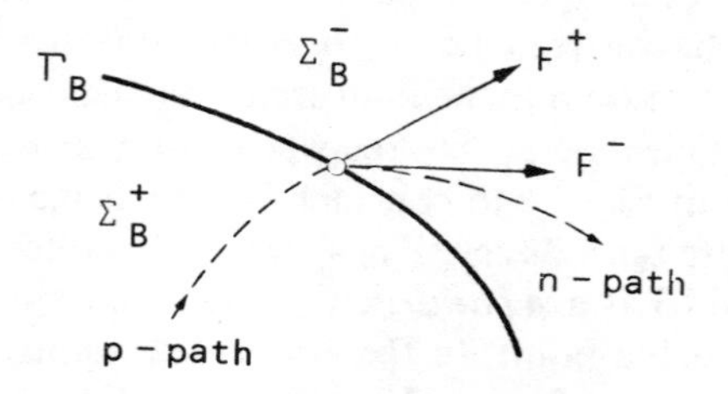

Fig. 3.4 *Regular switch point*

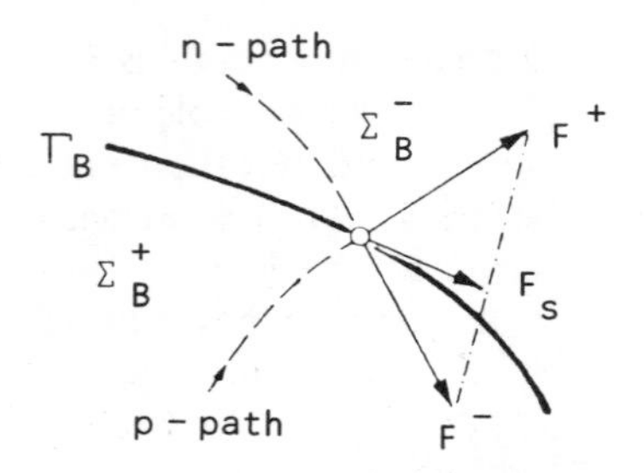

Fig. 3.5 *Sliding point*

and

$$\xi(x) \downarrow 0 \text{ as } x \text{ approaches } \Gamma_B \text{ from } \Sigma_B^- \Rightarrow \dot{\xi}(x)^- = \langle \nabla\xi(x), F^-(x) \rangle < 0 \tag{3.7b}$$

Clearly, on reaching Γ_B the state is constrained to remain on Γ_B, in which case the Filippov solution corresponds to the solution of the continuous equation $\dot{x} = F_s(x)$, where F_s is a continuous vector field on Γ defined as follows: $F_s(x)$ is the intersection of the tangent hyperplane (the set $\mathcal{N}$) to Γ at x and the line segment $\mu F^+(x) + (1 - \mu)F^-(x)$, $0 \leq \mu \leq 1$, joining $F^+(x)$ and $F^-(x)$ (the set $\cap_{\delta > 0} \overline{\text{cnvx}}\, F(B_\delta(x) \backslash \mathcal{N})$). In other words, the motion on Γ is defined to be the solution (in the usual sense) of the continuous differential equation

$$\dot{x} = \mu F^+(x) + (1 - \mu)F^-(x) = F_s(x) \tag{3.8a}$$

where μ is determined such that the vectors $F_s(x)$ and $\nabla\xi(x)$ are orthogonal, i.e.

$$\langle \nabla\xi(x), F_s(x) \rangle = \mu \langle F^+(x), \nabla\xi(x) \rangle + (1 - \mu)\langle F^-(x), \nabla\xi(x) \rangle = 0 \tag{3.8b}$$

giving

$$\mu = \frac{\langle \nabla\xi(x), F^-(x) \rangle}{\langle \nabla\xi(x), F^-(x) - F^+(x) \rangle} \tag{3.8c}$$

Hence, combining the above relations

$$\dot{x} = F_s(x) = \frac{\langle \nabla\xi(x), F^-(x) \rangle F^+(x) - \langle \nabla\xi(x), F^+(x) \rangle F^-(x)}{\langle \nabla\xi(x), F^-(x) - F^+(x) \rangle} \tag{3.9}$$

Thus, under conditions (3.7), the state is constrained to remain on the switching surface Γ and generates a trajectory governed by the equation (3.9). Such dynamic behaviour will be referred to as *sliding motion.* In the context of relay control systems this motion may be interpreted as idealized 'chattering' motion generated by an inertialess delay-free relay (see Utkin 1978). Finally, there is the nongeneric situation of Fig. 3.6 to consider for which the limit ($\delta \rightarrow 0$) direction vectors F^+ and F^- are both directed away from the switching surface Γ. Such a point will be referred to as a *separatrix.* Clearly a state trajectory can start but can never arrive at such a point. In the above discussion, solutions in a neighbourhood of an isolated surface of discontinuity only were considered; within

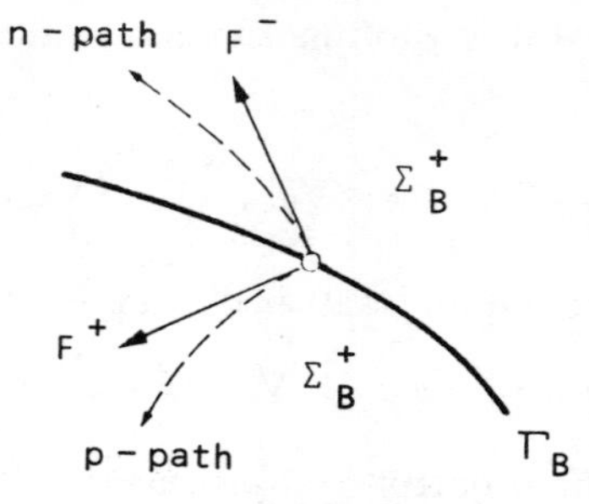

Fig. 3.6 *Separatrix*

B_δ the surface was assumed to be a continuously differentiable (n-1)-dimensional manifold. It is remarked that analysis of dynamic behaviour at corners of a surface of discontinuity or at the junction of one or more surfaces of discontinuity can also be treated by the theory.

3.3 Linear single-input systems: conditions for regular switching and sliding motion under discontinuous feedback control

Linear systems, with scalar saturable input, of the form

$$\dot{x} = Ax + bu; \qquad |u| \leq 1 \tag{3.10}$$

for which the control is specified as a discontinuous feedback†

$$u = u(x) = -\operatorname{sgn}(\xi(x)) \tag{3.11}$$

receive detailed analysis in later chapters. For this reason, the general results of Section 3.2 will now be interpreted for this particular class of systems (for related results see also Weissenberger, 1966).

The function ξ: $\mathbb{R}^n \rightarrow \mathbb{R}$ will be referred to as the switching function and is assumed to be continuously differentiable in the region of interest. The set $\Gamma = \{x\colon \xi(x) = 0\}$ constitutes the switching surface which is assumed to partition the state space (or the relevant region thereof) into two mutually exclusive regions: $\Sigma^+ = \{x\colon \xi(x) < 0\}$ and $\Sigma^- = \{x\colon \xi(x) > 0\}$. The feedback control is then defined, via the signum function, as

$$u = u(x) = -\operatorname{sgn}(\xi(x)) = \begin{cases} +1; & x \in \Sigma^+ \\ \sigma; & x \in \Gamma;\ -1 \leq \sigma \leq 1 \\ -1; & x \in \Sigma^- \end{cases} \tag{3.12}$$

† The convention of adopting the negative of the signum function in (3.11) is used here to conform with the notation of later chapters.

Now, the total time derivative $\dot{\xi}$ of the switching function ξ on a trajectory of (3.10)–(3.12) is given by

$$\dot{\xi} = \left\langle \frac{\partial \xi}{\partial x}, \dot{x} \right\rangle = \langle \nabla\xi, \dot{x} \rangle = \langle \nabla\xi, Ax \rangle - \langle \nabla\xi, b \rangle \operatorname{sgn}(\xi(x)) \tag{3.13}$$

Hence, if the following condition is satisfied at a point $x \in \Gamma$

(i) *regular switching condition:* $|\langle \nabla\xi, Ax \rangle| > |\langle \nabla\xi, b \rangle|$ (3.14)

then, by continuity of the trajectory $x(\cdot)$, $\dot{\xi}$ has the same sign in both Σ^+ and Σ^- in a neighbourhood of the point $x \in \Gamma$, i.e. the flow is as depicted in Fig. 3.4 and regular switching motion results. The set of points $x \in \Gamma$ satisfying (3.14) constitutes the set of regular switch points for the feedback system. The value of control σ on this set is defined as $+1$ or -1 according as the motion is from Σ^- to Σ^+ or Σ^+ to Σ^-, i.e. the value of control at a regular switch point is (arbitrarily) taken to be the control value defined by (3.12) *after* penetration of the switching surface Γ.

On the other hand, if the following conditions hold at a point $x \in \Gamma$

(ii) *sliding conditions:*
(a) $|\langle \nabla\xi(x), Ax \rangle| < |\langle \nabla\xi(x), b \rangle|$
(b) $\langle \nabla\xi(x), b \rangle > 0$
(3.15)

then, from (3.13), in a neighbourhood of the point $x \in \Gamma$, $\dot{\xi}$ and ξ are of opposite sign in both Σ^+ and Σ^-, i.e. the flow is as depicted in Fig. 3.5 and sliding motion results. The set of points $x \in \Gamma$ satisfying (3.15) constitutes the sliding set for the feedback system. In accordance with the Filippov theory, the differential equation of sliding motion on this set is determined by setting $F^+(x) = Ax + b$ and $F^-(x) = Ax - b$ in (3.9) to give

$$\dot{x} = F_s(x) = Ax - b\left[\frac{\langle \nabla\xi(x), Ax \rangle}{\langle \nabla\xi(x), b \rangle}\right] \tag{3.16}$$

or, equivalently,

$$\dot{x} = \left[I - \frac{b\nabla\xi(x)'}{\langle \nabla\xi(x), b \rangle}\right] Ax \tag{3.17}$$

Comparing (3.10) and (3.16), it may be seen that the effective value of control $u_s = \sigma \in [-1, 1]$ at a sliding point x is given by

$$u_s = -\frac{\langle \nabla\xi(x), Ax \rangle}{\langle \nabla\xi(x), b \rangle} \tag{3.18}$$

Finally, if the following conditions hold at a point $x \in \Gamma$

(iii) *separatrix conditions:*
(a) $|\langle \nabla\xi(x), Ax \rangle| < |\langle \nabla\xi(x), b \rangle|$
(b) $\langle \nabla\xi(x), b \rangle < 0$
(3.19)

then, from (3.13), it may be seen that, in a neighbourhood of $x \in \Gamma$, $\dot{\xi}$ and ξ have the same sign in both Σ^+ and Σ^-, i.e. the flow is as depicted in Fig. 3.6 and a separatrix results. The set of points $x \in \Gamma$ satisfying (3.19) constitutes the set of separatrices for the feedback system.

Applying the above results to Example 3.2.1, $\xi(x) = x_1 + x_2$ so that $\nabla\xi(x) = (1, 1)'$ and $\langle \nabla\xi(x), b \rangle = 1 > 0$ with $\langle \nabla\xi(x), Ax \rangle = x_2$. Hence, in view of (3.19), separatrices do not occur. On the other hand, from (3.15), the sliding set for the feedback system is the segment Γ_s of Γ characterised by

$$\Gamma_s = \{x \in \Gamma : |x_2| < 1\}$$

as previously noted; all other points of Γ are regular switch points. From (3.18), on the sliding set Γ_s, the effective control input is $u_s = \sigma = -x_2$ generating a solution curve (in Γ) of equation (3.17), viz.

$$\dot{x} = \begin{bmatrix} 0 & 1 \\ 0 & -1 \end{bmatrix} x$$

3.4 General and special invariance of multiple integrator systems

In the design of feedback controllers for saturating systems, Persson (1963) heuristically justified the adoption of a certain nonlinear feedback structure (determined up to a set of design parameters) on the basis that the system response should be 'independent of signal amplitude' in the sense that, if the design parameter values are calculated to give satisfactory performance for 'small' initial conditions, then these parameter values should also yield satisfactory performance for 'large' initial conditions, i.e. the system response should be invariant under changes of signal (and time) scale.

Fuller (1970a, 1971) clarified the theoretical basis of Persson's approach by showing that time-optimal multiple integrator systems with scalar saturable control exhibit such a property of invariance† which implies that, if a time-optimal trajectory to the origin from some initial state is known, then implicitly a family of 'invariant' or 'similar' trajectories is known, each member of the family being related to the original trajectory by an appropriate transformation of amplitude and time scales. Multiple integrator systems exhibit this invariance property, not only in the case of the time-optimal control problem, but also for a wide range of integral cost problems for which the integrand satisfies a certain 'homogeneity-type' condition, as will now be shown using the approach of Fuller (1970a, 1971).

† System invariance properties were also noted, in special cases, by Wonham (1963), Grensted and Fuller (1965), and Fuller (1966).

3.4.1 General invariance of optimal multiple integrator systems

Attention is now restricted to multiple integrator systems with a scalar saturable input:

$$\frac{dx}{dt} = Ax + bu; \qquad |u| \leq 1;$$

$$A = \begin{bmatrix} 0 & 1 & 0 & \cdots & 0 \\ 0 & 0 & 1 & \cdots & 0 \\ \vdots & & & \ddots & \\ 0 & 0 & 0 & \cdots & 1 \\ 0 & 0 & 0 & \cdots & 0 \end{bmatrix}; \qquad b = \begin{bmatrix} 0 \\ 0 \\ \vdots \\ 0 \\ \frac{1}{a} \end{bmatrix} \tag{3.20a}$$

or, equivalently,

$$\frac{dx_i}{dt} = x_{i+1}; \qquad i = 1,2, \ldots, n-1$$

$$\frac{dx_n}{dt} = \frac{u}{a}; \qquad |u| \leq 1 \tag{3.20b}$$

where $a > 0$ is a gain parameter. The following optimal regulator problem is considered:

$$\text{P1:} \begin{cases} \text{minimise the cost functional} \\ J(u) = \displaystyle\int_0^{t_f} L(x(t), u(t), a)\, dt; \qquad t_f \text{ free} & (3.21) \\ \text{subject to (3.20) and the boundary conditions} \\ x(0) = x^0; \qquad x(t_f) = 0 & (3.22) \end{cases}$$

i.e. a control $u(t)$, $0 \leq t \leq t_f$ is sought which transfers the multiple integrator system (3.20) from a given initial state x^0 to the origin while minimising the cost (3.21). The possible dependence of the cost integrand L on the gain parameter a has been indicated explicitly.

Now introduce the linear transformations (amplitude and time rescaling)

$$\left.\begin{aligned} \tilde{t} &= T(t) \triangleq \kappa t \\ \tilde{x}_i &= X_i(x_i) \triangleq \gamma \kappa^{1-i} x_i; \\ \tilde{u} &= u \\ \tilde{a} &= A(a) \triangleq \gamma^{-1} \kappa^n a \end{aligned}\right\} \quad i = 1, 2, \ldots, n; \qquad \gamma, \kappa > 0 \tag{3.23}$$

then

Property 3.4.1 General invariance of the optimal system: The optimal control problem P1 is invariant under transformation (3.23) if

$$L[(x_i), u, a] = \alpha(\gamma, \kappa, a)L[(X_i^{-1}(x_i)), u, A^{-1}(a)] \tag{3.24}$$

for all (x_i), for all $\gamma, \kappa, a > 0$, and for some $\alpha = \alpha(\gamma, \kappa, a) > 0$.

Proof
A straightforward calculation reveals that, under (3.23), problem P1 is transformed into problem P2:

P2:

minimise:

$$J(\tilde{u}(\cdot)) = \int_0^{t_f} L[(X_i^{-1}(\tilde{x}_i(T^{-1}(\tilde{t})))), \tilde{u}(T^{-1}(\tilde{t})), A^{-1}(\tilde{a})]\, d(T^{-1}(\tilde{t}))$$

$$= \int_0^{t_f} L[(X_i^{-1}(\tilde{x}_i(\tau))), \tilde{u}(\tau), A^{-1}(\tilde{a})]\, d\tau;\ t_f \text{ free} \tag{3.25}$$

subject to:

$$\frac{d\tilde{x}_i}{d\tilde{t}} = \tilde{x}_{i+1}; \qquad i = 1, 2, \ldots, n-1$$

$$\frac{d\tilde{x}_n}{d\tilde{t}} = \frac{\tilde{u}}{\tilde{a}}; \qquad |\tilde{u}| \le 1 \tag{3.26}$$

with boundary conditions

$$\tilde{x}(0) = \tilde{x}^0 = (X_i(x_i^0)); \qquad \tilde{x}(t_f) = 0; \qquad t_f \text{ free} \tag{3.27}$$

Noting that the cost functional can be multiplied by an arbitrary positive constant $\alpha > 0$ without affecting the cost minimising control (i.e. $\underset{u \in U}{\text{minimise}}\ J(u) \Leftrightarrow \underset{u \in U}{\text{minimise}}\ \alpha J(u)$ for all $\alpha > 0$), multiplying (3.25) by $\alpha > 0$ (which may depend on the parameters $\gamma, \kappa, \tilde{a}$) yields the equivalent minimisation problem:

$$\text{minimise: } J(\tilde{u}) = \int_0^{t_f} \alpha(\gamma, \kappa, \tilde{a})L[(X_i^{-1}(\tilde{x}_i)), \tilde{u}, A^{-1}(\tilde{a})]\, d\tau \tag{3.28}$$

subject to (3.26)–(3.27).

Comparing (3.20)–(3.22) and (3.26)–(3.28) reveals that problems P1 and P2 are equivalent if

$$\alpha(\gamma, \kappa, \tilde{a})L[(X_i^{-1}(\tilde{x}_i)), \tilde{u}, A^{-1}(\tilde{a})] = L[(\tilde{x}_i), \tilde{u}, \tilde{a}] \tag{3.29}$$

for all $(\tilde{x}_i)$, for all $\gamma, \kappa, \tilde{a} > 0$ and for some $\alpha = \alpha(\gamma, \kappa, \tilde{a}) > 0$, i.e. if the cost integrand

L satisfies (3.29), then the optimal control problem (3.20)–(3.22) is invariant under transformation (3.23). If an optimal control $u^*(t), 0 \le t \le t_f$, generates an optimal trajectory $(x_i(t))$, $0 \le t \le t_f$, taking system (3.20) from $x(0) = x^0$ to the origin $x(t_f) = 0$, then the control $\tilde{u}^*(\tilde{t}) = u^*(\kappa^{-1}\tilde{t})$, $0 \le \tilde{t} \le \kappa t_f$, generates an optimal trajectory $(\tilde{x}_i(\tilde{t})) = (\gamma\kappa^{1-i}x_i(\kappa^{-1}\tilde{t}))$, $0 \le \tilde{t} \le \kappa t_f$, taking system (3.26) from $\tilde{x}(0) = (\tilde{x}_i(0)) = (\gamma\kappa^{1-i}x_i^0)$ to $\tilde{x}(\kappa t_f) = 0$.

For example, the time-optimal control problem ($L \equiv 1$) satisfies (3.29) with $\alpha = 1$; on the other hand, the quadratic-cost problem with $L = \langle x, Qx \rangle + u^2$ (Q positive semidefinite) *fails* to satisfy (3.29) in general and consequently does *not* exhibit general invariance.

3.4.2 General invariance of the feedback system

Consider again the multiple integrator system (3.20), but now under discontinuous feedback control

$$u = -\text{sgn}\,(\Xi(x, a)) = -\text{sgn}\,(\Xi((x_i), a)) \tag{3.30}$$

where the dependence of the switching function $\Xi: \mathbb{R}^n \to \mathbb{R}$ on the gain parameter a has been indicated explicitly. Under what conditions does the feedback system exhibit general invariance?

Property 3.4.2 General invariance of the feedback system: The multiple integrator system (3.20) with feedback control (3.30) is invariant under transformation (3.23) if

$$\Xi((x_i), a) = \beta(\gamma, \kappa, a)\Xi((X_i^{-1}(x_i)), A^{-1}(a)) \tag{3.31}$$

for all $x = (x_i)$, for all $\gamma, \kappa, a > 0$ and for some $\beta = \beta(\gamma, \kappa, a) > 0$.

Proof

Under (3.23), the transformed system equations become

$$\left.\begin{aligned} \frac{d\tilde{x}_i}{d\tilde{t}} &= \tilde{x}_{i+1}; \quad i = 1, 2, \ldots, n-1 \\ \frac{d\tilde{x}_n}{d\tilde{t}} &= -\frac{1}{\tilde{a}}\,\text{sgn}\,[\Xi((X_i^{-1}(\tilde{x}_i)), A^{-1}(\tilde{a}))] \end{aligned}\right\} \tag{3.32}$$

Now, the argument of the signum function can be multiplied by an arbitrary positive constant $\beta > 0$ without changing the value of the function, i.e. $\text{sgn}\,(\Xi) = \text{sgn}\,(\beta\Xi)$, $\forall\, \beta > 0$ ($\Xi \neq 0$). Consequently, multiplying the argument of the signum function in (3.32) by $\beta > 0$ (where β may depend on the parameters $\gamma, \kappa, \tilde{a}$) yields the system:

$$\begin{aligned} \frac{d\tilde{x}_i}{d\tilde{t}} &= \tilde{x}_{i+1}; \quad i = 1, 2, \ldots, n-1 \\ \frac{d\tilde{x}_n}{d\tilde{t}} &= -\frac{1}{\tilde{a}}\,\text{sgn}\,[\beta(\gamma, \kappa, \tilde{a})\Xi((X_i^{-1}(\tilde{x}_i)), A^{-1}(\tilde{a}))] \end{aligned} \tag{3.33}$$

Comparison of (3.33) with (3.20) and (3.30) reveals that the feedback system exhibits general invariance under (3.23) if (3.31) is satisfied.

3.4.3 Special invariance of the optimal system

Suppose now that $\kappa > 0$ is no longer arbitrary but instead satisfies the relation

$$\gamma = \kappa^n; \qquad \kappa > 0 \tag{3.34}$$

then, from (3.23),

$$\tilde{a} = a \tag{3.35}$$

and the time and state transformations become

$$\begin{aligned} \tilde{t} &= \kappa t \\ \tilde{x}_i &= \kappa^{n+1-i} x_i; \qquad i = 1, 2, \ldots, n \end{aligned} \tag{3.36}$$

Property 3.4.1 of general invariance, when restricted through (3.34), will be referred to as *special invariance* of the optimal system and implies

Property 3.4.3 Special invariance of the optimal system: If

$$x(t) = (x_1(t), x_2(t), \ldots, x_n(t))', \qquad 0 \le t \le t_f \tag{3.37}$$

is an optimal trajectory of system (3.20), then so is

$$(\kappa^{-n} x_1(\kappa t), \kappa^{-(n-1)} x_2(\kappa t), \ldots, \kappa^{-1} x_n(\kappa t))', \qquad 0 \le t \le \kappa^{-1} t_f \tag{3.38}$$

for all $\kappa > 0$.

Proof

In view of (3.35), first note that, apart from notation, (3.20) and (3.26) define the *same* dynamical system. Hence, if

$$x = \theta(t) = (\theta_1(t), \theta_2(t), \ldots, \theta_n(t))', \qquad 0 \le t \le t_f$$

is an (optimal) trajectory of (3.20), then $\tilde{x} = \theta(\tilde{t})$, $0 \le \tilde{t} \le t_f$, is an (optimal) trajectory of (3.26). Application of transformations (3.23) and (3.34) to the latter trajectory of (3.26) yields an (optimal) trajectory of (3.20), whence the result that

$$(\kappa^{-n} \theta_1(\kappa t), \kappa^{-(n-1)} \theta_2(\kappa t), \ldots, \kappa^{-1} \theta_n(\kappa t))', \qquad 0 \le t \le \kappa^{-1} t_f$$

is an (optimal) trajectory of (3.20) for all $\kappa > 0$.

3.5 (Θ, Φ)-space: a reduced state space for invariant triple integrator systems

The property of special invariance (property 3.4.3) is naturally exhibited by multiple integrator systems optimised with respect to integral costs with integrands satisfying (3.29). Moreover, the invariance property is sometimes imposed on feedback control structures, via (3.31), to render the system response 'independent of signal amplitude' (Persson 1963). With this property, if a single trajectory of the system is known, then the invariant family of similar trajectories can be constructed by extension and contraction of the time and amplitude scales. Thus, in effect, a single known trajectory contains all information necessary to characterise each member of the invariant family. This observation suggests a certain redundance in approaches which analyse the system behaviour in the full n-dimensional state space $\mathbb{R}^n$. Because each family of invariant trajectories can be 'represented' by selecting an individual member of the family, a judicious choice of an $(n-1)$-dimensional *reduced state space* would seem beneficial, wherein each family of invariant trajectories in $\mathbb{R}^n$ corresponds uniquely to a representative trajectory in the reduced state space. Such an approach proves especially fruitful in the case of a triple integrator system exhibiting the requisite property of special invariance. Through the introduction of a nonlinear transformation of the state vector field in $\mathbb{R}^3$ and subsequent projection into a two-dimensional reduced state space, it is possible to depict and analyse the dynamic behaviour of the system. This reduced state space for triple integrator systems, adopted initially by Grensted and Fuller (1965) and described in detail by Fuller (1971), is now introduced and finds widespread application in later chapters.

3.5.1 z-space

The system to be considered in the controlled triple integrator

$$\dot{x}_1 = x_2; \quad \dot{x}_2 = x_3; \quad \dot{x}_3 = \frac{u}{a}; \quad |u| \leq 1; \quad a > 0 \tag{3.39}$$

It is assumed that any cost functional associated with the system, or any feedback structure imposed on the system, is such that the overall system exhibits the property of special invariance (property 3.4.3).

The following nonlinear state transformation is introduced:

$$\begin{aligned} z_1 &= |ax_1|^{1/3} \operatorname{sgn}(x_1) \\ z_2 &= |ax_2|^{1/2} \operatorname{sgn}(x_2) \\ z_3 &= ax_3 \end{aligned} \tag{3.40}$$

Now, in terms of these new variables, the property of special invariance implies that if

$$z(t) = (z_1(t), z_2(t), z_3(t))', \qquad 0 \leq t \leq t_f \tag{3.41}$$

is a trajectory of the system in z-space (generated by $u(t), 0 \leq t \leq t_f$), then, for all $\kappa > 0$,

$$\kappa^{-1} z(\kappa t), \qquad 0 \leq t \leq \kappa^{-1} t_f \tag{3.42}$$

is also a trajectory of the system (generated by $u(\kappa t), 0 \leq t \leq \kappa^{-1} t_f$). Note that, if P is a point of trajectory (3.41), then any point on the straight line (ray) from the origin through P is a point of a similar trajectory (3.42). As κ ranges over the positive reals, (3.42) generates a surface in z-space (composed of straight line generators) which constitutes a cone with vertex at the origin, as depicted in Fig. 3.7.

3.5.2 (Θ, Φ)-*space*

Referring to Fig. 3.8, consider a spherical surface S in z-space centred on the origin, such that the ray through P intersects S at the point Q. As P traces the trajectory $z(t)$, $0 \leq t \leq t_f$, in z-space Q describes a locus on the sphere S, i.e. the trajectory $z(\cdot)$ in z-space maps onto a locus on S; moreover, every similar trajectory, related to $z(\cdot)$ through (3.42), maps onto the *same* locus on S. If this construction is used for the entire set of families of similar trajectories (i.e. the complete vector field in z-space), then a set of nonintersecting loci on S is obtained. If the spherical coordinates (Θ, Φ) are adopted for points on S, then the required two-dimensional projection of the state vector field results. Specifically, Θ and Φ satisfy

$$\tan \Theta = \frac{z_2}{z_1}; \qquad \tan \Phi = \frac{z_3}{\sqrt{(z_1^2 + z_2^2)}} \tag{3.43}$$

or

$$\Theta = \begin{cases} \tan^{-1} \dfrac{z_2}{z_1}; & z_1 > 0 \\ \dfrac{\pi}{2}; & z_1 = 0,\ z_2 > 0 \\ -\dfrac{\pi}{2}; & z_1 = 0,\ z_2 < 0 \\ \pi + \tan^{-1} \dfrac{z_2}{z_1}; & z_1 < 0 \end{cases} \tag{3.44a}$$

$$\Phi = \begin{cases} \tan^{-1} \dfrac{z_3}{\sqrt{(z_1^2 + z_2^2)}}; & z_1^2 + z_2^2 \neq 0 \\ \dfrac{\pi}{2}; & z_1^2 + z_2^2 = 0, \quad z_3 > 0 \\ -\dfrac{\pi}{2}; & z_1^2 + z_2^2 = 0, \quad z_3 < 0 \end{cases} \tag{3.44b}$$

where the range of the function $\tan^{-1}(\cdot)$ is $(-\pi/2, \pi/2)$, i.e. $\tan^{-1}$ denotes principal values.

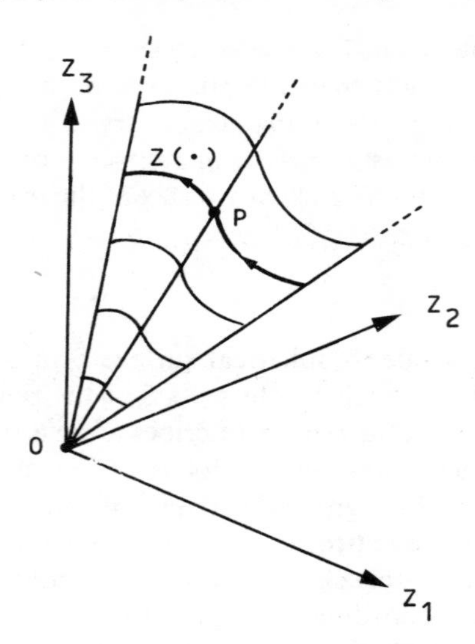

Fig. 3.7 *Cone of similar trajectories*

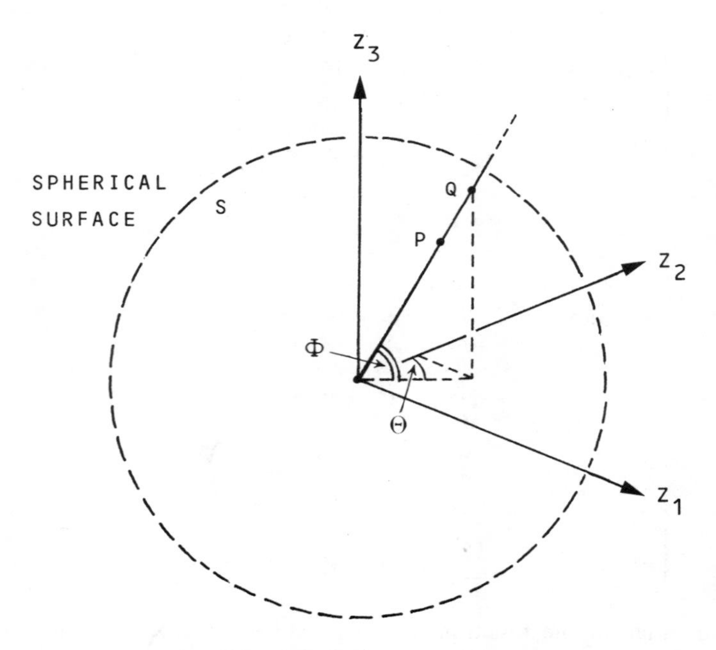

Fig. 3.8 *Spherical co-ordinates Θ, Φ*

3.5.3 p-paths and n-paths in (Θ, Φ)*-space*

Consider the set of p-paths, i.e. the set of state trajectories generated by a constant control $u \equiv +1$. The equations defining a member of this set can easily be found by integrating the state equations (3.39), viz.

$$x_1(t) = x_1^0 + x_2^0 t + \tfrac{1}{2}x_3^0 t^2 + \frac{t^3}{6a}$$

$$x_2(t) = x_2^0 + x_3^0 t + \frac{t^2}{2a} \qquad (3.45)$$

$$x_3(t) = x_3^0 + \frac{t}{a}$$

where $x^0 = (x_1^0, x_2^0, x_3^0)'$ is the trajectory starting point. The set of such p-paths projected into (Θ, Φ)-space is depicted in Fig. 3.9. Each p-path in (Θ, Φ)-space corresponds to a family of similar p-paths in the original state space. The portrait of p-paths exhibits two singular points in (Θ, Φ)-space: (i) the point C^+ in the fourth quadrant, away from which other p-paths diverge, corresponds to the unique p-path in the original state space which goes to the origin; (ii) the point D^+ in the first quadrant, towards which other p-paths converge, corresponds to the unique p-path in the original state space which leaves the origin.

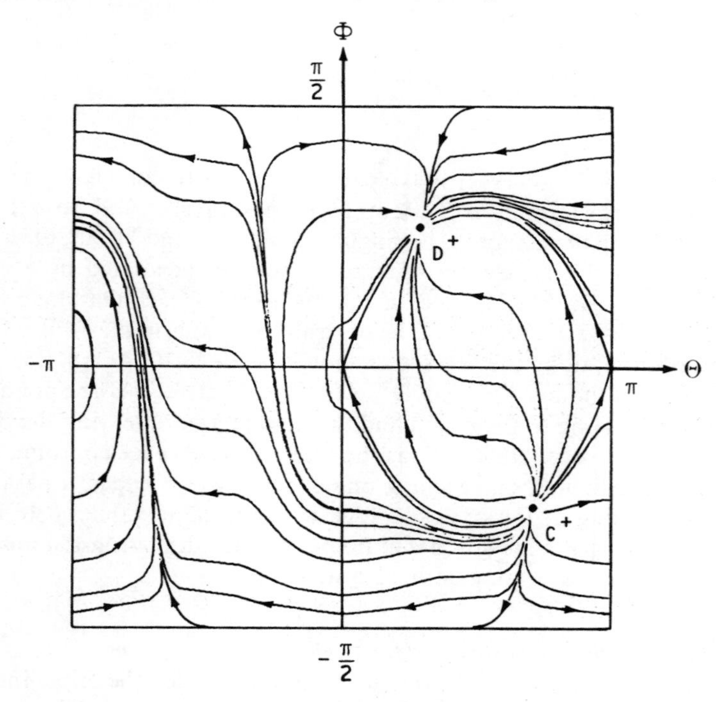

Fig. 3.9 p-*paths*

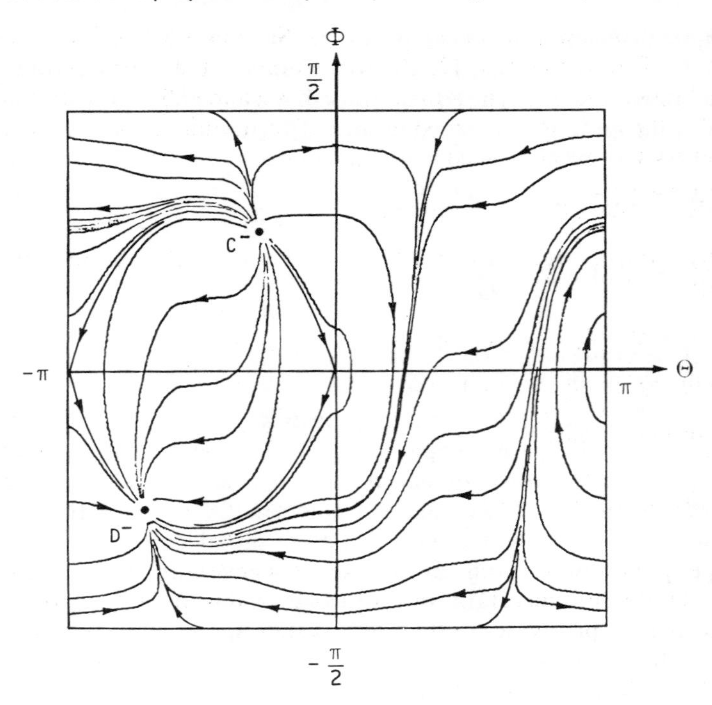

Fig. 3.10 n-*paths*

In effect, the p-paths (curves) in x-space which go to and leave the origin become, under transformation (3.40), straight lines (rays) to and from the origin in z-space which, in turn, map to the singular points C^+ and D^+ in (Θ, Φ)-space.

The set of n-paths, i.e. the set of state trajectories generated by a constant control $u \equiv -1$, can be depicted in (Θ, Φ)-space in a similar manner. The portrait of n-paths is a reflection of the portrait of p-paths in the Θ-axis followed by a π-translation along the Θ-axis, as shown in Fig. 3.10. Again, this portrait has two singular points C^- and D^- which, respectively, correspond to the unique n-path leading to the origin and the unique n-path leaving the origin in the original state space. Note that, as the reduced state space coordinates Θ, Φ are angular coordinates, on reaching one of the $\theta = \pm\pi$ edges, a path is continued by restarting at the opposite edge with the same value of Φ. In later Sections, the reduced state space will frequently be adopted in the analysis of certain third-order saturating control systems.

3.5.4 Time-optimal state portrait for the triple integrator system

By way of illustration, the time-optimal state portrait for the triple integrator system will now be constructed in (Θ, Φ)-space; the derivation of the associated feedback control law (previously stated in Section 1.3.2) is postponed until Chapter 6. From theorem 2.6.2, the optimal regulating control $u^*(t)$, $0 \leq t \leq t_f^*$ (assuming it exists) which transfers system (3.30) from an initial state $x(0) = x^0 = (x_1^0, x_2^0, x_3^0)'$ to the state origin in minimum time t_f^* is uniquely (almost everywhere) determined by the relation

$$u^*(t) = \text{sgn}\ \langle \eta(t), b \rangle = \text{sgn}\ (\eta_3(t)) \tag{3.46}$$

where $\eta(t) = (\eta_1(t), \eta_2(t), \eta_3(t))'$, $0 \leq t \leq t_f^*$, is a nontrivial solution of the adjoint equations

$$\begin{gathered} \dot{\eta}_1 = 0; \quad \dot{\eta}_2 = -\eta_1; \quad \dot{\eta}_3 = -\eta_2 \\ \eta(0) = \eta^0 = (\eta_1^0, \eta_2^0, \eta_3^0)' \neq 0 \end{gathered} \tag{3.47}$$

Integrating (3.47) yields

$$\eta_3(t) = \eta_3^0 - \eta_2^0 t + \eta_1^0 \frac{t^2}{2} \tag{3.48}$$

i.e. η_3 is a quadratic function of t and consequently can change sign at most twice on any finite interval. In view of (3.46), it immediately follows that the optimal control is piecewise constant with $|u^*(t)| = 1$ and with at most two sign changes on the interval $(0, t_f^*)$, i.e. with at most three intervals of control constancy. Hence the control sequences

$$\{\pm 1\}, \{\mp 1, \pm 1\}, \{\pm 1, \mp 1, \pm 1\}$$

only, are optimal control candidates, where $\{\pm 1, \mp 1, \pm 1\}$ denotes a control ± 1 on a finite interval followed by a control ∓ 1, etc.

Referring to Fig. 3.11, the time-optimal state portrait may now be constructed in (Θ, Φ)-space as follows.

(*i*) $\{\pm 1\}$ *control sequence trajectories:* Clearly these correspond to optimal switchless trajectories which go to the origin under constant control $u^* \equiv +1$ or $u^* \equiv -1$; in (Θ, Φ)-space, these trajectories map to the singular points C^+ $(u^* \equiv +1)$ and C^- $(u^* \equiv -1)$.

(ii) $\{\mp 1, \pm 1\}$ *control sequence trajectories:* These correspond to trajectories which contain one switch in control and which map to the continuous curve $\Gamma = \Gamma^+ \cup \Gamma^- \cup C^+ \cup C^-$ in (Θ, Φ)-space comprised of the singular points C^+, C^- together with the p-path (Γ^+) leading to C^- (an optimal $\{+1, -1\}$ control sequence path) and the n-path (Γ^-) leading to C^+ (an optimal $\{-1, +1\}$ control sequence path).

(*iii*) $\{\pm 1, \mp 1, \pm 1\}$ *control sequence trajectories:* These correspond to trajectories with the maximum number (two) of control switches and map to the set of paths in (Θ, Φ)-space comprised of a p-path (n-path) which goes to Γ^- (Γ^+) followed by an n-path (p-path) in Γ^- (Γ^+) to the singular point C^+ (C^-).

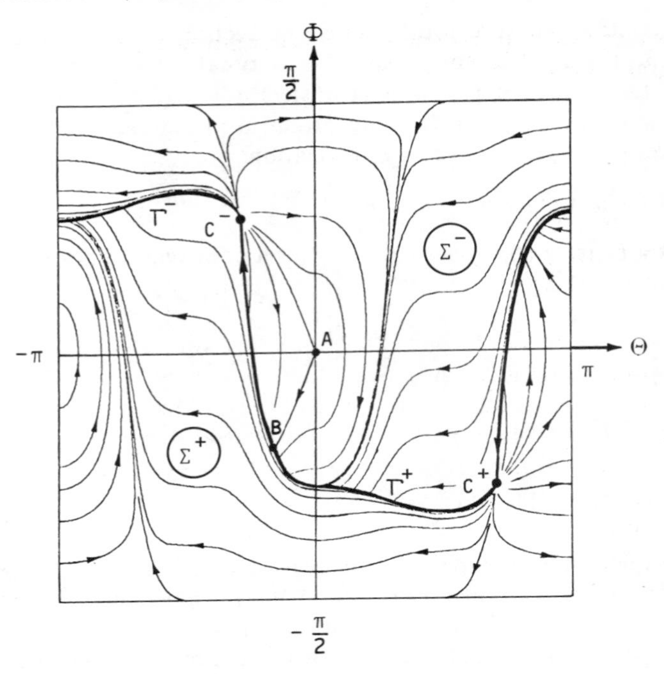

Fig. 3.11 *Time-optimal state portrait*

The curve $\Gamma = \Gamma^+ \cup \Gamma^- \cup C^+ \cup C^-$ corresponds to the time-optimal switching surface in $\mathbb{R}^3$ which partitions the state space into two mutually exclusive regions Σ^+ and Σ^- of positive $(u^* = +1)$ and negative $(u^* = -1)$ control, respectively. These control regions map to regions Σ^+ and Σ^- in (Θ, Φ)-space lying below and above the curve Γ as shown in Fig. 3.11. For example, consider the time-optimal trajectory to the origin from a starting point on the positive x_1 state semi-axis (equivalent to a step input to the system), i.e. $x(0) = x^0 = (x_1^0, 0, 0)'$, $x_1^0 > 0$. Under transformations (3.40) and (3.44), every such starting point maps to the point A of Fig. 3.11 (the origin in reduced state space). From A, the n-path AB is followed until the switching curve Γ is met at point B where the first control switch occurs. The p-path Γ^+ is then followed from B to the singular point C^- where the second control switch occurs. In (Θ, Φ)-space, the state point then remains stationary at C^-, while, in the original state space, the n-path to the state origin is followed. The time-optimal paths to the origin from other starting points can be traced in a similar manner. Note that a minimum-time path to the origin exists from each starting point $x^0 \in \mathbb{R}^3$, i.e. the system is time-optimally controllable to the origin from all points of the state space.

3.6 System equivalence: systems reducible to equivalent pure integrator form

In Sections 3.4 and 3.5, attention was restricted to multiple integrator systems. Special study of this restricted class of systems can be justified by the observation that, in a sufficiently small neighbourhood of the state origin, nth-order linear systems with bang-bang input essentially behave as a system with n integrating elements. Additional indication of the importance of multiple integrator systems is provided in this section by identifying a class of single-input systems, with real eigenvalues in simple ratio, for which the time-optimal control problem (and indeed other optimal control problems) are reducible to the corresponding (or related) problem for multiple integrator systems. This property of system equivalence was noted by Fuller (1973d) and discussed in detail in Fuller (1974a); furthermore, the property forms the basis of a quasi-time-optimal control technique for more general systems (Ryan 1975, 1978a) described in Chapter 10.

The saturating single-input system to be considered consists of n cascaded first-order elements with n associated real eigenvalues λ_i in the simple ratio $n:n-1:n-2:\cdots:2:1$, i.e. $\lambda_i = (n+1-i)\lambda$, λ constant, as shown in

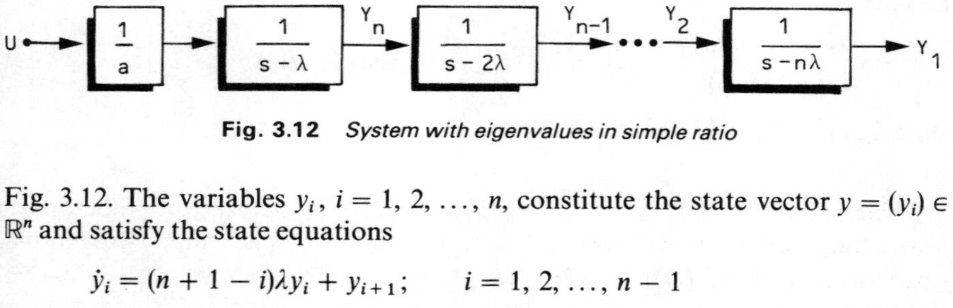

Fig. 3.12 *System with eigenvalues in simple ratio*

Fig. 3.12. The variables y_i, $i = 1, 2, \ldots, n$, constitute the state vector $y = (y_i) \in \mathbb{R}^n$ and satisfy the state equations

$$\dot{y}_i = (n+1-i)\lambda y_i + y_{i+1}; \qquad i = 1, 2, \ldots, n-1$$

$$\dot{y}_n = \lambda y_n + \frac{u}{a}; \quad |u| \leq 1; \quad \lambda = \text{constant} \neq 0; \quad a = \text{constant} > 0 \tag{3.49a}$$

with initial condition

$$y(0) = (y_i(0)) = (y_i^0) = y^0 \tag{3.49b}$$

Let $u(t)$, $0 \leq t \leq t_f$, be an admissible regulating control which generates a trajectory $y(t) = (y_i(t))$ of (3.49) from $y(0) = y^0 = (y_i^0)$ to $y(t_f) = 0$.

3.6.1 Equivalent multiple integrator system

The following transformation is first introduced

$$w_i(t) = \exp(-\lambda_i t) y_i(t) = \exp(-(n+1-i)\lambda t) y_i(t), \qquad i = 1, 2, \ldots, n \tag{3.50}$$

under which (3.49) becomes

$$\frac{dw_i}{dt} = \exp(-\lambda t)w_{i+1}; \qquad i = 1, 2, \ldots, n-1$$

$$\frac{dw_n}{dt} = \exp(-\lambda t)\frac{u}{a}; \quad |u| \leq 1, \quad a > 0 \tag{3.51a}$$

with the *same* initial condition

$$w(0) = (w_i(0)) = (y_i(0)) = (y_i^0) = y^0 \tag{3.51b}$$

Clearly, $u(\cdot)$ which transfers (3.49) from y^0 to the origin $y(t_f) = 0$ also transfers (3.51) from the same initial state $w(0) = y^0$ to the origin $w(t_f) = 0$.

Now introduce the additional transformation

$$s = \mathscr{T}(t) = \frac{1}{\lambda}[1 - \exp(-\lambda t)]; \qquad 0 \leq t \leq t_f \tag{3.52a}$$

with the inverse transformation

$$t = \mathscr{T}^{-1}(s) = \mathscr{S}(s) = -\frac{1}{\lambda}\ln(1 - \lambda s); \quad 0 \leq s \leq \mathscr{T}(t_f) \tag{3.52b}$$

and define the functions

$$\left.\begin{array}{ll} x_i(s) = w_i(\mathscr{S}(s)); & i = 1, 2, \ldots, n \\ v(s) = u(\mathscr{S}(s)) & \end{array}\right\} \; 0 \leq s \leq \mathscr{T}(t_f) \tag{3.53}$$

then, from (3.52) and (3.53),

$$\begin{aligned} \frac{dx_i}{ds} &= \frac{dw_i}{dt} \cdot \frac{dt}{ds} \\ &= \frac{dw_i}{dt}\exp(\lambda t); \qquad i = 1, 2, \ldots, n \end{aligned} \tag{3.54}$$

In consequence, (3.51) simplifies to the multiple integrator form

$$\frac{dx_i}{ds} = x_{i+1}; \qquad i = 1, 2, \ldots, n-1$$

$$\frac{dx_n}{ds} = \frac{v}{a}; \qquad |v| \leq 1; \quad a > 0 \tag{3.55a}$$

and, noting that $s = 0 \Leftrightarrow t = 0$, the initial state is unchanged, viz.

$$x(0) = (x_i(0)) = (y_i(0)) = (y_i^0) = y^0 \tag{3.55b}$$

Moreover, the control $v(s)$, $0 \leq s \leq \mathscr{T}(t_f)$, will transfer (3.55) from y^0 to the origin, $x(s_f) = 0$, in transformed time

$$s_f = \mathscr{T}(t_f) = \frac{1}{\lambda}[1 - \exp(-\lambda t_f)] \tag{3.56}$$

3.6.2 Transformation of the time-optimal control

Suppose now that $u^*(t)$, $0 \leq t \leq t_f^*$, is the time-optimal control which transfers system (3.49) from $y(0) = y^0$ to the origin $y(t_f^*) = 0$ in minimum time t_f^*. Then the control function $v^*(s) = u^*(\mathscr{S}(s))$, $0 \leq s \leq s_f^*$, transfers the multiple integrator system (3.55) from the same initial state $x(0) = y^0$ to the origin $x(s_f^*) = 0$ in time s_f^*, where

$$s_f^* = \mathscr{T}(t_f^*) = \frac{1}{\lambda}\,[1 - \exp\ (-\lambda t_f^*)] \tag{3.57}$$

That s_f^* is the *minimum time* to the origin for the multiple integrator system is easily established by contradiction as follows: suppose $v^*(\cdot)$ is *not* time-optimal, then there exists a control function $\hat{v}(s)$, $0 \leq s \leq \hat{s}_f$, which transfers system (3.55) from y^0 to the origin in time $\hat{s}_f < s_f^*$. Corresponding to the function $\hat{v}(\cdot)$ there exists a function $\hat{u}(t) = \hat{v}(\mathscr{T}(t))$, $0 \leq t \leq \hat{t}_f$, which transfers system (3.49) from y^0 to the origin in time $\hat{t}_f = \mathscr{S}(\hat{s}_f)$. But (within its domain of definition) $\mathscr{S}$ is a monotonically increasing function for all $\lambda \neq 0$, so that $\hat{s}_f < s_f^* \Rightarrow \hat{t}_f < t_f^*$, which clearly contradicts the optimality of t_f^*. Hence, s_f^* is the minimum time to the origin for system (3.55), i.e. the time-optimal control for the original system is mapped by the transformation into the time-optimal control for the multiple integrator system. Conversely, if $\lambda < 0$ (negative real eigenvalues, i.e. stable plant) then the time-optimal control for the multiple integrator system (3.55) maps to the time-optimal control for system (3.49). However, if $\lambda > 0$ (positive real eigenvalues, i.e. unstable plant) then it may be seen from (3.52b) that the inverse transformation $\mathscr{T}^{-1}(s) = \mathscr{S}(s)$ is well defined only for $0 \leq s < 1/\lambda$. Hence, not every time-optimal control function for the multiple integrator system (3.55) can be related to a time-optimal control function for system (3.49) when $\lambda > 0$ but only those multiple integrator controls corresponding to the set of initial states of (3.55) from which the origin can be attained in a time $s_f^* < 1/\lambda$.

In summary, if the time-optimal control u^* exists for the original system (3.49), then this control is mapped by the transformation into the time-optimal control v^* for the equivalent multiple integrator system. Conversely, the time-optimal control v^* for the multiple integrator system (3.55) is mapped by the transformation into the time-optimal control u^* for system (3.49) if $\lambda < 0$; however, if $\lambda > 0$, then time-optimal controls v^* associated with initial states y^0 sufficiently close to the origin, only, can be mapped to corresponding optimal controls u^* for the original system (3.49). This latter observation proves useful in characterising the domain of null controllability $\mathscr{C}$ for unstable plants, as will be seen later.

Finally, suppose that the optimal control (when it exists) for each of systems (3.49) and (3.55) is known as an explicit function of the state, viz.

$$\begin{aligned} u^*(t) &= \Psi_1(y(t)) \\ v^*(s) &= \Psi_2(x(s)) \end{aligned} \tag{3.58}$$

where $\Psi_1(y) = -\text{sgn}\,(\Xi_1(y))$ and $\Psi_2(x) = -\text{sgn}\,(\Xi_2(x))$, Ξ_1 and Ξ_2 being time-optimal switching functions. It readily follows that the functions Ψ_1 and Ψ_2 are, in fact, identical as the following argument shows. Setting $t = 0 \Leftrightarrow s = 0$, then

$$u^*(0) = \Psi_1(y(0)) = \Psi_1(y^0)$$

and

$$v^*(0) = \Psi_2(x(0)) = \Psi_2(y^0)$$

But, $u^*(0) = v^*(\mathscr{S}(0)) = v^*(0)$ and hence

$$\Psi_1(y^0) = \Psi_2(y^0) \tag{3.59}$$

Since y^0 is arbitrary (within the domain of null controllability $\mathscr{C}$ for (3.49)) the two functions must be identical. Note that this result implies that the time-optimal feedback synthesis for the original system (3.49) is *independent of the parameter* λ.

Example 3.6.1 : In Example 2.6.1, it was shown that the time-optimal feedback control for the double integrator system

$$\dot{x}_1 = x_2; \quad \dot{x}_2 = u; \quad |u| \leq 1; \quad x = (x_1, x_2)' \in \mathbb{R}^2$$

is given by

$$u^* = u^*(x) = -\text{sgn}\,(\Xi(x))$$

where the time-optimal switching function $\Xi\colon \mathbb{R}^2 \to \mathbb{R}$ is defined as

$$\Xi(x) = \begin{cases} \xi(x) = x_1 + \tfrac{1}{2}x_2|x_2|; & \xi(x) \neq 0 \\ x_2; & \xi(x) = 0 \end{cases}$$

From the preceding analysis, it immediately follows that the time-optimal feedback control for the system

$$\dot{y}_1 = 2\lambda y_1 + y_2$$

$$\dot{y}_2 = \lambda y_2 + v; \quad |v| \leq 1; \quad \lambda < 0; \quad y = (y_1, y_2)' \in \mathbb{R}^2$$

with negative real eigenvalues in the ratio 2 : 1, is given by

$$v^* = v^*(y) = -\text{sgn}\,(\Xi(y))$$

where the time-optimal switching function $\Xi\colon \mathbb{R}^2 \to \mathbb{R}$ is again defined as above. In Chapter 5 this example is also treated for the case of $\lambda > 0$ (unstable plant); in this case the above synthesis is again valid when restricted to the domain of null controllability $\mathscr{C}$ which can be characterised via the domain of definition of the inverse transformation $\mathscr{T}^{-1} = \mathscr{S}$ of (3.52b).

Chapter 4

Linear autonomous time-optimal regulators

Chapter 2 provided the mathematical framework for analysis of optimal saturating control systems; the fundamental structure and properties of these systems have been developed in Chapter 3, with particular emphasis on feedback synthesis. In the present chapter, the study of time-optimal systems is further developed; additional properties of time-optimal regulating systems are derived. Finally, a predictive control strategy for single-input systems is described and its time-optimality is established. This strategy forms the basis of an efficient method for obtaining explicit closed-form expressions for the time-optimal feedback control laws for many of the specific systems, of up to fourth order, to be considered in Chapters 5, 6 and 7.

4.1 Normal system: theorem on number of switchings in time-optimal control

The *proper*, linear, autonomous, time-optimal regulator problem is assumed throughout this chapter; the system equations are

$$\dot{x} = Ax + Bu; \qquad x(0) = x^0; \qquad x(t) \in \mathbb{R}^n; \qquad u(t) \in \Omega \subset \mathbb{R}^m \tag{4.1a}$$

with

$$\text{Rank}\ [B : AB : \cdots : A^{n-1}B] = n \tag{4.1b}$$

and where the control restraint set Ω is taken as the unit cube in $\mathbb{R}^m$, viz.

$$|u_i(t)| \leq 1; \qquad i = 1, 2, \ldots, m \tag{4.2}$$

In this section, the system is further restricted to satisfy the *normality* condition (2.68), i.e.

$$\text{Rank}\ [b^j : Ab^j : \cdots : A^{n-1}b^j] = n \tag{4.3}$$

for all column vectors b^j, $j = 1, 2, \ldots, m$, of the $n \times m$ matrix B. It is required to determine an admissible control $u^* \in U$ which generates a trajectory emanating from x^0 and reaching the origin $x(t_f^*) = 0$ in minimum time t_f^*. From theorem 2.6.2, the optimal control (when it exists) is piecewise constant and is uniquely determined (almost everywhere) by the relations

$$\begin{aligned} u_j^*(t) &= \operatorname{sgn} \langle \eta(t), b^j \rangle \\ &= \operatorname{sgn} \langle \exp(-A^T t)\eta^0, b^j \rangle; \qquad j = 1, 2, \ldots, m \end{aligned} \tag{4.4}$$

for some $\eta^0 \neq 0$.

The above information proved sufficient to determine the time-optimal control synthesis in a number of earlier examples. A double integrator system (for which the eigenvalues of the system A matrix are both zero) was treated in Example 2.6.1 and the treatment was extended to a system with real non-zero eigenvalues in simple ratio in Example 3.6.1, using the system equivalence property of Section 3.6. In each of these cases, the optimal control was found to contain at most *one* switch on $(0, t_f^*)$, i.e. at most two intervals of constant control on $[0, t_f^*]$; moreover, a time-optimal path to the origin exists from all starting points $x^0 \in \mathbb{R}^2$. A triple integrator system was considered in Sections 1.3.2 and 3.5.4 and a time-optimal path to the state origin was found to exist for all starting points $x^0 \in \mathbb{R}^3$; for this third-order system the time-optimal piecewise constant control was shown to contain at most *two* switches on $(0, t_f^*)$, i.e. three intervals of control constancy on $[0, t_f^*]$. Note that, in each of the above examples, the eigenvalues of the system A matrix are real and the (maximum) number of intervals of control constancy is equal to the dimension of the system. By way of contrast to this observation, Example 2.6.2 (harmonic oscillator) considered a second-order system for which the eigenvalues of the system A matrix are *imaginary* and for which the number of intervals of control constancy on the time-optimal control is *unbounded*. These examples serve to illustrate the following theorems on the number of control switchings for linear time-optimal systems (see also Pontryagin *et al.*, 1962).

Theorem 4.1.1: For the linear, autonomous, normal system in $\mathbb{R}^n$, defined by (4.1)–(4.3), if every eigenvalue of the system A matrix is real, then each piecewise constant component $u_j^*(t)$, $0 \leq t \leq t_f^*$, $j = 1, 2, \ldots, m$ of the time-optimal control can contain at most $(n - 1)$ switches on $(0, t_f^*)$, i.e. at most n intervals of constant control on $[0, t_f^*]$.

Proof

Suppose that the matrix A has a set of r *distinct* real eigenvalues $\{\lambda_1, \lambda_2, \ldots, \lambda_r\}$, the jth eigenvalue having multiplicity m_j, so that $\sum_{j=1}^r m_j = n$ (the order of the system). Noting that the eigenvalues of the adjoint matrix $-A^T$ are simply the negatives of those of A, then the components $\eta_j(\cdot)$ (and all linear combinations thereof) of every nontrivial solution $\eta(\cdot)$ of the adjoint equation $\dot{\eta} = -A^T\eta$ are all of the form

$$p_1(t) \exp(-\lambda_1 t) + p_2(t) \exp(-\lambda_2 t) + \cdots + p_r(t) \exp(-\lambda_r t) \tag{4.5}$$

where $p_j(t)$ are polynomials (not all zero), the degree of which does not exceed $m_j - 1$, where m_j is the multiplicity of eigenvalue λ_j. For example, if the eigenvalues are all distinct (i.e. $m_j = 1, j = 1, 2, \ldots, m$) then $p_j = c_j$ (constant).

It will now be shown (by induction) that the function (4.5) can have at most $(n - 1)$ real roots. The result is obvious for $r = 1$, since $p_1(t) \exp(-\lambda_1 t)$ has the same roots as $p_1(t)$. Suppose now that the result is true for $r = R - 1 (\geq 1)$ and false for $r = R$, i.e. suppose that the function

$$p_1(t) \exp(-\lambda_1 t) + p_2(t) \exp(-\lambda_2 t) + \cdots + p_R(t) \exp(-\lambda_R t) \tag{4.6}$$

has at least $n = \sum_{j=1}^{R} m_j$ real roots. Multiplication of (4.6) by $\exp(\lambda_R t) > 0$ does not effect its roots and yields the function

$$p_1(t) \exp((\lambda_R - \lambda_1)t) + p_2(t) \exp((\lambda_R - \lambda_2)t) + \cdots + p_R(t) \tag{4.7}$$

with at least $\sum_{j=1}^{R} m_j$ real roots, identical to those of (4.6). Noting that the derivative of a smooth function has at least one real root between each pair of real roots of the function itself, it readily follows that the m_R-th derivative of (4.7) must contain at least $\sum_{j=1}^{R} m_j - m_R = \sum_{j=1}^{R-1} m_j$ real roots. Now, the m_R-th derivative of the polynomial $p_R(t)$ (of degree $\leq m_R - 1$) is zero and hence the m_R-th derivative of the function (4.7) is of the form

$$q_1(t) \exp((\lambda_R - \lambda_1)t) + q_2(t) \exp((\lambda_R - \lambda_2)t) + \cdots + q_{R-1} \exp((\lambda_R - \lambda_{R-1})t) \tag{4.8}$$

where $q_j(t)$ are polynomials again of degree $m_j - 1$ at most, and where the terms $(\lambda_R - \lambda_j), j = 1, 2, \ldots, R - 1$, are real and distinct. Now the original supposition that the general result is true for $r = R - 1$ can be applied to conclude that (4.8) has at most $\sum_{j=1}^{R-1} m_j - 1$ real roots which contradicts the earlier result that (4.8) has at least $\sum_{j=1}^{R-1} m_j$ real roots. Hence, the second supposition that the result does not hold for $r = R$ is false, i.e. the result is true for $r = R$ if true for $r = R - 1$. Since the result is true for $r = 1$, the inductive argument can be invoked to establish the general result that the function (4.5) has at most $(n - 1)$ real roots.

From (4.4), the optimal control is given by

$$\begin{aligned} u_j^*(t) &= \operatorname{sgn} \langle \eta(t), b^j \rangle \\ &= \operatorname{sgn} (\eta_1(t) b_1^j + \eta_2(t) b_2^j + \cdots + \eta_n(t) b_n^j); \end{aligned}$$

$$j = 1, 2, \ldots, m \tag{4.9}$$

where

$$b^j = \begin{bmatrix} b_1^j \\ b_2^j \\ \vdots \\ b_n^j \end{bmatrix} \text{ is the } j\text{th column of the matrix } B.$$

The normality condition (4.3) ensures that $\langle \eta(t), b^j \rangle$ does not vanish identically.

Finally, $\langle \eta(t), b^j \rangle$, being a (nontrivial) linear combination of the functions $\eta_i(t)$, is of the form (4.5) and hence has at most $(n-1)$ real roots corresponding to the control discontinuities in (4.9). This completes the proof.

The above theorem establishes an upper bound on the number of time-optimal control switches when the eigenvalues of the system A matrix are real. Example 2.6.2 illustrates, by counterexample, that this upper bound is *invalid* when A has *complex eigenvalues*; the general result in this case is stated below.

Theorem 4.1.2: If A has one or more conjugate pairs of complex eigenvalues, then the number of time-optimal control switches is unbounded in the sense that, given an arbitrary integer $N > 0$, there exists an initial state $x^0 \in \mathbb{R}^n$ for which the corresponding time-optimal regulating control $u^*(t)$, $0 \le t \le t_f^*$, has at least one component $u_j^*(\cdot)$ containing more than N discontinuities.

4.2 Proper system: isochronal hypersurfaces

Let Υ_t denote the set of all states x for which there exists an optimal control transferring the *proper* system (4.1)–(4.2) from x to the origin in the *same minimum time* $t \ge 0$, i.e. Υ_t is the *t-isochronal* set or *t-isochrone* in $\mathbb{R}^n$. It will now be shown that, for each $t > 0$, Υ_t forms a closed convex hypersurface (shell) and, as t increases continuously over the positive reals, the corresponding isochronal hypersurfaces expand continuously from the origin.

Referring to definition 2.5.1 (see also Section 2.6), consider the attainable set

$$K_0(-t) = \left\{ x\colon x = \int_0^{-t} \exp\left(-A(t+s)\right) B u(s)\, ds; \quad u(s),\ -t \le s \le 0 \text{ admissible} \right\}$$

or, equivalently, setting $\sigma = t + s$,

$$K_0(-t) = \left\{ x\colon x = -\int_0^{t} \exp\left(-A\sigma\right) B u[\sigma]\, d\sigma; \quad u[\sigma] = u(\sigma - t),\ 0 \le \sigma \le t, \text{ admissible} \right\}$$

i.e. $K_0(-t)$ is the set of states attainable from the origin in *backwards* or *reverse time* $-t$ under an admissible control $u \in U$. The t-isochronal set Υ_t is precisely the boundary $\partial K_0(-t)$ of the convex, compact set $K_0(-t)$ as the following argument demonstrates.

Suppose initially that $x^1 \in \Upsilon_t$ so that there exists an optimal control $u^*(s)$, $0 \le s \le t$, driving x^1 to the origin in minimum time t. Moreover, by theorem 2.5.3, u^* is *extremal*.† Clearly, the control $u^*[\sigma] = u^*(\sigma - t)$, $0 \le \sigma \le t$, is also

† Recall definition 2.5.2 which states that a control is extremal if it drives the initial state to the boundary of the attainable set; moreover, by theorem 2.5.2, a control is extremal if and only if it satisfies the maximum principle.

extremal and drives the system from the origin to x^1. Hence, by the above definition of the set $K_0(-t)$ and theorem 2.5.2, it follows that x^1 is a boundary point of $K_0(-t)$, giving the result $x^1 \in \Upsilon_t \Rightarrow x^1 \in \partial K_0(-t)$.

Now suppose that $x^1 \in \partial K_0(-t)$ so that there exists an extremal control $u^*[\sigma], 0 \leq \sigma \leq t$, such that

$$x^1 = -\int_0^t \exp(-A\sigma)Bu^*[\sigma]\, d\sigma \in \partial K_0(-t)$$

Clearly, the control $u^*(s) = u^*[t-s], 0 \leq s \leq t$, is also *extremal* (and, by theorem 2.5.2, satisfies the maximum principle) and generates a trajectory $x(s)$, $0 \leq s \leq t$, from $x(0) = x^1$ to the origin $x(t) = 0$. Finally, as the system is *proper* it may be concluded from theorem 2.6.1 that u^* is a *time-optimal* control which drives x^1 to the state origin in minimum time t; this demonstrates the reverse implication $x^1 \in \partial K_0(-t) \Rightarrow x^1 \in \Upsilon_t$.

Combining the above results establishes the equivalence of the sets $\partial K_0(-t)$ and Υ_t. Consequently, the t-isochrone has the characterisation

$$\Upsilon_t = \left\{x\colon x = -\int_0^t \exp(-A\sigma)Bu^*[\sigma]\, d\sigma; \qquad u^* \text{ extremal}\right\} \tag{4.10}$$

That Υ_t $(t > 0)$ is a closed convex hypersurface (shell) in $\mathbb{R}^n$ follows from the convexity and compactness (theorem 2.5.1) of the set $K_0(-t)$. For $t = 0$, $\Upsilon_t = \Upsilon_0 = \{0\}$ and, as t increases continuously over the positive reals, Υ_t expands continuously from the origin (since the set $K_0(-t)$ expands likewise by theorem 2.5.1 and lemma 2.6.1) forming the family $\bigcup_{t\geq 0} \Upsilon_t$ of nonintersecting shells; this family defines the domain of null controllability $\mathscr{C}$ for the system, as will be discussed in Section 4.3.

Example 4.2.1 : Consider again the double integrator system

$$\dot{x}_1 = x_2; \qquad \dot{x}_2 = u; \qquad |u| \leq 1$$

Each extremal control u^* for this second-order system (with real nonpositive eigenvalues) satisfies $|u^*(t)| = 1$ for a.a. t, with at most two intervals of control constancy, i.e. an extremal control defined on the interval $[0, t_f^*]$ can be expressed in the form

$$u^*[t] = \begin{cases} C; & 0 \leq t < t_s \\ -C; & t_s \leq t \leq t_f^* \end{cases}; \qquad C = \pm 1$$

where t_s denotes the instant of control switching. Now, for the example under consideration,

$$A = \begin{bmatrix} 0 & 1 \\ 0 & 0 \end{bmatrix}; \qquad B = b = \begin{bmatrix} 0 \\ 1 \end{bmatrix}; \qquad \text{and} \quad A^k = 0, k = 2, 3, \ldots$$

Hence,

$$\exp(-A\sigma) = I - A\sigma + \tfrac{1}{2}A^2\sigma^2 - \cdots + \frac{(-1)^i}{i!}A^i\sigma^i + \cdots$$
$$= I - A\sigma$$
$$= \begin{bmatrix} 1 & -\sigma \\ 0 & 1 \end{bmatrix}$$

so that

$$\int_0^{t_f*} \exp(-A\sigma)Bu^*[\sigma]\, d\sigma = \int_0^{t_f*} (I - A\sigma)Bu^*[\sigma]\, d\sigma$$
$$= C\left[\int_0^{t_s}\begin{bmatrix} -\sigma \\ 1 \end{bmatrix} d\sigma - \int_{t_s}^{t_f*}\begin{bmatrix} -\sigma \\ 1 \end{bmatrix} d\sigma\right]$$
$$= C\begin{bmatrix} \tfrac{1}{2}[(t_f^*)^2 - 2t_s^2] \\ -(t_f^* - 2t_s) \end{bmatrix}$$

By (4.10), the t_f^*-isochrone Υ_{t_f*} is given by

$$\Upsilon_{t_f*} = \{x = (x_1, x_2)' : x_1 = \tfrac{1}{2}C[(t_f^*)^2 - 2t_s^2];$$
$$x_2 = -C(t_f^* - 2t_s);\ C = \pm 1;\ 0 < t_s \leq t_f^*\}$$

Eliminating t_s from the above expressions for x_1 and x_2 yields

$$\Upsilon_{t_f*} = \left\{x = (x_1, x_2)' : x_1 - \tfrac{1}{2}C(t_f^*)^2 = -\frac{C}{4}(x_2 + Ct_f^*)^2;\right.$$
$$\left. C = \pm 1;\ |x_2| \leq t_f^*\right\} \tag{4.11}$$

i.e. the t_f^*-isochrone is comprised of two parabolic arcs (see also Rang, 1963), bounded by $|x_2| \leq t_f^*$, viz. (i) with vertex at $(\tfrac{1}{2}(t_f^*)^2, -t_f^*)$ and opening to the left, corresponding to $C = +1$; and (ii) with vertex at $(-\tfrac{1}{2}(t_f^*)^2, t_f^*)$ and opening to the right, corresponding to $C = -1$. It is straightforward to verify that the t_f^*-isochrone is identical to the constant cost surface $V(x) = t_f^*$, previously derived for the double integrator system in Example 2.6.1 and depicted in Fig. 2.5, viz.

$$\Upsilon_{t_f*} = \{x : V(x) = t_f^*\} \tag{4.12a}$$

where, from (2.84),

$$V(x) = x_2 \operatorname{sgn}(\Xi) + 2\sqrt{x_1 \operatorname{sgn}(\Xi) + \tfrac{1}{2}x_2^2} \tag{4.12b}$$

with

$$\Xi = \Xi(x) = \begin{cases} \xi(x) = x_1 + \tfrac{1}{2}x_2|x_2|; & \xi(x) \neq 0 \\ x_2; & \xi(x) = 0 \end{cases} \tag{4.12c}$$

As t_f^* increases from zero, a family of isochrones is generated which expands to fill the state space $\mathbb{R}^2$, i.e.

$$\bigcup_{t_f^* \geq 0} \Upsilon_{t_f*} = \mathbb{R}^2$$

4.3 Proper system: domain of null controllability

The set $\mathscr{C}$ of all initial states $x^0 \in \mathbb{R}^n$ which can be transferred to the state origin in finite time by an admissible control is termed the *domain of null controllability* for the linear, autonomous, proper system (4.1)–(4.2).† By theorem 2.5.3, it follows that, for each $x^0 \in \mathscr{C}$, there exists an optimal (and hence extremal) control which transfers x^0 to the origin in minimum time. Moreover, as the system is proper, it may be concluded from theorems 2.5.2, 2.5.3 and 2.6.1 that a state x^0 belongs to $\mathscr{C}$ if, and only if, there exists a time $0 \leq t_f^* < \infty$ and an *extremal* control $u^*(t)$, $0 \leq t \leq t_f^*$ such that

$$x(t_f^*) = \phi(t_f^*, x^0, u^*) = \exp(At_f^*)x^0 + \int_0^{t_f*} \exp(A(t_f^* - s))Bu^*(s)\,ds = 0$$

In other words,

$$x^0 \in \mathscr{C} \Leftrightarrow \exp(At_f^*)\left[x^0 + \int_0^{t_f*} \exp(-As)Bu^*(s)\,ds\right] = 0$$

$$\Leftrightarrow x^0 = -\int_0^{t_f*} \exp(-As)Bu^*(s)\,ds \qquad (4.13)$$

for some time $t_f^* \in [0, \infty)$ and some extremal control u^* defined on $[0, t_f^*]$. Recalling the characterisation (4.10) of the t_f^*-isochronal set Υ_{t_f*}, (4.13) states that x^0 belongs to the set $\mathscr{C}$ if, and only if, x^0 belongs to the set Υ_{t_f*} for some $t_f^* \in [0, \infty)$. Consequently, the domain of null controllability $\mathscr{C}$ is the collection of all isochronal sets Υ_{t_f*} generated as t_f^* ranges over the non-negative reals, viz.

$$\mathscr{C} = \bigcup_{t_f* \geq 0} \Upsilon_{t_f*} \qquad (4.14)$$

The following theorem establishes certain properties of the domain of null controllability $\mathscr{C}$.

Theorem 4.3.1: The domain of null controllability $\mathscr{C}$ for the linear, autonomous, proper system (4.1)–(4.2) is an *open, convex* set in $\mathbb{R}^n$. If *no* eigenvalue of the system A matrix has a positive real part, then $\mathscr{C}$ coincides with the whole state space $\mathbb{R}^n$, i.e. all initial states are null controllable. If *all* eigenvalues of A have positive real parts, then $\mathscr{C}$ is an open, convex and *bounded* set in $\mathbb{R}^n$.

† For results on null controllability for *time-varying* linear systems with constrained controls, see Barmish and Schmitendorf (1980) and Schmitendorf and Barmish (1980).

Proof
Convexity follows from the convexity of the sets $K_0(-t_f^*)$ which comprise the domain of null controllability. To see that $\mathscr{C}$ is open note that, since the t-isochronal sets Υ_t expand with increasing t, given a point $x^1 \in \Upsilon_{t_1*}$ there always exists a time $t_2^* > t_1^*$ such that x^1 lies inside the t_2^*-isochrone Υ_{t_2*}.

To prove that $\mathscr{C} = \mathbb{R}^n$ when *no* eigenvalue of A has a positive real part, use is made of the following property of convex sets.

> *Property of convex sets:* If K is a convex set in $\mathbb{R}^n$ with the property that, given any real number $k > 0$ and any vector $\eta^0 \in \mathbb{R}^n$, there exists a vector $y \in K$ such that $\langle \eta^0, y \rangle > k$, then $K = \mathbb{R}^n$. In other words, if for every vector $\eta^0 \in \mathbb{R}^n$ a vector y can be found in the convex set K such that the scalar product $\langle \eta^0, y \rangle$ takes an arbitrarily large value, then K coincides with the whole space $\mathbb{R}^n$. For example, the convex set K of Fig. 4.1*a* clearly fails to meet this requirement for every choice of η^0; on the other hand, the convex set K of Fig. 4.1*b* satisfies the requirement for η^0 in the direction y but not in the direction z. If the requirement can be met for every choice of η^0, then $K = \mathbb{R}^n$.

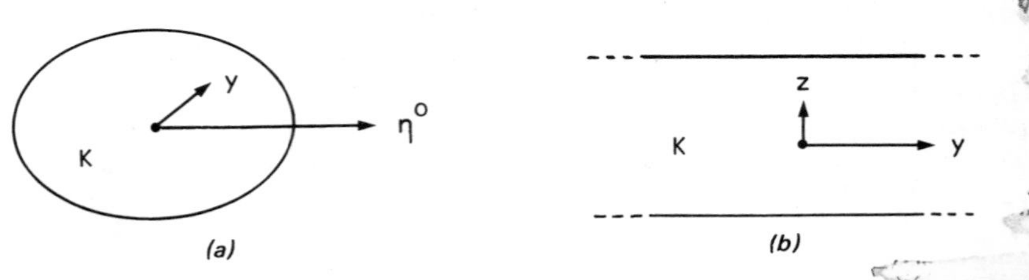

Fig. 4.1

Returning to the proof of the theorem, since the system is *proper*, for any $\eta^0 \neq 0$ at least one component of the $\mathbb{R}^m$-valued function $v(s) = B^T \exp(-A^T s)\eta^0$ does not vanish identically on any finite interval I, i.e. $v_j(s) = \langle \exp(-As)b^j, \eta^0 \rangle \not\equiv 0$ on any finite interval I for some $1 \leq j \leq m$, as otherwise the (proper) condition (4.1b) is violated (this can be established as in the proof of lemma 2.6.1). Suppose $v_j \not\equiv 0$ is such that

$$\int_0^\infty |v_j(s)| \, ds < \infty \tag{4.15}$$

Under this supposition, the integral $\int_0^\infty v_j(s)\, ds$ converges and hence the real function

$$z(t) = \int_t^\infty v_j(s)\, ds; \qquad 0 \leq t < \infty \tag{4.16}$$

is well defined. Now, the eigenvalues λ_i of A are the roots of the characteristic equation $\Delta(\lambda) = |\lambda I - A| = 0$ and, by the Cayley–Hamilton theorem, the matrix A satisfies its own characteristic equation, i.e. $\Delta(A) = 0$ which, in turn, implies that

$$\begin{aligned}\Delta(-D)v_j &= \Delta(-D)\langle \eta^0, \exp(-As)b^j\rangle \\ &= \langle \eta^0, \Delta(A)\exp(-As)b^j\rangle \\ &= 0\end{aligned} \tag{4.17}$$

where D denotes the differential operator. From (4.16),

$$Dz = \frac{d}{dt}z = -v_j$$

which, when combined with (4.17), yields a linear, homogeneous differential equation for z, viz.

$$D\Delta(-D)z = 0 \tag{4.18a}$$

with characteristic equation

$$\Delta_1(\lambda) = \lambda\,\Delta(-\lambda) = 0 \tag{4.18b}$$

Clearly, the roots of (4.18b) have *non-negative* real parts (since the roots λ_i of $\Delta(\lambda) = 0$ have nonpositive real parts) so that (4.18a) is *not* asymptotically stable, i.e.

$$z(t) \nrightarrow 0 \quad \text{as} \quad t \to \infty$$

which yields the required contradiction as, from its definition in (4.16), $z(t) \to 0$ as $t \to \infty$. Hence, supposition (4.15) is false so that

$$\int_0^t |v_j(s)|\, ds \to \infty \quad \text{as} \quad t \to \infty$$

Finally, selecting the admissible control vector function $u = (u_1, u_2, \ldots, u_m)'$ as follows

$$\left.\begin{aligned} u_j(t) &= \operatorname{sgn}(v_j(t)) = \operatorname{sgn}\langle \eta^0, \exp(-At)b^j\rangle \\ u_i(t) &= 0; \qquad i \neq j \end{aligned}\right\} 0 \le t < \infty$$

and defining the vector function

$$\begin{aligned} y(t) &= \int_0^t \exp(-As)Bu(s)\, ds \\ &= \int_0^t \exp(-As)b^j u_j(s)\, ds \in \mathscr{C} \end{aligned}$$

then

$$\langle \eta^0, y(t) \rangle = \left\langle \eta^0, \int_0^t \exp(-As) b^j u_j(s) \, ds \right\rangle$$

$$= \int_0^t |\langle \eta^0, \exp(-As) b^j \rangle| \, ds$$

$$= \int_0^t |v_j(s)| \, ds \to \infty \qquad \text{as} \qquad t \to \infty$$

Hence, by the earlier property of convex sets, the result is established, i.e.

$$\mathscr{C} = \mathbb{R}^n \qquad \text{if} \qquad \operatorname{Re}(\lambda_i) \leq 0; \qquad i = 1, 2, \ldots, n$$

Consider now the unstable case in which the eigenvalues of A all have positive real parts ($\operatorname{Re}(\lambda_i) > 0$, $i = 1, 2, \ldots, n$), then the eigenvalues of $-A$ all have negative real parts and, consequently, there exist constants $M > 0$ and $\omega > 0$ such that $\|\exp(-As)\| \leq M \exp(-\omega s)$. Hence, for any admissible control $u \in U$,

$$\left\| \int_0^t \exp(-As) Bu(s) \, ds \right\|$$

$$\leq \int_0^t \|\exp(-As)\| \; \|B\| \; \|u(s)\| \, ds$$

$$\leq \sqrt{m}\, M \|B\| \int_0^t \exp(-\omega s) \, ds$$

$$(\text{as } |u_i(s)| \leq 1,\ i = 1, 2, \ldots, m \Rightarrow \|u(s)\| \leq \sqrt{m})$$

$$= C[1 - \exp(-\omega t)]; \qquad 0 < C = \sqrt{m}\, M \|B\| \omega^{-1} < \infty$$

In particular,

$$\|y(t)\| \leq \sqrt{m}\, M \|B\| \omega^{-1} \qquad \text{for all} \qquad 0 \leq t < \infty$$

from which the boundedness of the domain of null controllability $\mathscr{C}$ immediately follows. This concludes the proof of the theorem.

In summary, the open, convex domain of null controllability $\mathscr{C}$ coincides with the whole state space $\mathbb{R}^n$ when the eigenvalues of the system A matrix have non-positive real parts (note that this includes the case of eigenvalues lying on the imaginary axis in the complex plane), while $\mathscr{C}$ is an open, convex and bounded set in $\mathbb{R}^n$ when *all* eigenvalues of A have positive real parts. What if some, but not all, eigenvalues have positive real parts? Loosely speaking, in this case the open, convex domain of null controllability $\mathscr{C}$ is unbounded in certain directions (associated with the eigenvectors corresponding to the stable eigenvalues of A) and bounded in other directions (associated with the eigenvectors corresponding to the unstable eigenvalues of A), as will be seen in a number of specific cases in Chapters 5, 6, 7.

Example 4.3.1 System with positive eigenvalues in simple ratio: Consider the system

$$\dot{x}_1 = 2x_1 + x_2; \qquad \dot{x}_2 = x_2 + u; \qquad |u| \le 1 \tag{4.19}$$

The eigenvalues of the system matrix

$$A = \begin{bmatrix} 2 & 1 \\ 0 & 1 \end{bmatrix} \quad \text{are} \quad \lambda_1 = 2, \lambda_2 = 1.$$

Hence, by theorem 4.3.1, the domain of null controllability $\mathscr{C}$ is an open, convex and *bounded* set in $\mathbb{R}^n$. Noting that the system eigenvalues are in the simple ratio 2 : 1, the system equivalence property of Section 3.6 can be invoked to reduce the system to the equivalent double integrator form of Example 4.2.1. In particular, by virtue of the time transformation (3.52b), it may be concluded that, if the t_f^*-isochrone Υ_{t_f*} for the double integrator is given (as in Example 4.2.1) by

$$\Upsilon_{t_f*} = \{x\colon T(x) = V(x) = t_f^*\} \tag{4.20a}$$

where

$$T(x) = x_2 \operatorname{sgn}(\Xi) + 2\sqrt{x_1 \operatorname{sgn}(\Xi) + \tfrac{1}{2}x_2^2} \tag{4.20b}$$

$$\Xi = \Xi(x) = \begin{cases} \xi(x) = x_1 + \tfrac{1}{2}x_2|x_2|; & \xi(x) \neq 0 \\ x_2; & \xi(x) = 0 \end{cases} \tag{4.20c}$$

then the t_f^*-isochrone for system (4.19) may be expressed as

$$\Upsilon_{t_f*} = \{x\colon \ln[1 - T(x)]^{-1} = t_f^*\} \tag{4.21}$$

Consequently, the open, convex and bounded domain of null controllability may be explicitly characterised as

$$\mathscr{C} = \bigcup_{t_f^* \ge 0} \Upsilon_{t_f*} = \{x\colon 1 - T(x) > 0\} \tag{4.22}$$

with $T(x)$ defined by (4.20), i.e. $\mathscr{C}$ corresponds to the region interior to the closed curve $\partial\mathscr{C} = \{x\colon T(x) = 1\}$ as depicted in Fig. 4.2.

4.4 Single-input, nonoscillatory systems: time-optimal control on state axes and in regions where all state co-ordinates have same sign

Consider the system of Fig. 4.3 decomposed serially into n cascaded first-order elements, the respective outputs of which (the variables $(x_1(t), x_2(t), \ldots, x_n(t))'$) constitute the state vector $x(t) \in \mathbb{R}^n$ and satisfy the state equations:

$$\dot{x}_i = \lambda_i x_i + x_{i+1}; \qquad i = 1, 2, \ldots, n-1 \tag{4.23a}$$

$$\dot{x}_n = \lambda_n x_n + \frac{u}{a}; \qquad |u| \le 1; \qquad a > 0 \tag{4.23b}$$

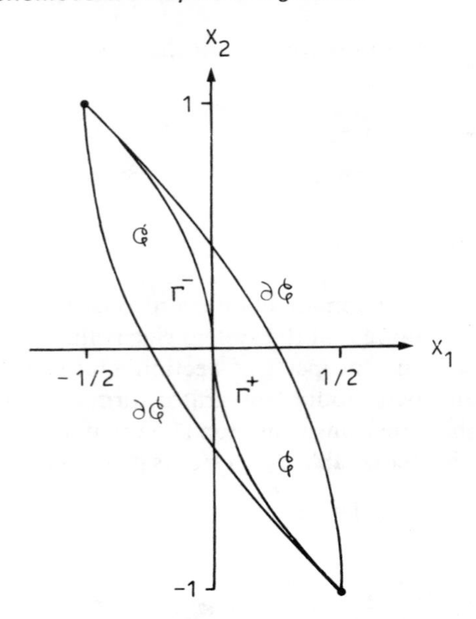

Fig. 4.2 *Example 4.3.1: domain of null controllability $\mathscr{C}$ and time-optimal switching curve* Γ

It is remarked that the real eigenvalues $\lambda_1, \lambda_2, \ldots, \lambda_n$ need *not* be distinct and can take positive, zero or negative values. It will be shown that, at a null controllable state $x^0 \in \mathscr{C}$ in the region where all state coordinates are non-negative, the time-optimal control u^* takes the value -1; while, by symmetry, if $x^0 \in \mathscr{C}$ lies in the region where all state coordinates are nonpositive, then the time-optimal control u^* takes the value $+1$. This result is proved for multiple integrator systems in Fuller (1974d).

Denoting, by Q^p, the set of null controllable states in the positive orthant, i.e.

$$Q^p = \{x \in \mathscr{C}: x_i \geq 0;\ i = 1, 2, \ldots, n;\ x \neq 0\} \tag{4.24}$$

and, by Q^n, the set of null controllable states in the negative orthant, i.e.

$$Q^n = \{x \in \mathscr{C}: x_i \leq 0;\ i = 1, 2, \ldots, n;\ x \neq 0\} \tag{4.25}$$

then,

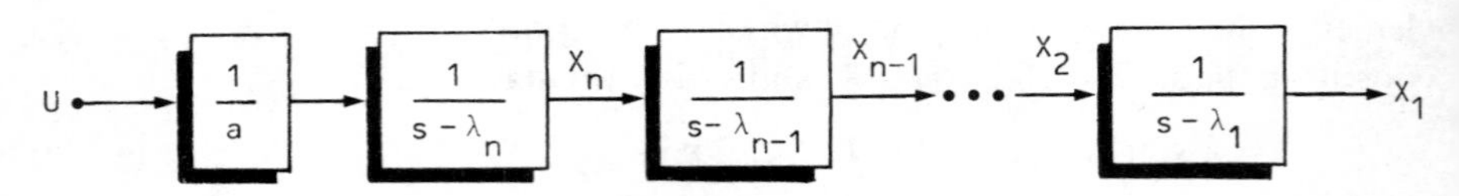

Fig. 4.3 *Serially decomposed system*

Lemma 4.4.1: The time-optimal control u^* takes the value -1 at every point of Q^p and the value $+1$ at every point of Q^n, i.e.

$$u^* = u^*(x) = \begin{cases} -1; & x \in Q^p \\ +1; & x \in Q^n \end{cases} \tag{4.26}$$

Proof
From the earlier theorems, for an arbitrary null controllable state $x^0 \in \mathscr{C}$, a unique optimal control $u^*(t)$, $0 \le t \le t_f^*$, exists which generates a trajectory $x(t) = (x_1(t), x_2(t), \ldots, x_n(t))'$ emanating from $x(0) = x^0$ and terminating at the origin $x(t_f^*) = 0$ in minimum time t_f^*; moreover, the control function $u^*(\cdot)$ is piecewise constant with $|u^*(t)| = 1$ and contains N intervals of constancy, i.e. N-1 discontinuities on $(0, t_f^*)$ occurring at times $t = \tau_i \in (0, t_f^*)$, $i = 1, 2, \ldots, N-1$, where

$$1 \le N \le n \tag{4.27}$$

Denote, by $u^0(= \pm 1)$, the value of control on the first interval, i.e.

$$u^*(t) = u^0; \qquad 0 \le t < \tau_1 \tag{4.28}$$

then, on the final or Nth interval,

$$u^*(t) = (-1)^{N+1}u^0; \qquad \tau_{n-1} \le t < t_f^* \tag{4.29}$$

Let $I_i \ge 0$, $i = 1, 2, \ldots, n$ denote the number of zeros of the continuous function $x_i(\cdot)$ which occur on the open interval $(0, t_f^*)$; i.e. interior zeros only are counted, the zero end value $x_n(t_f^*) = 0$ and a possible zero initial value $x_i(0) = 0$ are excluded. It will first be established that

$$N - 1 \ge I_n \ge \cdots \ge I_{i+1} \ge I_i \ge \cdots \ge I_1 \tag{4.30}$$

i.e. the number of zeros of the function $x_i(\cdot)$ cannot exceed the number of zeros of the function $x_{i+1}(\cdot)$ or the number of control function discontinuities.† Consider initially the continuous piecewise differentiable function $x_n(t)$, $0 \le t \le t_f^*$, with endpoints $x_n(0) = x_n^0$ and $x_n(t_f^*) = 0$, generated by $u^*(\cdot)$ through the linear first-order differential equation (4.23b). To every zero of $x_n(\cdot)$ there must correspond one or two intervals of control constancy, as depicted in Fig. 4.4*a*, *b*; moreover, on the final (Nth) interval of control constancy, no interior zeros occur (Fig. 4.4*c*). Hence, the number I_n of zeros of the function $x_n(\cdot)$ which occur between its endpoints cannot exceed the number $(N-1)$ of discontinuities in the piecewise constant control function u^*, i.e.

$$N - 1 \ge I_n \ge 0$$

Consider now the continuously differentiable function $x_{n-1}(t)$, $0 \le t \le t_f^*$, with endpoints $x_{n-1}(0) = x_{n-1}^0$ and $x_{n-1}(t_f^*) = 0$, which is generated through the

† This result is not surprising; for example, with $\lambda < 0$, each successive block of Fig. 4.3 acts as a lowpass filter, smoothing the output of the preceding block.

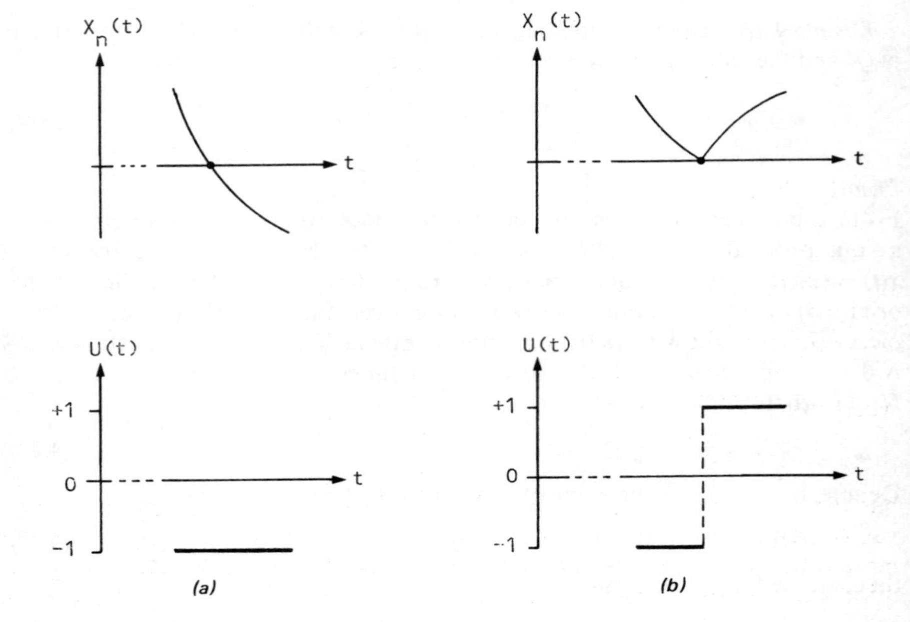

(a) (b)

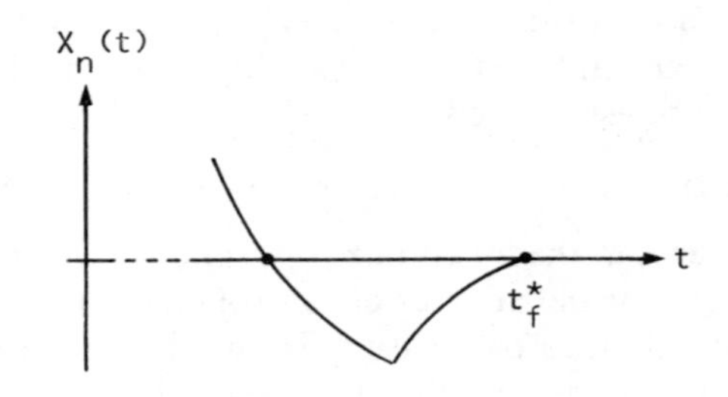

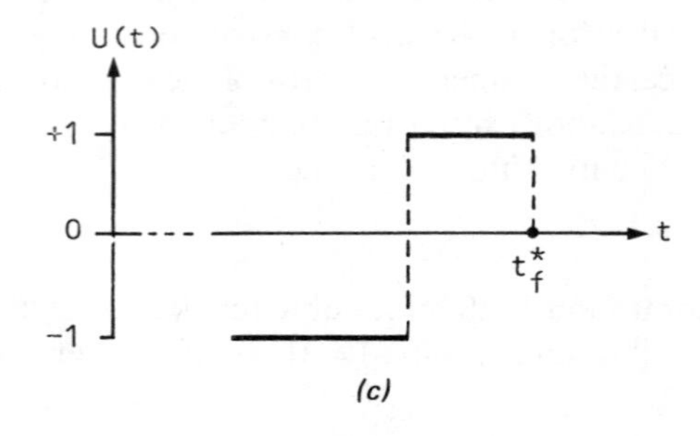

(c)

Fig. 4.4

first-order differential equation (4.23a) with $i = n - 1$, viz.

$$x_{n-1}(t) = \exp(\lambda_{n-1}t)\left[x_{n-1}^0 + \int_0^t \exp(-\lambda_{n-1}s)x_n(s)\,ds\right] \quad (4.31)$$

Clearly, the function $x_{n-1}(t)$ has the same zeros as the function $g_{n-1}(t) = \exp(-\lambda_{n-1}t)x_{n-1}(t)$, $0 \le t \le t_f^*$. Consequently, from (4.31), the number I_{n-1} of interior zeros of $x_{n-1}(\cdot)$ cannot exceed the number of interior zeros of the continuously differentiable function

$$g_{n-1}(t) = x_{n-1}^0 + \int_0^t f_n(s)\,ds; \qquad 0 \le t \le t_f^* \quad (4.32a)$$

where the continuous, piecewise differentiable function

$$f_n(t) = \exp(-\lambda_{n-1}t)x_n(t); \qquad 0 \le t \le t_f^* \quad (4.32b)$$

has the same zeros as $x_n(\cdot)$. But, the derivative f_n of the function g_{n-1} must vanish at least once between each pair of successive interior zeros of g_{n-1}; moreover, as $g_{n-1}(t_f^*) = \exp(-\lambda_{n-1}t_f^*)x_{n-1}(t_f^*) = 0$, the derivative f_n of g_{n-1} must also vanish at least once between the final interior zero and the zero endpoint $(g_{n-1}(t_f^*) = 0)$. Hence, the number I_{n-1} of interior zeros of g_{n-1} (or x_{n-1}) cannot exceed the number I_n of interior zeros of f_n (or x_n), i.e.

$$N - 1 \ge I_n \ge I_{n-1} \ge 0$$

The above argument can be applied in a similar manner to the remaining functions $x_i(\cdot)$, $i = n-2, n-3, \ldots, 2, 1$, to establish (4.30).

Now let x^0 lie in the region Q^p, i.e.

$$x_i^0 \ge 0; \qquad i = 1, 2, \ldots, n \quad (4.33)$$

with at least one non-zero component $x_j^0 \ne 0$. Suppose, contrary to the lemma, that:

$$u^0 = +1 = \operatorname{sgn}(x_j^0) \quad (4.34)$$

then it is easily verified that, at $t = 0^+$,

$$\operatorname{sgn}(x_i(0^+)) = +1; \qquad i = 1, 2, \ldots, n \quad (4.35)$$

while, at $t = t_f^{*-}$,

$$\operatorname{sgn}(x_i(t_f^{*-})) = (-1)^{N+n-i}; \qquad i = 1, 2, \ldots, n \quad (4.36)$$

i.e. all components of the state vector are positive in a (sufficiently small) neighbourhood of the starting point, while the components have an alternating sequence of signs immediately prior to attaining the origin. This latter property (4.36) can be verified by integrating the state equations (4.23), in *reverse time*, backwards from the origin under the final-interval control value $(-1)^{N+1}$.

Clearly, if $\operatorname{sgn}(x_i(0^+)) = -\operatorname{sgn}(x_i(t_f^{*-}))$, then the function $x_i(\cdot)$ must have an *odd* number I_i of zeros on $(0, t_f^*)$; while, if $\operatorname{sgn}(x_i(0^+)) = \operatorname{sgn}(x_i(t_f^{*-}))$, then I_i must be *even* valued (or *zero*). From (4.35) and (4.36), all components of $x(0^+)$ have the same sign, while the components of $x(t_f^{*-})$ have alternating signs, and hence

$$I_{i+1}\begin{Bmatrix}\text{even (or zero)}\\ \text{odd}\end{Bmatrix} \Leftrightarrow I_i\begin{Bmatrix}\text{odd}\\ \text{even (or zero)}\end{Bmatrix}; \qquad i = 1, 2, \ldots, n-1 \tag{4.37}$$

Moreover, from (4.36),

$$N-1\begin{Bmatrix}\text{even (or zero)}\\ \text{odd}\end{Bmatrix} \Leftrightarrow I_n\begin{Bmatrix}\text{odd}\\ \text{even (or zero)}\end{Bmatrix} \tag{4.38}$$

In view of (4.37) and (4.38), it may be concluded that equality cannot hold in (4.30), so that the integer I_{i+1} must exceed I_i in value by at least 1 and integer $(N-1)$ must exceed I_n by at least 1, i.e.

$$\begin{aligned} I_{i+1} &\geq I_i + 1; \qquad i = 1, 2, \ldots, n-1 \\ N - 1 &\geq I_n + 1 \end{aligned} \tag{4.39}$$

Combining (4.30) and (4.39) yields the result

$$N - 1 \geq I_n + 1 \geq I_{n-1} + 2 \geq I_{n-2} + 3 \geq \cdots \geq I_1 + n \geq n \tag{4.40}$$

But, by theorem 4.4.1, the number N of intervals of control constancy cannot exceed the order n of the system. Consequently, (4.40) leads to the required contradiction. Hence, supposition (4.34) is false and it may be concluded that the optimal control takes the value -1 for states $x \in Q^p$ in the positive orthant; by symmetry, the optimal control takes the value $+1$ for states $x \in Q^n$ in the negative orthant. This establishes the lemma, a particular case of which is the following corollary (see also Fuller, 1980d) which is used in the next section in proving the time-optimality of a predictive control strategy.

Corollary 4.4.1 : The time-optimal control at a point $x = (0, 0, \ldots, x_i, \ldots, 0)'$, $x_i \neq 0$, on the x_i state axis is given by

$$u^* = -\operatorname{sgn}(x_i) \tag{4.41}$$

4.5 Time-optimal strategy of Gulko *et al.*

The concept of predictive control is well established; the underlying idea is to determine the controlling input to a plant by comparing predicted responses to a range of candidate inputs, these predictions being generated via a model of the plant operating on a fast time scale. Coales and Noton (1956) introduced one such predictive strategy for second-order relay control systems. This strategy was subsequently generalised to third and higher order single-input systems by Gulko *et al.* (Gulko, 1963; Gulko and Kogan, 1963; Gulko *et al.*, 1964). Fuller (1973b) established that, for linear plants, the idealised strategy gives exact time-optimal control to the origin, where, by idealised, is meant that the model of the plant operates on an infinitely fast time scale so that the predicted responses can be generated instantaneously.

In the present context, it is the *time-optimality* of the idealised predictive strategy that is of interest as this provides the basis of an efficient method for obtaining closed-form expressions for the time-optimal feedback controls and associated switching hypersurfaces in a variety of cases to be investigated in the following three chapters. Discussion on the practicality of, and modifications to, the strategy in its original context as a suboptimal predictive control technique is deferred until Chapter 10.

For single-input systems with real eigenvalues, the idealised strategy of Gulko *et al.* may be interpreted as follows.

4.5.1 Time-optimal strategy for serially decomposed systems

The time-optimal regulator problem is studied in the case of the serially decomposed single-input system of Fig. 4.3, viz.

$$\dot{x}_i = \lambda_i x_i + x_{i+1}; \qquad i = 1, 2, \ldots, n-1$$

$$\dot{x}_n = \lambda_n x_n + \frac{u}{a}; \qquad |u| \le 1; \qquad a > 0 \tag{4.42a}$$

or, equivalently,

$$\dot{x} = Ax + bu \tag{4.42b}$$

where

$$A = \begin{bmatrix} \lambda_1 & 1 & 0 & \cdots & 0 & 0 \\ 0 & \lambda_2 & 1 & \cdots & 0 & 0 \\ \vdots & & \vdots & \ddots & \vdots & \vdots \\ 0 & 0 & 0 & \cdots & \lambda_{n-1} & 1 \\ 0 & 0 & 0 & \cdots & 0 & \lambda_n \end{bmatrix}; \qquad b = \begin{bmatrix} 0 \\ 0 \\ \vdots \\ 0 \\ \dfrac{1}{a} \end{bmatrix} \tag{4.42c}$$

with a null controllable initial state

$$x(0) = x^\alpha = (x_1^\alpha, x_2^\alpha, \ldots, x_n^\alpha)' \in \mathscr{C} \subseteq \mathbb{R}^n \tag{4.42d}$$

Consider now the $(n-1)$th-order subsystem formed by deleting the first member of equation set (4.42a); the time-optimal *subsystem* control $u^s(t)$, $0 \le t \le t_{n-1}^s$ which transfers this subsystem from its initial state

$x^s(0) = x^{s\alpha} = (x_2^\alpha, x_3^\alpha, \ldots, x_n^\alpha)'$ to the *subspace* ($\mathbb{R}^{n-1}$) origin $x^s(t_{n-1}^s) = 0$ in minimum time t_{n-1}^s will clearly transfer the overall nth-order system from $x(0) = x^\alpha$ to a point on the x_1 state axis, i.e.

$$x(t_{n-1}^s) = \phi(t_{n-1}^s, x^\alpha, u^s) = x^\beta = (x_1^\beta, 0, 0, \ldots, 0)' \tag{4.43}$$

The value of control (at x^α) for time-optimal control to the *origin* for the overall nth-order system can now be expressed (as will be proved) as

$$u^*(0) = \begin{cases} -\operatorname{sgn}(x_1^\beta); & x_1^\beta \neq 0 \\ u^s(0); & x_1^\beta = 0 \end{cases} \tag{4.44}$$

In its original predictive control context, this result is used as follows: the quantity x_1^β is calculated by using a fast model to generate the trajectory from x^α to x^β under the time-optimal subsystem control u^s, the input to the actual plant is then determined by (4.44).

Proof of time-optimality: Denote, by Γ, the $(n-1)$-dimensional hypersurface in state space composed of the origin and the set of state trajectories which go to the origin under a piecewise constant control $u^*(\cdot)$, with $|u(t)| = 1$ and at most $(n-1)$ intervals of control constancy (which, by theorem 4.1.1, is one less than the maximum number n). From theorems 2.6.2, 4.1.1 and 4.3.1 it may be concluded that Γ partitions the domain of null controllability $\mathscr{C} \subseteq \mathbb{R}^n$ into two mutually exclusive regions, Σ^+ and Σ^-, wherein the time-optimal control takes the values $+1$ and -1, respectively; Γ is, in fact, the time-optimal switching hypersurface.

Consider now the state path $\alpha\beta$ from x^α to x^β, generated by the time-optimal *subsystem* control u^s, i.e.

$$\alpha\beta = \{x(t) = \phi(t, x^\alpha, u^s), 0 \leq t \leq t_{n-1}^s; \quad x(t_{n-1}^s) = x^\beta = (x_1^\beta, 0, 0, \ldots, 0)'\}$$

For definiteness, let

$$x_1^\beta > 0 \tag{4.45}$$

then it will be established that the trajectory $x(t), 0 \leq t \leq t_{n-1}^s$, lies entirely in the region Σ^-. By corollary 4.4.1, the optimal control at β is given by $u^* = -\operatorname{sgn}(x_1^\beta)$ so that the endpoint x^β of path $\alpha\beta$ certainly lies in the region Σ^-. Now assume that the trajectory $x(\cdot)$, i.e. path $\alpha\beta$, does *not* lie entirely in Σ^-, then it must intersect the time-optimal switching hypersurface Γ at some point x^γ. From x^γ, a path $\gamma 0$ to the origin can be generated in Γ by an admissible piecewise constant control $\tilde{u}$, with $|\tilde{u}(t)| = 1$ and at most $(n-1)$ intervals of constancy. Then, along $\gamma 0$, the $(n-1)$th-order subsystem is transferred time-optimally from $x^{s\gamma} = (x_2^\gamma, x_3^\gamma, \ldots, x_n^\gamma)'$ to the subspace origin (since the control $\tilde{u}$ satisfies the sufficient conditions for subsystem time-optimality). But the path $\gamma\beta$

also corresponds to a time-optimal path from $x^{s\gamma}$ to the subspace origin. Hence, by uniqueness of time-optimal control, paths $\gamma 0$ and $\gamma\beta$ must coincide, implying $x^{\beta} = 0$ which contradicts (4.45). Therefore, the supposition that $\alpha\beta$ does not lie entirely in Σ^- is false so that, in particular, the initial state x^{α} lies in Σ^-, whence the result

$$x^{\alpha} \in \Sigma^- \Rightarrow u^*(0) = -1 = -\operatorname{sgn}(x_1^{\beta})$$

Thus, (4.44) has been established for $x_1^{\beta} > 0$ and, by symmetry, it must also hold for $x_1^{\beta} < 0$. It now remains to establish the result for $x_1^{\beta} = 0$. In this case, the subsystem control function u^s not only gives time-optimal control to the subspace ($\mathbb{R}^{n-1}$) origin, but also to the origin of the overall state space $\mathbb{R}^n$, so that $u^* \equiv u^s$ and, in particular,

$$u^*(0) = u^s(0); \qquad \text{if} \qquad x_1^{\beta} = 0$$

In fact, in the latter case, the starting point x^{α} is a point of the overall time-optimal switching hypersurface Γ. This completes the proof.

4.5.2 Application of the time-optimal strategy to the control synthesis problem

Time-optimal control synthesis for serially decomposed systems of the form (4.42), only, will be studied here. However, it is remarked that the approach can easily be modified to apply to single-input diagonalised systems (using the time-optimality of the predictive strategy for diagonalised systems as proved by Fuller (1973b)). It is assumed that the time-optimal *feedback* control $u^s(x^s) = -\operatorname{sgn}(\Xi^s(x^s))$ is explicitly known for the $(n-1)$th-order subsystem obtained on deletion of the first member of equation set (4.42). Moreover, as the $(n-1)$ subsystem eigenvalues (λ_i, $i = 2, 3, \ldots, n$) are real, the optimal control can contain at most $(n-2)$ switches occurring at times t_i^s, $i = 1, 2, \ldots, n-2$, the subsystem origin being reached at time t_{n-1}^s. It is assumed that the dependence, $t_i^s = T_i^s(x^s)$, of these times on the subsystem starting point x^s is known explicitly. Summarising, it is assumed that the time-optimal control problem for the $(n-1)$th order subsystem has been completely solved in the sense that closed-form expressions are available for the functions Ξ^s: $\mathscr{C}^s \subseteq \mathbb{R}^{n-1} \to \mathbb{R}$ and T_i^s: $\mathscr{C}^s \subseteq \mathbb{R}^{n-1} \to [0, \infty)$, $i = 1, 2, \ldots, n-1$ (where $\mathscr{C}^s$ denotes the subsystem domain of null controllability). This knowledge of the subsystem enables the optimal feedback control $u^*(x) = -\operatorname{sgn}(\Xi(x))$, Ξ: $\mathscr{C} \subseteq \mathbb{R}^n \to \mathbb{R}$, for the full nth order system to be derived from the strategy as follows. Starting from an arbitrary null-controllable state $x = \left[\begin{smallmatrix} x_1 \\ x^s \end{smallmatrix}\right] \in \mathscr{C}$ at time $t = 0$, the endpoint x^{β} of the (fictitious) optimal trajectory to the x_1 state axis, i.e. subspace origin,† is easily characterised (as an explicit function of x) by integrating the full state equations

† Note that $x \in \mathscr{C} \Rightarrow x^s \in \mathscr{C}^s$ (but the converse is not necessarily true) so that a trajectory to the subspace origin exists for all $x \in \mathscr{C}$.

from $t = 0$ to $t = t^s_{n-1} = T^s_{n-1}(x^s)$, viz

$$x^\beta = x(t^s_{n-1}) = \phi(t^s_{n-1}, x, u^s) = \exp\left(AT^s_{n-1}(x^s)\right)$$

$$\times \left[x - \operatorname{sgn}\left(\Xi^s(x^s)\right) \sum_{i=0}^{n-2} (-1)^i \int_{T^s_i(x^s)}^{T^s_{i+1}(x^s)} \exp(-A\tau) b \, d\tau \right]$$

$$= \begin{bmatrix} \beta(x) \\ 0 \\ \vdots \\ 0 \end{bmatrix} \tag{4.46}$$

where $T^s_0 \triangleq 0$ and A, b are of the form (4.42*c*). Eqn. (4.46) defines a mapping

$$(x_1, x_2, \ldots, x_n)' \mapsto (\beta(x), 0, \ldots, 0)'$$

which transforms an arbitrary initial state $x \in \mathscr{C}$ into the trajectory endpoint x^β on the x_1 state axis. In accordance with the strategy, the optimal control value at x for minimum-time transition to the origin is then determined by the real-valued vector function β, viz.

$$u^*(x) = \begin{cases} +1; & \text{if } \beta(x) < 0 \\ u^s(x^s); & \text{if } \beta(x) = 0 \\ -1; & \text{if } \beta(x) > 0 \end{cases} \tag{4.47}$$

Note that the upper triangular form (4.42c) of the system A matrix for the serially decomposed plant (4.42) implies that the matrix exponential

$$\exp(At) = I + At + \tfrac{1}{2}A^2t^2 + \cdots + \frac{1}{i!} A^i t^i + \cdots$$

$$= \begin{bmatrix} \exp(\lambda_1 t) & \cdots & \cdots & \cdots \\ & \exp(\lambda_2 t) & \cdots & \cdots \\ & & \ddots & \vdots \\ 0 & & & \exp(\lambda_n t) \end{bmatrix}$$

is also an upper-triangular matrix with $\exp(\lambda_i t) > 0$, $i = 1, 2, \ldots, n$, as its diagonal elements. Consequently, multiplying both sides of (4.46) by the positive scalar $\exp(-\lambda_1 T^s_{n-1}(x^s)) > 0$, gives

$$\exp(-\lambda_1 T^s_{n-1}(x^s)) x^\beta = \begin{bmatrix} \exp(-\lambda_1 T^s_{n-1}(x_1))\beta(x) \\ 0 \\ \vdots \\ 0 \end{bmatrix}$$

where

$$\exp(-\lambda_1 T^s_{n-1}(x^s))\beta(x) \triangleq \xi(x) = x_1 + F(x^s) \tag{4.48}$$

i.e. the above procedure yields a standard form consisting of the x_1 component plus some function F of the subsystem state vector x^s. Finally, as $\exp(-\lambda_1 T^s_{n-1}(x^s)) > 0$,

$$\text{sgn}\,(\beta(x)) = \text{sgn}\,[\exp\,(-\lambda_1 T^s_{n-1}(x^s)\beta(x))] = \text{sgn}\,(\xi(x))$$

and the optimal control (4.47) may be written as

$$u^*(x) = -\text{sgn}\,(\Xi(x)) \tag{4.49a}$$

where

$$\Xi(x) = \begin{cases} \xi(x); & \xi(x) \neq 0 \\ \Xi^s(x^s); & \xi(x) = 0 \end{cases} \tag{4.49b}$$

with $\xi(x)\colon \mathscr{C} \to \mathbb{R}$ in the standard form (4.48). This approach to optimal control synthesis finds widespread application in the following three chapters.

Chapter 5

First- and second-order time-optimal control system synthesis

Time-optimal feedback regulators for first- and second-order saturating control systems are derived in this chapter. In addition, time-optimal isochrones and domains of null controllability are explicitly characterised in a number of cases. As time-optimal quantities are assumed throughout (and also in Chapters 6 and 7), the superscript *, hitherto used to denote optimality, is omitted for notational convenience.

5.1 First-order linear system

The first-order autonomous system

$$\dot{x} = \lambda x + \frac{u}{a}; \qquad \lambda = \text{constant}; \qquad a = \text{constant} > 0;$$
$$|u(t)| \leq 1; \qquad x(t) \in \mathbb{R} \tag{5.1}$$

is studied initially; the three cases for which the parameter λ is negative, zero and positive valued are considered separately.

Case (*i*) $\lambda < 0$: In this case it may be concluded from theorem 4.3.1 that a time-optimal solution exists for all starting points $x \in \mathbb{R}$. From (4.41) it follows immediately that the *constant* time-optimal control may be synthesised by the feedback law

$$u(x) = -\text{sgn}\,(x); \qquad x \neq 0 \tag{5.2}$$

with $u = 0$ at $x = 0$. By integrating the state equation (5.1) under the constant optimal control (5.2) from $t = 0$ to $t = t_1$, the minimum time $t_1 = T_1(x)$ to the origin from an arbitrary non-zero initial state $x(0) = x \neq 0$ is easily calculated as

$$t_1 = T_1(x) = -\frac{1}{\lambda} \ln\,(1 - a\lambda|x|) \tag{5.3}$$

Case (*ii*) $\lambda = 0$: Again a time-optimal control exists for all initial states $x(0) = x \in \mathbb{R}$ and is synthesised by (5.2). The time to the origin $t_1 = T_1(x)$ in this case is easily calculated as

$$t_1 = T_1(x) = a|x| \tag{5.4}$$

Case (*iii*) $\lambda > 0$: In this case the uncontrolled plant $\dot{x} = \lambda x$ is *unstable.* From theorem 4.3.1 a time-optimal control exists only for starting points $x(0) = x \in \mathscr{C}$ in the domain of null controllability $\mathscr{C}$; in such cases the time to the origin is again given by (5.3), viz.

$$t_1 = T_1(x) = \frac{1}{\lambda} \ln (1 - a\lambda|x|)^{-1}; \qquad \lambda > 0 \tag{5.5}$$

Note that $t_1 \to \infty$ as $|x| \to 1/a\lambda$, so that the open convex domain of null controllability may be explicitly characterised as

$$\mathscr{C} = \bigcup_{t_1 \geq 0} \Upsilon_{t_1} = \bigcup_{t_1 \geq 0} \{x\colon T_1(x) = t_1\} = \left\{x\colon |x| < \frac{1}{a\lambda}\right\} \tag{5.6}$$

5.2 Single-input second-order systems

Linear, autonomous, single-input systems of the general form

$$\dot{x} = Ax + bu; \qquad x(t) \in \mathbb{R}^2; \qquad |u(t)| \leq 1 \tag{5.7}$$

will now be considered. The pair (A, b) is assumed to be a controllable pair (rank $[b\colon Ab] = 2$) and hence the time-optimal control problem is normal. The 2×2 matrix A has eigenvalues λ_1, λ_2.

5.2.1 Systems with real eigenvalues

The case of *real* eigenvalues λ_1, λ_2 is treated initially. Without loss of generality, the serially decomposed system representation is assumed, i.e.

$$\begin{aligned} \dot{x}_1 &= \lambda_1 x_1 + x_2 \\ \dot{x}_2 &= \lambda_2 x_2 + \frac{u}{a}; \qquad |u| \leq 1; \qquad a > 0 \end{aligned} \tag{5.8}$$

The time-optimality of the strategy of Gulko *et al.* will be exploited in the derivation of the optimal feedback control synthesis, as outlined in Section 4.5.2. For the case at hand, the $(n-1)$th-order subsystem is defined by the scalar equation

$$\dot{x}_2 = \lambda_2 x_2 + \frac{u}{a}; \qquad |u| \leq 1; \qquad a > 0 \tag{5.9}$$

of the form studied in Section 5.1. Hence, the optimal subsystem feedback control is

$$u^s(x_2) = -\text{sgn}\,(\Xi^s(x_2)) = -\text{sgn}\,(x_2); \qquad x_2 \neq 0 \tag{5.10}$$

(with $u^s(0) = 0$) and the minimum time to the subspace origin is given by

$$t_1^s = T_1^s(x_2) \triangleq \begin{cases} -\dfrac{1}{\lambda}\ln\,(1 - a\lambda|x_2|); & \lambda < 0 \\ a|x_2|; & \lambda = 0 \\ \dfrac{1}{\lambda}\ln\,(1 - a\lambda|x_2|)^{-1}; & \lambda > 0; \quad |x_2| < \dfrac{1}{a\lambda} \end{cases} \tag{5.11}$$

Applied to the full system, the control (5.10) generates a trajectory with endpoint $x^\beta = (x_1^\beta, 0)' = (\beta(x), 0)'$ given by (4.46), viz.

$$\begin{bmatrix} \beta(x) \\ 0 \end{bmatrix} = \exp\,(AT_1^s(x_2))\left(x - \text{sgn}\,(x_2)\int_0^{T_1^s(x_2)} \exp\,(-As)\begin{bmatrix} 0 \\ 1/a \end{bmatrix} ds\right) \tag{5.12}$$

For the serially decomposed system under consideration, the function β is easily calculated as

$$\beta(x) = \exp\,(\lambda_1 T_1^s(x_2))\xi(x) \tag{5.13a}$$

where

$$\xi(x) = \begin{cases} x_1 - \dfrac{\text{sgn}\,(x_2)}{a\lambda_2}\displaystyle\int_0^{T_1^s(x_2)} \\ \quad \times \exp\,(-\lambda_1 s)[(1 - a\lambda_2|x_2|)\exp\,(\lambda_2 s) - 1]\,ds; & \lambda_2 \neq 0 \\ x_1 - \dfrac{\text{sgn}\,(x_2)}{a}\displaystyle\int_0^{T_1^s(x_2)} \exp\,(-\lambda_1 s)[s - a|x_2|]\,ds; & \lambda_2 = 0 \end{cases} \tag{5.13b}$$

Noting that $\text{sgn}\,(\beta(x)) = \text{sgn}\,(\xi(x))$, the time-optimal feedback control law may be expressed as

$$u = -\text{sgn}\,(\Xi(x)); \qquad \Xi(x) = \begin{cases} \xi(x); & \xi(x) \neq 0 \\ x_2; & \xi(x) = 0 \end{cases}; \qquad x \in \mathscr{C} \tag{5.14}$$

with $u = 0$ at $x = 0$. Several specific systems will now be considered.

5.2.1 (i) *Double integrator system* $\lambda_1 = 0 = \lambda_2$

From theorem 4.3.1, $\mathscr{C} = \mathbb{R}^2$, i.e. all states are controllable to the origin. The minimum time to the subspace origin is $T_1^s(x_2) = a|x_2|$. Evaluating the right-hand side of (5.13b) the switching function $\xi(x)$ is given by

$$\xi(x) = x_1 + \tfrac{1}{2}ax_2|x_2| \tag{5.15}$$

which is in agreement with result (2.81) of Example 2.6.1 where the double integrator system with $a = 1$ was considered.

The set $\Gamma = \{x: \xi(x) = 0\}$ constitutes the optimal switching curve. For any initial state x (at time $t = 0$) the optimal control contains at most one discontinuity occurring at time $t_1 = T_1(x) \geq 0$, the origin being subsequently attained in minimum time $t_2 = T_2(x) \geq 0$. The functions T_i: $\mathbb{R}^2 \rightarrow [0, \infty)$ are easily characterised as follows:

First consider an initial state $x \notin \Gamma$ (i.e. *not* on the optimal switching curve), the optimal control then contains precisely *one* discontinuity at some time $t_1 > 0$ with $t_2 > t_1 > 0$. Integrating the state equation along the optimal path from x to the origin yields the relations

$$x_1 + x_2 t_2 - \frac{\text{sgn}\ (\xi(x))}{a}\left[\frac{t_2^2}{2} - (t_2 - t_1)^2\right] = 0$$

$$x_2 - \frac{\text{sgn}\ (\xi(x))}{a}[t_2 - 2(t_2 - t_1)] = 0$$

which may be solved for t_1 and t_2 to give the required results

$$t_1 = ax_2\ \text{sgn}\ (\xi(x)) + \sqrt{ax_1\ \text{sgn}\ (\xi(x)) + \tfrac{1}{2}(ax_2)^2}$$

$$t_2 = ax_2\ \text{sgn}\ (\xi(x)) + 2\sqrt{ax_1\ \text{sgn}\ (\xi(x)) + \tfrac{1}{2}(ax_2)^2}; \qquad x \notin \Gamma \Leftrightarrow \xi(x) \neq 0.$$

On the other hand, if $x \in \Gamma$ ($\Leftrightarrow \xi(x) = 0$), the origin is reached in minimum time $t_2 = a|x_2|$ via a *switchless* trajectory in Γ generated by the optimal (subsystem) control $u = -\text{sgn}\ (x_2)$. Combining these results, the dependence of the switching and minimum times on the initial state x may be expressed as

$$t_1 = T_1(x) = ax_2\ \text{sgn}\ (\Xi) + \sqrt{ax_1\ \text{sgn}\ (\Xi) + \tfrac{1}{2}(ax_2)^2} \tag{5.16a}$$

$$t_2 = T_2(x) = ax_2\ \text{sgn}\ (\Xi) + 2\sqrt{ax_1\ \text{sgn}\ (\Xi) + \tfrac{1}{2}(ax_2)^2} \tag{5.16b}$$

where

$$\Xi = \Xi(x) = \begin{cases} \xi(x); & \xi(x) \neq 0 \\ x_2; & \xi(x) = 0 \end{cases}$$

It is remarked that the function T_2: $\mathbb{R}^2 \rightarrow [0, \infty)$ is continuous whereas the function T_1: $\mathbb{R}^2 \rightarrow [0, \infty)$ exhibits a discontinuity at the switching curve Γ. Strictly speaking, for $x \in \Gamma$ a switching time t_1 does not exist as the origin is attained via a switchless trajectory in Γ. However, note that, if $x \in \Gamma \Leftrightarrow \xi(x) = 0$, then $ax_1\ \text{sgn}\ (\Xi) = ax_1\ \text{sgn}\ (x_2) = -\tfrac{1}{2}(ax_2)^2$ so that (5.16) yields the result $t_1 = t_2 = a|x_2|$. Hence, for $x \in \Gamma$, (5.16) should be interpreted as simply defining the minimum time to the origin, a distinction between the terms switching and final times being inappropriate.

Finally, note that,

$$\mathscr{C} = \bigcup_{t_2 \geq 0} \Upsilon_{t_2} = \bigcup_{t_2 \geq 0} \{x: T_2(x) = t_2\} = \mathbb{R}^2$$

i.e. as t_2 increases from zero, the family of isochronal surfaces expand to fill the state space. Again these results are in agreement with the earlier treatment in Example 2.6.1 for which the value function $V(x)$ is synonymous with the 'isochronal' function $T_2(x)$.

5.2.1 (ii) Integrator-plus-lag system : $\lambda_1 = 0; \lambda_2 = \lambda < 0$
Again, by theorem 4.3.1, $\mathscr{C} = \mathbb{R}^2$ (all states null controllable). In this case, with $\lambda_2 = \lambda < 0$,

$$T_1^s(x_2) = -\frac{1}{\lambda} \ln (1 - a\lambda |x_2|)$$

and the switching function $\xi: \mathbb{R}^2 \to \mathbb{R}$ becomes, via (5.13b),

$$\xi(x) = x_1 - \frac{x_2}{\lambda} - \frac{\operatorname{sgn}(x_2)}{a\lambda^2} \ln (1 - a\lambda |x_2|) \tag{5.17}$$

The set $\Gamma = \{x: \xi(x) = 0\}$ constitutes the optimal switching curve which partitions the state space into two mutually exclusive regions Σ^+ and Σ^- of positive and negative control effort, respectively. For parameter values $\lambda = -1$ and $a = 1$, the optimal vector field is depicted in Fig. 5.1. As in the previous example, for an arbitrary trajectory starting point $x \in \mathbb{R}^2$ the optimal control contains at most one discontinuity occurring at time $t_1 = T_1(x)$, the state origin being attained in minimum time $t_2 = T_2(x)$. The functions $T_i: \mathbb{R}^2 \to [0, \infty)$ may be

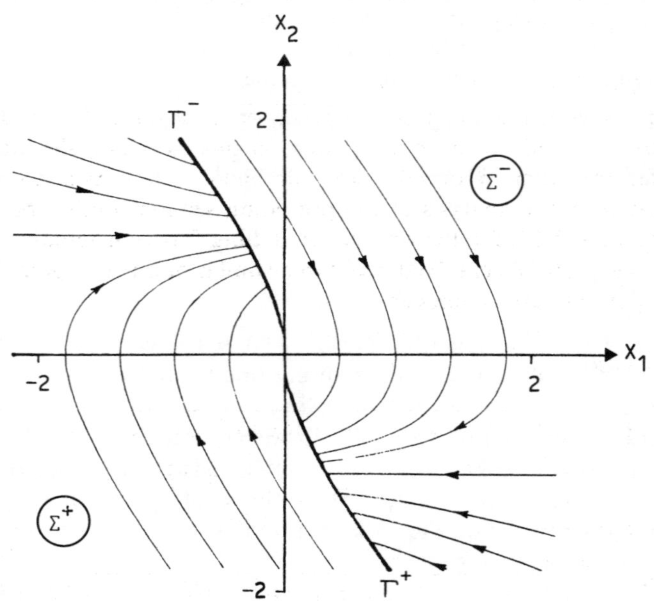

Fig. 5.1 *Integrator-plus-lag*
Time-optimal synthesis for parameter values $a = 1$, $\lambda = -1$

explicitly characterised by integrating the state equations along the optimal path from x to the origin. This straightforward procedure yields the result:

$$t_1 = T_1(x) = -\frac{1}{\lambda}$$
$$\times \ln\left[1 + \sqrt{1 - [1 - a\lambda x_2 \operatorname{sgn}(\Xi)] \exp[-a\lambda(\lambda x_1 - x_2)\operatorname{sgn}(\Xi)]}\right] - a(\lambda x_1 - x_2)\operatorname{sgn}(\Xi) \qquad (5.18a)$$

$$t_2 = T_2(x) = -\frac{2}{\lambda}$$
$$\times \ln\left[1 + \sqrt{1 - [1 - a\lambda x_2 \operatorname{sgn}(\Xi)] \exp[-a\lambda(\lambda x_1 - x_2)\operatorname{sgn}(\Xi)]}\right] - a(\lambda x_1 - x_2)\operatorname{sgn}(\Xi) \qquad (5.18b)$$

where

$$\Xi = \Xi(x) = \begin{cases} \xi(x); & \xi(x) \neq 0 \\ x_2; & \xi(x) = 0 \end{cases}$$

As in the previous example, for $x \in \Gamma$ the above expressions are interpreted as equivalently defining the minimum time

$$T_1^s(x_2) = T_1(x) = T_2(x) = -\frac{1}{\lambda}\ln(1 - a\lambda|x_2|); \qquad x \in \Gamma$$

to the origin via the *switchless* trajectory in Γ generated by the optimal (subsystem) control $u \equiv -\operatorname{sgn}(x_2)$.

5.2.1 (iii) Unstable system I : $\lambda_1 = 0;\ \lambda_2 = \lambda > 0$

The plant to be controlled is now composed of an unstable first-order element ($\lambda_2 = \lambda > 0$) in cascade with an integrator. In this case, not all states are null controllable; the domain of null controllability is an open convex subset $\mathscr{C} \subset \mathbb{R}^2$. Within $\mathscr{C}$ the analysis of the preceding section holds; specifically for $x \in \mathscr{C}$ expression (5.17) for the optimal switching function remains valid. The functions T_i: $\mathscr{C} \to [0, \infty)$ can be derived in a similar manner to give the optimal switching (t_1) and final (t_2) times as

$$t_1 = T_1(x) = \frac{1}{\lambda}$$
$$\times \ln\left[1 - \sqrt{1 - [1 - a\lambda x_2 \operatorname{sgn}(\Xi)] \exp[-a\lambda(\lambda x_1 - x_2)\operatorname{sgn}(\Xi)]}\right]^{-1} - a(\lambda x_1 - x_2)\operatorname{sgn}(\Xi) \qquad (5.19a)$$

$$t_2 = T_2(x) = \frac{2}{\lambda}$$
$$\times \ln\left[1 - \sqrt{1 - [1 - a\lambda x_2 \operatorname{sgn}(\Xi)] \exp[-a\lambda(\lambda x_1 - x_2)\operatorname{sgn}(\Xi)]}\right]^{-1} - a(\lambda x_1 - x_2)\operatorname{sgn}(\Xi) \qquad (5.19b)$$

Note that, for all x such that $|x_2| < 1/a\lambda$, $T_2(x)$ is a well defined non-negative real number, with $T_2(x) \to \infty$ as $|x_2| \to 1/a\lambda$. Conversely, for $|x_2| > 1/a\lambda$ a time-optimal path to the origin does not exist. These observations yield the explicit characterisation of the open, convex domain of null controllability

$$\mathscr{C} = \bigcup_{t_2 \geq 0} \Upsilon_{t_2} = \bigcup_{t_2 \geq 0} \{x\colon T_2(x) = t_2\} = \left\{x\colon |x_2| < \frac{1}{a\lambda}\right\} \tag{5.20}$$

i.e. a strip of width $2/a\lambda$ in the state plane, centred on the x_1 state axis. The optimal switching curve $\Gamma = \{x : \xi(x) = 0\}$ lies entirely in $\mathscr{C}$ and partitions $\mathscr{C}$ into the two mutually exclusive regions Σ^+ and Σ^-. For parameter values $\lambda = 1 = a$, the optimal vector field in $\mathscr{C}$ is depicted in Fig. 5.2. Note that the eigenvector $(1, 0)'$ associated with the nonpositive eigenvalue $(\lambda_1 = 0)$ is directed along the x_1 state axis corresponding to the unbounded direction of $\mathscr{C}$. This example is also treated in Vakilzadeh (1974).

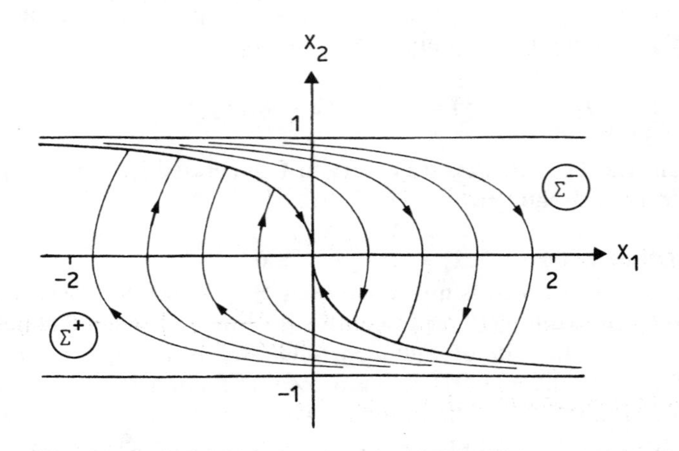

Fig. 5.2 *Unstable system I*
Domain of null controllability and time-optimal synthesis for parameter values $a = 1, \lambda = 1$

5.2.1 (iv) System with negative real and distinct eigenvalues: $\lambda_1 < 0$, $\lambda_2 < 0$, $\lambda_1 \neq \lambda_2$

In this case $\mathscr{C} = \mathbb{R}^2$ and no loss in generality is incurred in assuming the eigenvalue ordering

$$\lambda_1 < \lambda_2 < 0 \qquad \text{with} \qquad \alpha \triangleq \frac{\lambda_1}{\lambda_2} > 1 \tag{5.21}$$

A straightforward application of (5.13b) yields the optimal switching function

$$\xi(x) = x_1 + \frac{x_2}{\lambda_1 - \lambda_2} - \frac{\operatorname{sgn}(x_2)}{a\lambda_1(\lambda_1 - \lambda_2)}[1 - (1 - a\lambda_2|x_2|)^{\alpha}];$$

$$\alpha = \frac{\lambda_1}{\lambda_2} > 1 \tag{5.22}$$

Note that, under the linear state transformation,

$$w = \begin{bmatrix} w_1 \\ w_2 \end{bmatrix} = \begin{bmatrix} a\lambda_1(\lambda_1 - \lambda_2) & a\lambda_1 \\ 0 & a\lambda_2 \end{bmatrix} \begin{bmatrix} x_1 \\ x_2 \end{bmatrix} = Px \tag{5.23}$$

the original system equations are reduced to the convenient diagonalised form

$$\dot{w} = \begin{bmatrix} \dot{w}_1 \\ \dot{w}_2 \end{bmatrix} = \begin{bmatrix} \lambda_1 & 0 \\ 0 & \lambda_2 \end{bmatrix} \begin{bmatrix} w_1 \\ w_2 \end{bmatrix} + \begin{bmatrix} \lambda_1 \\ \lambda_2 \end{bmatrix} u; \qquad |u| \le 1 \tag{5.24}$$

In terms of the new state vector w, the switching function (5.22) becomes

$$\xi(P^{-1}w) = \frac{1}{a\lambda_1(\lambda_1 - \lambda_2)}\{w_1 + \operatorname{sgn}(w_2)[1 - (1 + |w_2|)^{\alpha}]\} \tag{5.25}$$

Defining the function $\tilde{\xi}$ as

$$\tilde{\xi}(w) = w_1 + \operatorname{sgn}(w_2)[1 - (1 + |w_2|)^{\alpha}]; \qquad \alpha = \frac{\lambda_1}{\lambda_2} > 1$$

and noting that

$$\operatorname{sgn}(\xi(x)) = \operatorname{sgn}(\xi(P^{-1}w)) = \operatorname{sgn}\left(\frac{\tilde{\xi}(w)}{a\lambda_1(\lambda_1 - \lambda_2)}\right) = \operatorname{sgn}(\tilde{\xi}(w))$$

and

$$\operatorname{sgn}(w_2) = \operatorname{sgn}(a\lambda_2 x_2) = -\operatorname{sgn}(x_2)$$

the optimal feedback control law (5.14) may be rewritten in the form

$$u = -\operatorname{sgn}(\tilde{\Xi}(w)); \qquad \tilde{\Xi}(w) = \begin{cases} \tilde{\xi}(w); & \tilde{\xi}(w) \ne 0 \\ -w_2; & \tilde{\xi}(w) = 0 \end{cases} \tag{5.26}$$

For all $\alpha > 1$, the portrait of p-paths (i.e. the state vector field generated with $u \equiv +1$) for system (5.24) exhibits a stable node N^+ at the point $(-1, -1)$ towards which all p-paths converge; the portrait of n-paths exhibits a stable node N^- at $(+1, +1)$. Fig. 5.3 depicts the p- and n-paths corresponding to the parameter value $\alpha = 3$. The optimal switch curve $\tilde{\Gamma} = \{w\colon \tilde{\xi}(w) = 0\}$ is comprised of the origin together with the p-path ($\tilde{\Gamma}^+$) and n-path ($\tilde{\Gamma}^-$) which lead to the origin. For the illustrative value $\alpha = 3$, the time-optimal state portrait is depicted in Fig. 5.4.

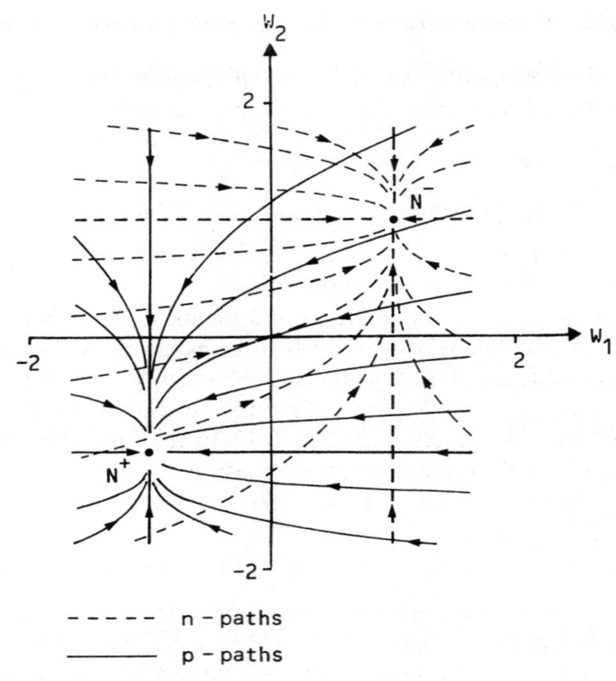

Fig. 5.3 p- *and* n- *paths for* $\alpha = 3$

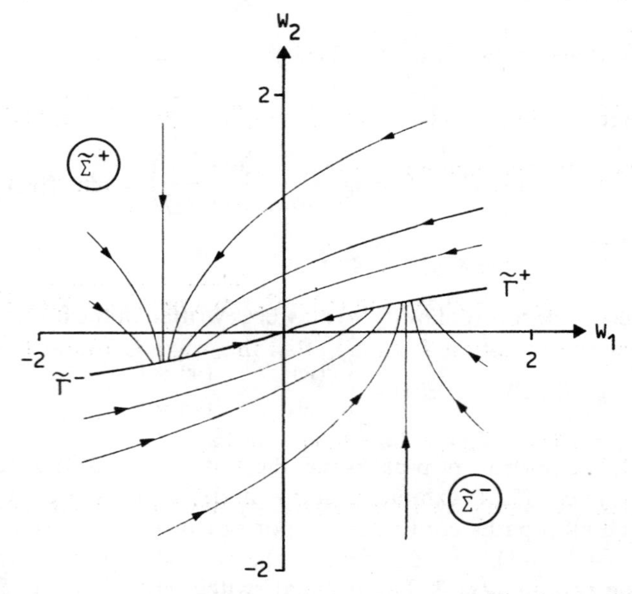

Fig. 5.4 *Time-optimal synthesis for* $\alpha = 3$

5.2.1 (v) System with negative real eigenvalues in simple ratio $: \lambda_1/\lambda_2 = \alpha = 2$
Consider now the above system when the eigenvalues satisfy

$$\lambda_1 = 2\lambda; \qquad \lambda_2 = \lambda; \qquad \lambda = \text{constant} < 0 \tag{5.27}$$

In this case $\alpha = 2$ and the switching function (5.22) reduces to

$$\xi(x) = x_1 + \tfrac{1}{2}ax_2\,|x_2| \tag{5.28}$$

which is *independent* of the parameter λ and identical to (5.15) for the double integrator system. Of course, this result is to be expected as the special system ($\alpha = 2$) under consideration is of the form (3.49) which was shown in Section 3.6 to be reducible *ab initio* to an equivalent multiple integrator form. Recalling the analysis of Section 3.6, reduction of form (3.49) to the equivalent multiple integrator form (3.55) involved the time transformation

$$t \mapsto \frac{1}{\lambda}\,[1 - \exp(-\lambda t)] \tag{5.29}$$

For the case at hand, it immediately follows (from the associated results (5.16) for the double integrator system and invoking the inverse transformation) that, for an initial state $x \in \mathbb{R}^2$ the optimal control contains at most one discontinuity occurring at time

$$t_1 = T_1(x) = -\frac{1}{\lambda} \times \ln\left[1 - \lambda[ax_2 \operatorname{sgn}(\Xi) + \sqrt{ax_1 \operatorname{sgn}(\Xi) + \tfrac{1}{2}(ax_2)^2}]\right] \tag{5.30a}$$

the origin being attained in minimum time

$$t_2 = T_2(x) = -\frac{1}{\lambda} \times \ln\left[1 - \lambda[ax_2 \operatorname{sgn}(\Xi) + 2\sqrt{ax_1 \operatorname{sgn}(\Xi) + \tfrac{1}{2}(ax_2)^2}]\right] \tag{5.30b}$$

with the usual interpretation, for $x \in \Gamma$, that the origin is attained in minimum time

$$T_1^s(x_2) = T_2(x) = T_1(x) = -\frac{1}{\lambda}\ln(1 - a\lambda\,|x_2|) \tag{5.31}$$

via a switchless trajectory (in Γ) generated by the optimal subsystem control $u \equiv -\operatorname{sgn}(x_2)$.

5.2.1 (vi) Unstable system II $: \lambda_1 > \lambda_2 > 0$
The plant to be controlled is now composed of two unstable first-order elements. In view of theorem 4.3.1, not all states are null controllable; time-optimal trajectories to the origin exist only for starting points in the domain of null

controllability $\mathscr{C} \subset \mathbb{R}^2$. Noting that

$$x = \begin{bmatrix} x_1 \\ x_2 \end{bmatrix} \in \mathscr{C} \Rightarrow x_2 \in \mathscr{C}^s = \left\{ x_2 \colon |x_2| < \frac{1}{a\lambda} \right\} \tag{5.32}$$

where $\mathscr{C}_s$ is the subsystem domain of null controllability (cf. (5.6)), it may be verified that (5.22), *restricted* to $\mathscr{C}$, is again a well-defined, real-valued time-optimal switching function. As in the stable case of the previous section, the analysis is facilitated by the introduction of the state transformation (5.23) which, again, transforms the system equations into the convenient diagonalised form (5.24), where now $\lambda_1 > \lambda_2 > 0$ (with $\alpha = \lambda_1/\lambda_2 > 1$ as before). In terms of the w state vector, the optimal switching function (5.22) becomes

$$\xi(P^{-1}w) = \frac{1}{a\lambda_1(\lambda_1 - \lambda_2)} [w_1 - \text{sgn}(w_2)[1 - (1 - |w_2|)^\alpha]]; \qquad w \in \tilde{\mathscr{C}} \quad (\Rightarrow |w_2| < 1) \tag{5.33}$$

where $\tilde{\mathscr{C}} = \{w \colon P^{-1}w = x \in \mathscr{C}\}$ is the domain of null controllability in w state space. Defining $\tilde{\xi} \colon \tilde{\mathscr{C}} \to \mathbb{R}$ as

$$\tilde{\xi}(w) = w_1 - \text{sgn}(w_2)[1 - (1 - |w_2|)^\alpha]; \qquad \alpha = \frac{\lambda_1}{\lambda_2} > 1 \tag{5.34}$$

and noting that $\text{sgn}(\xi(x)) = \text{sgn}(\xi(P^{-1}w)) = \text{sgn}(\tilde{\xi}(w))$ and $\text{sgn}(x_2) = \text{sgn}(a\lambda_2 x_2) = \text{sgn}(w_2)$ for all $x \in \mathscr{C}$ ($w \in \tilde{\mathscr{C}}$), the time-optimal feedback control law (within $\tilde{\mathscr{C}}$) may be expressed as

$$u = -\text{sgn}(\tilde{\Xi}(w)); \qquad \tilde{\Xi}(w) = \begin{cases} \tilde{\xi}(w); & \tilde{\xi}(w) \neq 0 \\ w_2; & \tilde{\xi}(w) = 0 \end{cases}; \qquad w \in \tilde{\mathscr{C}} \tag{5.35}$$

This control generates a time-optimal path to the origin from each point within the domain of null controllability $\tilde{\mathscr{C}}$ which will now be characterised.

For all $\alpha > 1$, the portrait of p-paths ($u \equiv +1$) for system (5.24) with positive real eigenvalues $\lambda_1 > \lambda_2 > 0$ exhibits an unstable node N^+ at the point $(-1, -1)$, away from which all p-paths diverge; by symmetry, the portrait of n-paths ($u \equiv -1$) exhibits an unstable node N^- at the point $(1, 1)$, away from which all n-paths diverge. For example, for the illustrative parameter value $\alpha = 3$, the fields of p- and n-paths are similar in shape to those of Fig. 5.3 with the essential difference that the direction of motion is reversed, N^+ and N^- becoming unstable nodes, away from which the p- and n-paths diverge. The optimal switching curve $\tilde{\Gamma} = \{w \in \tilde{\mathscr{C}} \colon \tilde{\xi}(w) = 0\}$ consists of the origin together with the p-path ($\tilde{\Gamma}^+$) and n-path ($\tilde{\Gamma}^-$) leading to the origin from N^+ and N^-, respectively; the optimal curve $\tilde{\Gamma}$ is depicted in Fig. 5.5 for the parameter value $\alpha = 3$. Consider now the closed curve $\partial\tilde{\mathscr{C}} = \partial\tilde{\mathscr{C}}^+ \cup \partial\tilde{\mathscr{C}}^-$ comprised of the p-path ($\partial\tilde{\mathscr{C}}^+$) from N^+ to N^- and the n-path ($\partial\tilde{\mathscr{C}}^-$) from N^- to N^+. It is easily shown that $\partial\tilde{\mathscr{C}}$ has the characterisation

$$\partial\tilde{\mathscr{C}} = \{w \colon w_1 \, \text{sgn}(\tilde{\Xi}) + 2^{1-\alpha}[1 - w_2 \, \text{sgn}(\tilde{\Xi})]^\alpha = 1\} \tag{5.36}$$

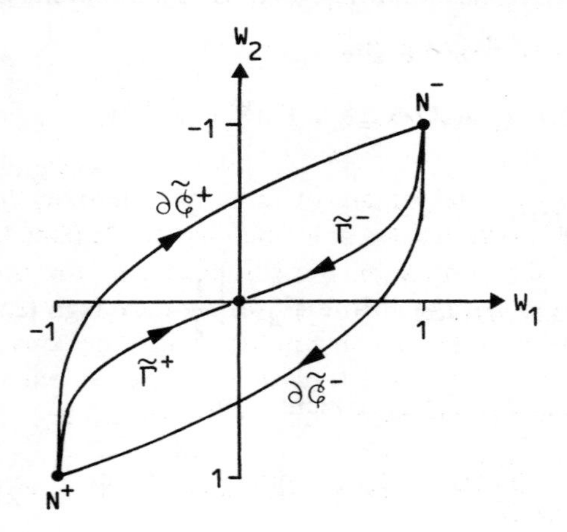

Fig. 5.5 *Unstable system II*
Domain of null controllability and time-optimal switching curve for $\alpha = 3$

and is depicted in Fig. 5.5 for the parameter value $\alpha = 3$. It is readily seen that a minimum-time path to the origin exists from each point in the *interior* of the region bounded by $\partial\tilde{\mathscr{C}}$; on the other hand, no such paths exist from points on and exterior to the curve $\partial\tilde{\mathscr{C}}$. Thus, the open, convex domain of null controllability $\tilde{\mathscr{C}}$ can be expressed as

$$\tilde{\mathscr{C}} = \{w\colon |w_1| < 1; \quad |w_2| < 1; \quad w_1 \operatorname{sgn}(\tilde{\Xi}) + 2^{1-\alpha}[1 - w_2 \operatorname{sgn}(\tilde{\Xi})]^\alpha < 1\} \tag{5.37}$$

For $\alpha = 3$, the domain of null controllability $\tilde{\mathscr{C}}$ and the time-optimal switching curve $\tilde{\Gamma}$ are depicted in Fig. 5.5.

5.2.1 (vii) System with positive real eigenvalues in simple ratio: $\lambda_1/\lambda_2 = \alpha = 2$
Consider now the previous system when the eigenvalues are in the ratio 2 : 1, viz.

$$\lambda_1 = 2\lambda; \qquad \lambda_2 = \lambda; \qquad \lambda = \text{constant} > 0 \tag{5.38}$$

As in the stable case of Section 5.2.1(v), the optimal switching function (5.22) simplifies to

$$\xi(x) = x_1 + \tfrac{1}{2}ax_2|x_2|; \qquad x \in \mathscr{C} \tag{5.39}$$

but now restricted to the domain of null controllability $\mathscr{C}$; again, note that (5.39) is independent of the parameter λ and is identical to the optimal switching function (5.15) for the double integrator system. Furthermore, from (5.23) and (5.37) with $\alpha = 2$, the domain of null controllability in x state space has the characterisation

$$\mathscr{C} = \{x: (2a\lambda^2 x_1 + 2a\lambda x_2)\ \mathrm{sgn}\ (\Xi) + \tfrac{1}{2}(1 - 2a\lambda x_2\ \mathrm{sgn}\ (\Xi) + a^2\lambda^2 x_2^2) < 1\}$$

or, on rearranging terms,

$$\mathscr{C} = \left\{x: 4[ax_1\ \mathrm{sgn}\ (\Xi) + \tfrac{1}{2}(ax_2)^2] < \left[\frac{1}{\lambda} - ax_2\ \mathrm{sgn}\ (\Xi)\right]^2\right\}$$

which can be expressed in the equivalent form

$$\mathscr{C} = \left\{x: ax_2\ \mathrm{sgn}\ (\Xi) + 2\sqrt{ax_1\ \mathrm{sgn}\ (\Xi) + \tfrac{1}{2}(ax_2)^2} < \frac{1}{\lambda}\right\} \tag{5.40}$$

where

$$\Xi = \Xi(x) = \begin{cases} \xi(x); & \xi(x) \neq 0 \\ x_2; & \xi(x) = 0 \end{cases}$$

This result is to be expected from the equivalence property of Section 3.6 by which the system is reducible to equivalent double integrator form, this reduction involving the time transformation

$$t \mapsto \frac{1}{\lambda}[1 - \exp(-\lambda t)]$$

As in Section 5.2.1(v), it follows from the equivalence property and results (5.16) for the equivalent double integrator system (when carried over to the present case via the inverse transformation $t \mapsto 1/\lambda \ln (1 - \lambda t)^{-1}$) that an initial state $x \in \mathscr{C}$ can be transferred to the origin in minimum time

$$t_2 = T_2(x) = \frac{1}{\lambda} \ln \left[1 - \lambda[ax_2\ \mathrm{sgn}\ (\Xi) + 2\sqrt{ax_1\ \mathrm{sgn}\ (\Xi) + \tfrac{1}{2}(ax_2)^2}]\right]^{-1} \tag{5.41}$$

Noting that $t_2 \to \infty$ as $ax_2\ \mathrm{sgn}\ (\Xi) + 2\sqrt{ax_1\ \mathrm{sgn}\ (\Xi) + \tfrac{1}{2}(ax_2)^2} \to 1/\lambda$ it follows that the domain of null controllability may be expressed as

$$\mathscr{C} = \bigcup_{t_2 \geq 0} \Upsilon_{t_2} = \bigcup_{t_2 \geq 0} \{x: T_2(x) = t_2\}$$

$$= \left\{x: ax_2\ \mathrm{sgn}\ (\Xi) + 2\sqrt{ax_1\ \mathrm{sgn}\ (\Xi) + \tfrac{1}{2}(ax_2)^2} < \frac{1}{\lambda}\right\}$$

which is in agreement with the earlier result (5.40) (see also Ryan, 1976b).

5.2.1(viii) Unstable system III: $\lambda_2 > 0; \lambda_1 < 0$

The plant to be controlled now consists of an unstable first-order element in cascade with a stable first-order element. In this case, $\alpha = \lambda_1/\lambda_2 < 0$ and it is readily verified (details omitted) that (5.22) again defines the time-optimal switching function but now restricted to the domain of null controllability (see also Flügge-Lotz, 1968; Vakilzadeh, 1978):

$$\mathscr{C} = \left\{x : |x_2| < \frac{1}{a\lambda_2}\right\} \tag{5.42}$$

i.e. a strip of width $2/a\lambda_2$ centred on the x_1 state axis, so that

$$\xi(x) = x_1 + \frac{x_2}{\lambda_1 - \lambda_2} - \frac{\text{sgn}\,(x_2)}{a\lambda_1(\lambda_1 - \lambda_2)}\,[1 - (1 - a\lambda_2|x_2|)^{\alpha}];$$
$$x \in \mathscr{C}; \qquad \alpha = \frac{\lambda_1}{\lambda_2} < 0 \tag{5.43}$$

For example, consider the case of

$$\lambda_1 = \lambda; \qquad \lambda_2 = -\lambda; \qquad \lambda = \text{constant} < 0$$

for which $\alpha = -1$ and, from (5.43),

$$\begin{aligned}\xi(x) &= x_1 + \frac{x_2}{2\lambda} - \frac{x_2}{2\lambda}\,(1 + a\lambda|x_2|)^{-1}; \qquad x \in \mathscr{C} \\ &= (1 + a\lambda|x_2|)^{-1}(x_1 + \tfrac{1}{2}ax_2|x_2| + a\lambda x_1|x_2|)\end{aligned}$$

Writing

$$\tilde{\xi}(x) = x_1 + \tfrac{1}{2}ax_2|x_2| + a\lambda x_1|x_2|$$

and noting that

$$x \in \mathscr{C} \Rightarrow (1 + a\lambda|x_2|)^{-1} > 0$$

the time-optimal feedback control may be expressed as

$$u = -\text{sgn}\,(\Xi(x)) = -\text{sgn}\,(\tilde{\Xi}(x))$$

where $\tilde{\Xi} : \mathscr{C} \to \mathbb{R}$ is given by

$$\tilde{\Xi}(x) = \begin{cases} \tilde{\xi}(x); & \tilde{\xi}(x) \neq 0 \\ x_2; & \tilde{\xi}(x) = 0 \end{cases}; \qquad x \in \mathscr{C}$$

Note that, as $\lambda \to 0$, $\mathscr{C} \to \mathbb{R}^2$ and $\tilde{\xi}(x) \to x_1 + \frac{1}{2}ax_2|x_2|$, so that, in the limit, the optimal solution for the double integrator system is recovered.

5.2.1(ix) System with repeated negative eigenvalues : $\lambda_1 = \lambda_2 = \lambda < 0$

In this case of coincident negative eigenvalues $\lambda_1 = \lambda_2 = \lambda < 0$, all states are null controllable, i.e. $\mathscr{C} = \mathbb{R}^2$, and a straightforward calculation yields the optimal switching function

$$\xi(x) = x_1 + \frac{x_2}{\lambda} + \frac{\text{sgn}\,(x_2)}{a\lambda^2}(1 - a\lambda|x_2|)\,\ln\,(1 - a\lambda|x_2|) \tag{5.44}$$

5.2.1(x) Unstable system IV: repeated positive eigenvalues: $\lambda_1 = \lambda_2 = \lambda > 0$
Finally, the case of coincident positive real eigenvalues

$$\lambda_1 = \lambda_2 = \lambda > 0$$

will be treated. In this case, the portrait of p-paths ($u \equiv +1$) exhibits an unstable node N^+ at the point $(1/a\lambda^2, -1/a\lambda)$, away from which the p-paths diverge; by symmetry, the portrait of n-paths ($u \equiv -1$) exhibits an unstable node N^- at the point $(-1/a\lambda^2, 1/a\lambda)$, away from which the n-paths diverge.

The time-optimal switching curve Γ is composed of the origin together with the p-path (Γ^+) leading to the origin from N^+ and the n-path (Γ^-) leading to the origin from N^-; Γ has the characterisation

$$\Gamma = \{x \in \mathscr{C}: \xi(x) = 0\}$$

where the switching function ξ is defined by (5.44) but now restricted to the domain of null controllability $\mathscr{C}$. The latter restriction implies that $x_2 \in \mathscr{C}^s = \{x: |x_2| < 1/a\lambda\}$ which, in turn, ensures that the argument of the logarithmic function of (5.44) is positive so that (5.44) is a well defined real-valued function on $\mathscr{C}$. As in Section 5.2.1(vi), it may be seen that the boundary $\partial\mathscr{C}$ of the domain of null controllability $\mathscr{C}$ coincides with the closed curve $\partial\mathscr{C} = \partial\mathscr{C}^+ \cup \partial\mathscr{C}^-$, composed of the p-path ($\partial\mathscr{C}^+$) from N^+ to N^- and the n-path ($\partial\mathscr{C}^-$) from N^- to N^+, which, on further analysis, may be expressed as

$$\partial\mathscr{C}^+ = \left\{x: -\frac{1}{a\lambda} \le x_2 < \frac{1}{a\lambda}; \quad x_1 + \frac{x_2}{\lambda} - \frac{1}{a\lambda^2}(1 + a\lambda x_2) \ln\left[\frac{1 + a\lambda x_2}{2}\right] = 0\right\}$$

$$\partial\mathscr{C}^- = \left\{x: -\frac{1}{a\lambda} < x_2 \le \frac{1}{a\lambda}; \quad x_1 + \frac{x_2}{\lambda} + \frac{1}{a\lambda^2}(1 - a\lambda x_2) \ln\left[\frac{1 - a\lambda x_2}{2}\right] = 0\right\} \tag{5.45}$$

For parameter values $a = 1 = \lambda$, the time-optimal state portrait within the domain of null controllability is depicted in Fig. 5.6.

5.2.2 Systems with complex eigenvalues
The time-optimal regulator problem for second-order, single-input oscillatory systems of the general form

$$\ddot{y} + 2\zeta\omega\dot{y} + \omega^2 y = \frac{u}{a}; \qquad |u| \le 1; \qquad |\zeta| < 1; \qquad a > 0 \tag{5.46}$$

will be treated in this section; this problem was solved initially by Bushaw (1953, 1958). To avoid excessive notation, the dimensionless form of (5.46) will

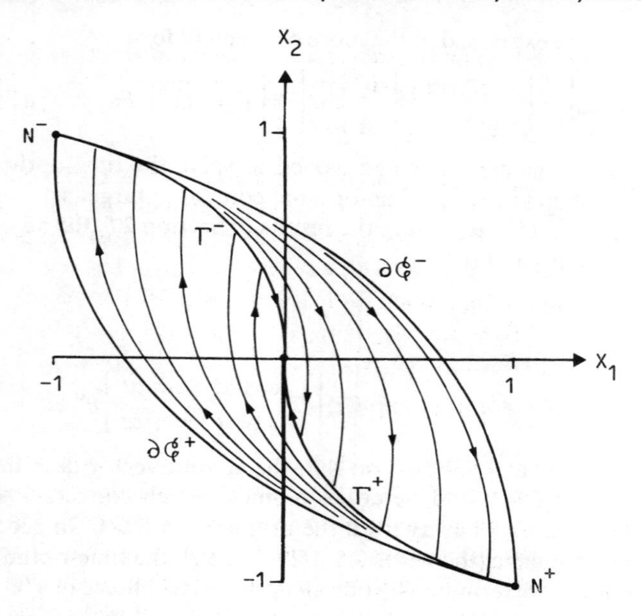

Fig. 5.6 *Unstable system IV*
Domain of null controllability and time-optimal synthesis for parameter values $a = 1, \lambda = 1$

be adopted (corresponding to an appropriate time and signal scale change), viz.

$$\ddot{y} + 2\zeta\dot{y} + y = u; \qquad |u| \leq 1; \qquad |\zeta| < 1 \tag{5.47}$$

The results obtained can be immediately carried over to the general form (5.46) via the transformations

$$y \mapsto a\omega^2 y; \qquad t \mapsto \omega t \text{ (so that } \dot{y} \mapsto a\omega\dot{y} \text{ and } \ddot{y} \mapsto a\ddot{y}) \tag{5.48}$$

Defining the state variables $y_1 = y$, $y_2 = \dot{y}$, (5.47) may be written in the form

$$\dot{y} = \begin{bmatrix} \dot{y}_1 \\ \dot{y}_2 \end{bmatrix} = \begin{bmatrix} 0 & 1 \\ -1 & -2\zeta \end{bmatrix} \begin{bmatrix} y_1 \\ y_2 \end{bmatrix} + \begin{bmatrix} 0 \\ 1 \end{bmatrix} u = \tilde{A}y + \tilde{b}u; \qquad |u| \leq 1;$$

$$|\zeta| < 1 \tag{5.49}$$

where the eigenvalues of the system matrix $\tilde{A}$ form the complex conjugate pair:

$$\lambda_1 = -\zeta + i\sqrt{1-\zeta^2}; \qquad \lambda_2 = -\zeta - i\sqrt{1-\zeta^2}$$

Writing $\alpha = \sqrt{1-\zeta^2}$ and introducing the additional linear transformation:

$$x = \begin{bmatrix} x_1 \\ x_2 \end{bmatrix} = \begin{bmatrix} 1 & \zeta \\ 0 & \alpha \end{bmatrix} \begin{bmatrix} y_1 \\ y_2 \end{bmatrix} = Py; \qquad \alpha = \sqrt{1-\zeta^2} \tag{5.50}$$

then (5.49) may be expressed in the more convenient form

$$\dot{x} = \begin{bmatrix} \dot{x}_1 \\ \dot{x}_2 \end{bmatrix} = \begin{bmatrix} -\zeta & \alpha \\ -\alpha & -\zeta \end{bmatrix} \begin{bmatrix} x_1 \\ x_2 \end{bmatrix} + \begin{bmatrix} \zeta \\ \alpha \end{bmatrix} u = Ax + bu; \qquad |u| \leq 1 \qquad (5.51)$$

The maximum principle will be employed to solve the time-optimal regulator problem for system (5.51) (for time-optimal control to target sets other than the origin, see Pinch, 1979). Recalling the results of Section 2.5, the adjoint equation for the problem at hand is

$$\dot{\eta} = -A^T\eta; \qquad \eta(0) = \eta^0$$

with solution

$$\eta(t) = \exp(-A^T t)\eta^0 = \exp(\zeta t) \begin{bmatrix} \cos \alpha t & \sin \alpha t \\ -\sin \alpha t & \cos \alpha t \end{bmatrix} \eta^0 \qquad (5.52)$$

Note that $\| \eta(t) \| = \exp(\zeta t) \| \eta^0 \|$ so that the adjoint vector describes a circle in the η-plane when $\zeta = 0$ and describes a spiral which converges to the origin when $\zeta < 0$ and diverges away from the origin when $\zeta > 0$. In accordance with the maximum principle (theorems 2.5.2, 2.5.3, 2.6.2), the time-optimal regulating control is uniquely determined by (for an appropriate choice of η^0):

$$\begin{aligned} u(t) &= \operatorname{sgn} < \eta(t), b > \\ &= \operatorname{sgn} \{\exp(\zeta t)[(\zeta\eta_1^0 + \alpha\eta_2^0) \cos \alpha t + (-\alpha\eta_1^0 + \zeta\eta_2^0) \sin \alpha t]\} \\ &= \operatorname{sgn} [\exp(\zeta t) \sin(\alpha t + \delta)]; \qquad 0 \leq t \leq t_f \end{aligned} \qquad (5.53)$$

Hence, the optimal control is piecewise constant, taking the values ± 1 only, with discontinuities separated π/α units in time; moreover, there is no upper bound on the number of intervals of control constancy (in contrast to the results of Section 5.2.1 for systems with real eigenvalues).

On integrating the state equations, the p- and n-paths for the system are found to satisfy

$$\left.\begin{aligned} x_1(t) &= \exp(-\zeta t)[(x_1^0 - u) \cos \alpha t + x_2^0 \sin \alpha t] + u \\ x_2(t) &= \exp(-\zeta t)[x_2^0 \cos \alpha t - (x_1^0 - u) \sin \alpha t] \end{aligned}\right\} \qquad (5.54)$$

passing through an arbitrary point $x^0 = (x_1^0, x_2^0)'$, with $u = +1$ for a p-path and $u = -1$ for an n-path. The time-optimal state portrait is composed of the appropriate 'piecing together' of such p- and n-paths as will now be shown. The three cases of (i) $\zeta = 0$, (ii) $0 < \zeta < 1$, and (iii) $-1 < \zeta < 0$ will be treated separately. From theorem 4.3.1, all states are null controllable in cases (i) and (ii), while in case (iii) only those states lying in the bounded, open, convex domain of null controllability $\mathscr{C}$ are controllable to the origin.

5.2.2(i) Harmonic oscillator: $\zeta = 0$

In this case $\alpha = 1$ and the p- and n-paths form families of concentric circles (traced in a clockwise sense with increasing time) in the state plane, centred on

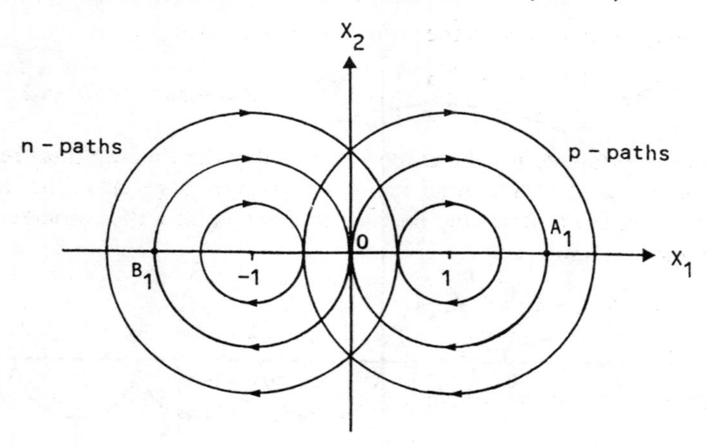

Fig. 5.7 *Harmonic oscillator*

the points with coordinates (1, 0) and (−1, 0), respectively, as depicted in Fig. 5.7. The time taken to complete a circuit of any such path is 2π (the period of oscillation). Since the origin is to be attained in finite time, every optimal trajectory must terminate with an arc of the p- or n-path leading to the origin. Moreover, from (5.53) with $\zeta = 0 \Rightarrow \alpha = 1$, it follows that, on an optimal trajectory, the control can remain constant on a time interval of duration π at most; hence, every optimal trajectory must terminate with one or other of the semicircular arcs A_1O, B_1O (see Fig. 5.8), or some subarc thereof, leading to the origin. Suppose, for definiteness, that a particular optimal trajectory switches at point P_1 onto the n-path leading to the origin, as depicted in Fig. 5.8. Prior to this switch, the trajectory must have followed an arc (of duration π) of the p-path leading to P_1; hence, the optimal switch prior to P_1 must have occurred at the diametrically opposite point P_2 of Fig. 5.8. Repeating this construction, as the point P_1 sweeps out the semicircular arc B_1O, the associated point P_2 sweeps out the semicircular arc A_2A_1. Every optimal switch point on B_1O maps uniquely to a preceding optimal switch point (occurring π time units earlier) on the semicircular arc A_2A_1, i.e. A_2A_1 is obtained from B_1O by the transformation $x_1 \mapsto x_1 + 4$, $x_2 \mapsto -x_2$. By symmetry, a similar construction yields the semicircular arc B_2B_1 of optimal switch points for trajectories terminating with an arc of the p-path A_1O leading to the origin. Consider again the optimal trajectory P_2P_1O; if this trajectory is traced back further, then, prior to the switch at P_2, the trajectory must have followed an n-path of duration π so that the optimal switch point prior to P_2 is located at the diametrically opposite point P_3. Again, as P_1 sweeps out the arc B_1O, P_2 sweeps out the arc A_2A_1 and P_3 sweeps out the semicircular arc B_3B_2 of Fig. 5.8. By symmetry, the arc A_3A_2 is obtained. Continuing this construction, the complete locus of optimal switch points, i.e. the time-optimal switching curve Γ, is seen to consist of the semi-

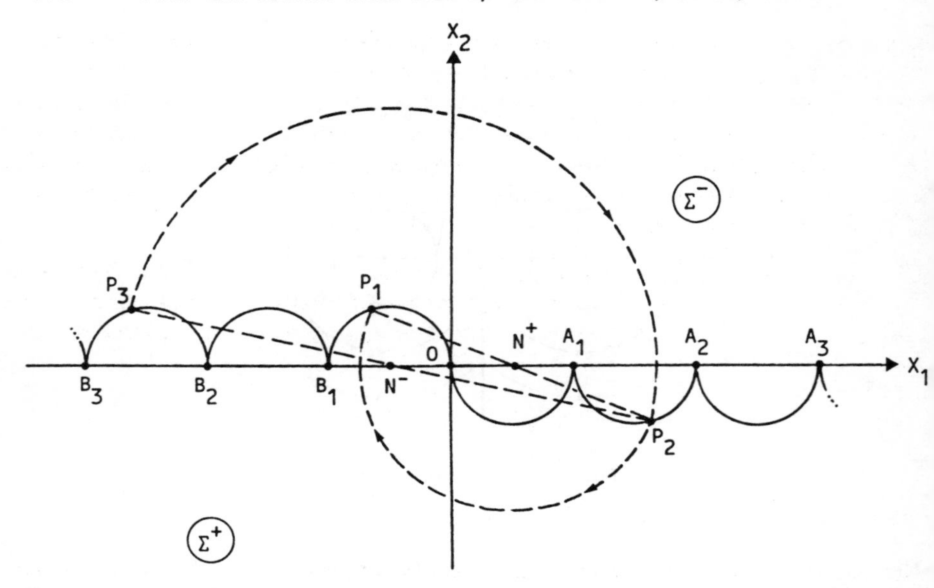

Fig. 5.8 *Harmonic oscillator*
Time-optimal switching curve

circular arcs A_1O, B_1O together with the sequences of semicircular arcs $\{A_{i+1}A_i\}_{i=1}^{\infty}$ and $\{B_{i+1}B_i\}_{i=1}^{\infty}$ obtained from A_1O and B_1O by translation along the x_1 state axis, as shown in Fig. 5.8. Note that the semicircular arc $A_{i+1}A_i$ consists of those points which can be steered (optimally) to the arc $B_i B_{i-1}$ in exactly π time units. The switching curve Γ may be expressed as

$$\Gamma = \{x\colon \xi(x) = 0\} \tag{5.55a}$$

where

$$\xi(x) = x_2 + \sqrt{1 - (|x_1| - 2N - 1)^2}\ \mathrm{sgn}\,(x_1);$$
$$2N < |x_1| \le 2N + 2, \qquad N = 0, 1, 2, \ldots \tag{5.55b}$$

This curve partitions the state plane into the two mutually exclusive regions

$$\Sigma^+ = \{x\colon \xi(x) < 0\} \qquad \text{and} \qquad \Sigma^- = \{x\colon \xi(x) > 0\}$$

wherein the optimal control takes the values $+1$ and -1, respectively. The optimal feedback control is now given by

$$u = -\mathrm{sgn}\,(\Xi(x)); \qquad \Xi(x) = \begin{cases} \xi(x); & \xi(x) \neq 0 \\ -x_1; & \xi(x) = 0 \end{cases} \tag{5.56}$$

Owing to the transcendental nature of the underlying equations, it is not possible in general to obtain closed-form expressions for the dependence $t_i = T_i(x)$ of the optimal switching and final times on the initial state x. Athans and Falb

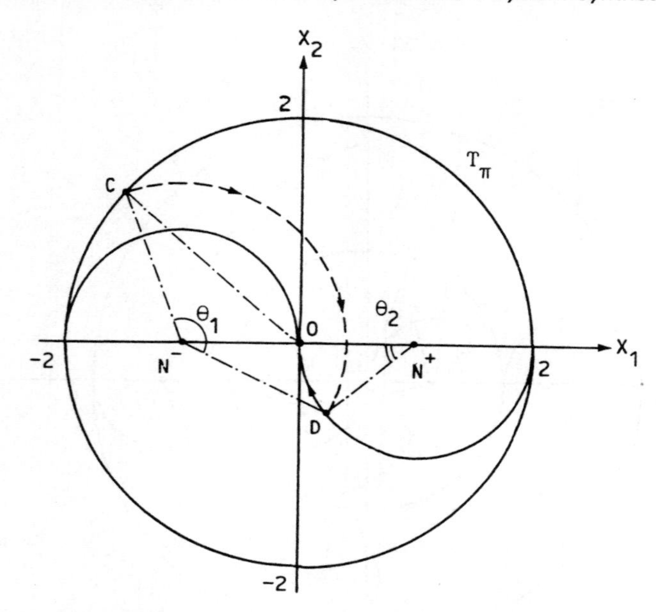

Fig. 5.9 *Harmonic oscillator* π-isochrone (Υ_π)

(1966) have given a geometric method for constructing the minimum-time isochrones. However, it is possible to obtain explicit results in some special cases, as will now be shown (see also Parks 1976).

Initially, consider the set

$$\Upsilon_\pi = \{x: \|x\| = 2\}$$

i.e. a circle of radius 2 centred on the origin in the state plane. It will be established that, for all starting points $x \in \Upsilon_\pi$, the origin is attained in the same minimum time $t_f = \pi$, i.e. Υ_π is the π-isochrone for the problem. Referring to Fig. 5.9, let C be an arbitrary starting point on Υ_π, then the first (and only) optimal switch occurs at point D on the semicircular arc A_1O; the duration of the arc CD is equal to the angle θ_1 subtended at N^-. From D, the p-path DO is followed to the origin of duration equal to the angle θ_2 subtended at N^+. Thus, the total time to the origin is $t_f = \theta_1 + \theta_2$. By simple geometry, it may be seen that triangles CN^-O and N^-DN^+ are congruent, from which it immediately follows that

$$t_f = \theta_1 + \theta_2 = \pi$$

Hence, the minimum time to the origin from an arbitrary starting point on Υ_π is π, i.e. Υ_π is the π-isochrone for the problem.

Consider now the set $\Upsilon_{2\pi} = \{x: \|x\| = 4\}$, i.e. a circle of radius 4 centred on

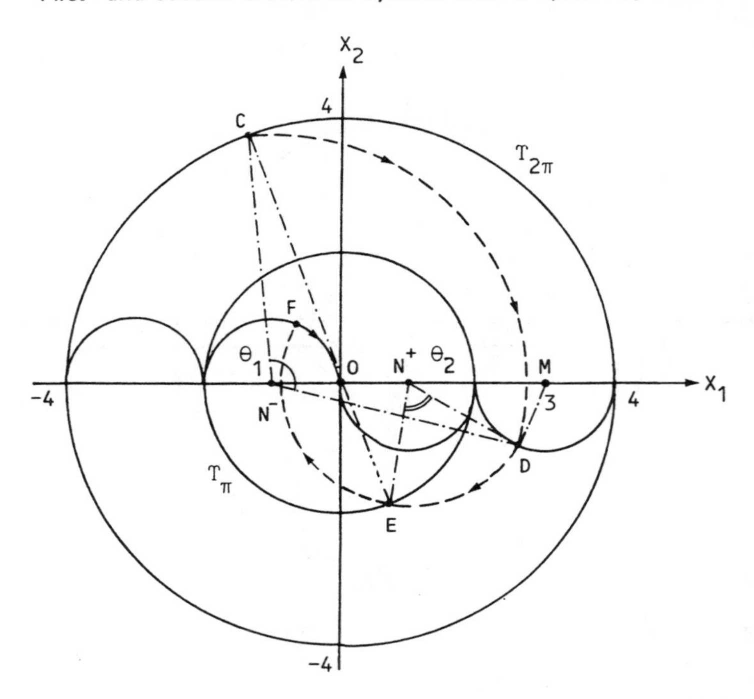

Fig. 5.10 *Harmonic oscillator* 2π-isochrone ($\Upsilon_{2\pi}$)

the origin in the state plane. That $\Upsilon_{2\pi}$ is, in fact, the 2π-isochrone can be established as follows.

Referring to Fig. 5.10, let C be an arbitrary starting point on $\Upsilon_{2\pi}$. From C, an n-path CD is followed to the first optimal switch point D; a p-path is then followed which intersects the π-isochrone Υ_π at the point E. From the above results it is clear that the minimum-time path EFO from E to the origin has duration π. It remains to show that path CDE also is of duration π (and hence the total duration t_f of the path CDEFO is 2π as required). Clearly, the duration t_1 of path CDE is given by $t_1 = \theta_1 + \theta_2$, where θ_1 is the duration of the n-path CD and θ_2 is the duration of the p-path DE. Again, by simple geometry, it may be seen that triangles CN^-O and N^-DM are congruent as are triangles ON^+E and DMN^+, from which the required result $\theta_1 + \theta_2 = \pi$ easily follows. Thus, the duration of the full path CDEFO to the origin is 2π, i.e. the set $\Upsilon_{2\pi}$ is the 2π-isochrone for the problem.

A more detailed analysis reveals the general result that the set

$$\Upsilon_{i\pi} = \{x\colon \|x\| = 2i\}, \qquad i = 1, 2, \ldots$$

constitutes the $i\pi$-isochronal set.

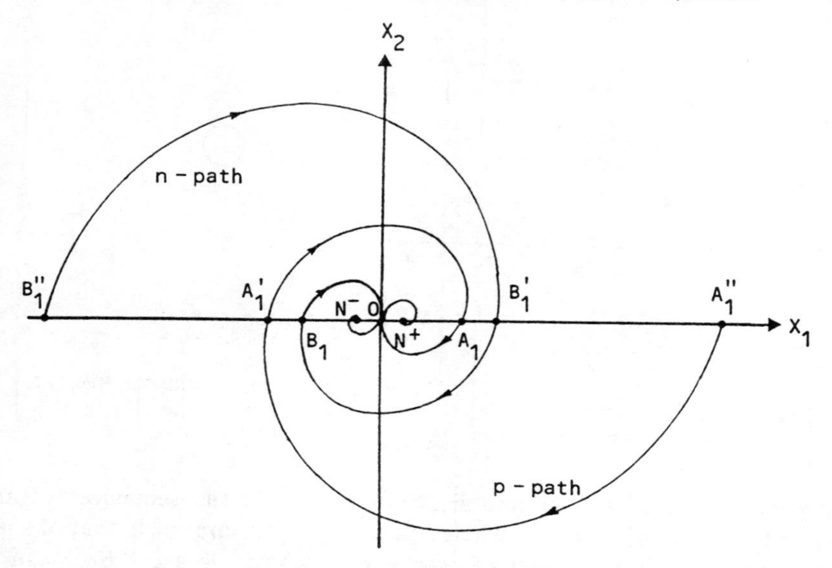

Fig. 5.11 *Damped oscillator*

5.2.2(ii) Damped oscillator: $0 < \zeta < 1$

From (5.54), it may be seen that the p- and n-paths form families of spirals, traced in a clockwise sense for increasing time, which converge to the points N^+ and N^- in the state plane with co-ordinates (1, 0) and (−1, 0), respectively. Fig. 5.11 depicts two important members of these families, viz. the p- and n-spirals which pass through the origin. Since the origin is to be attained in finite time, every optimal trajectory must terminate with an arc of these spirals leading to the origin. Again, in accordance with the maximum principle, on an optimal trajectory the control can remain constant on a time interval of duration π/α at most (see (5.53)); hence, every optimal trajectory must terminate with one or other of the arcs A_1O, B_1O of Fig. 5.11 (each of duration π/α) or some subarc thereof (of duration $< \pi/\alpha$) leading to the origin. It is readily verified from (5.54) that the coordinates of A_1 and B_1 are given by

$$A_1: \quad \begin{cases} x_1 = \exp(\zeta\pi/\alpha) + 1 \\ x_2 = 0 \end{cases}$$
$$B_1: \quad \begin{cases} x_1 = -\exp(\zeta\pi/\alpha) - 1 \\ x_2 = 0 \end{cases} \qquad (5.58)$$

As in the case of the harmonic oscillator, the complete switching curve Γ may be constructed from the basic arcs A_1O and B_1O and using the property that successive switch points are separated π/α units in time. An analogous construction to that adopted for the harmonic oscillator yields a switching curve Γ

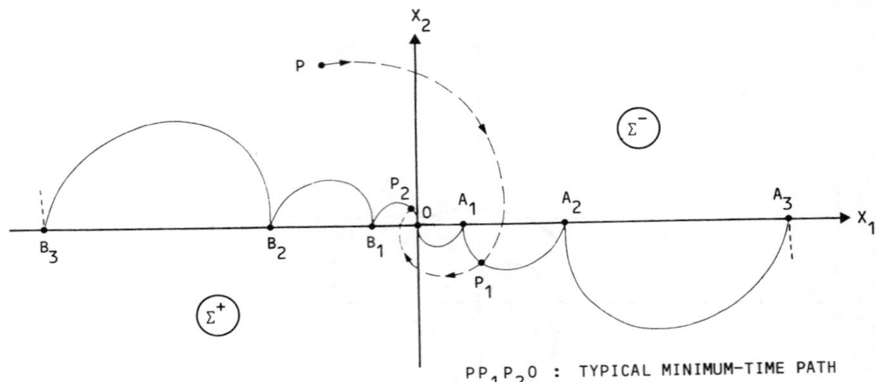

Fig. 5.12 *Damped oscillator*
Time-optimal switching curve

composed of the arcs A_1O and B_1O together with the sequences of arcs $\{A_{i+1}A_i\}_{i=1}^{\infty}$ and $\{B_{i+1}B_i\}_{i=1}^{\infty}$ where these sequences are such that the arc $A_{i+1}A_i$ corresponds to the set of points from which the arc B_iB_{i-1} can be attained (optimally) in a time π/α under a constant control $u = +1$, and conversely, $B_{i+1}B_i$ corresponds to the set of points from which A_iA_{i-1} can be attained (optimally) in a time π/α under a constant control $u = -1$. In the present case of a damped oscillator ($0 < \zeta < 1$), the sequences $\{A_{i+1}A_i\}$ and $\{B_{i+1}B_i\}$ are not composed of semicircular arcs (as in the case of $\zeta = 0$) but instead correspond (via translations and rotations) to the sequences of semiarcs (i.e. lying above and below the x_1 state axis) of the spirals of Fig.5.11 generated as these spirals are traced in the reverse time (anticlockwise) sense. In other words, arcs A_2A_1, $A_3A_2, \ldots$ (B_2B_1, $B_3B_2, \ldots$), which extend end-to-end from $A_1O(B_1O)$ along the positive (negative) x_1 state axis, are obtained from arcs $A_1'A_1$, $A_1''A_1'$, ... ($B_1'B_1$, $B_1''B_1'$, ...) by the appropriate reflections (in the x_1 state axis) and translations (along the x_1 state axis). For a damping value of $\zeta = \frac{1}{4}$, Fig. 5.12 depicts the optimal switching curve $\Gamma = \Gamma^+ \cup \Gamma^-$ which consists of the arc A_1O and the sequence of expanding spiral arcs $\{A_{i+1}A_i\}_{i=1}^{\infty}$ along the positive x_1 state axis (i.e. Γ^+) together with the arc B_1O and the sequence of expanding arcs $\{B_{i+1}B_i\}_{i=1}^{\infty}$ along the negative x_1 state axis (i.e. Γ^-). The curve Γ touches the x_1 axis at the points A_i, B_i, $i = 1, 2, \ldots$, the coordinates of which may be shown to satisfy (generalising (5.58))

$$A_i: \left\{ \begin{aligned} x_1 &= \exp(i\zeta\pi/\alpha) + 1 + 2\sum_{j=1}^{i-1} \exp(j\zeta\pi/\alpha) \\ x_2 &= 0 \end{aligned} \right\} \quad i = 1, 2, \ldots \tag{5.59a}$$

$$B_i: \left\{ \begin{aligned} x_1 &= -\left[\exp(i\zeta\pi/\alpha) + 1 + 2\sum_{j=1}^{i-1} \exp(j\zeta\pi/\alpha)\right] \\ x_2 &= 0 \end{aligned} \right\} \quad i = 1, 2, \ldots \tag{5.59b}$$

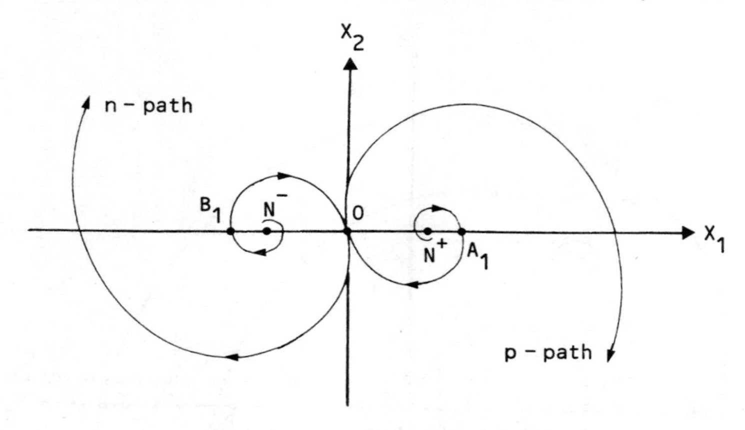

Fig. 5.13 *Unstable oscillator*

5.2.2(iii) Unstable oscillator : $-1 < \zeta < 0$

Finally, the case of an unstable oscillator ($-1 < \zeta < 0$) will be considered. Again, the p- and n-paths form families of logarithmic spirals (with nodes N^+, N^- at (1, 0) and (-1, 0) as before) traversed in a clockwise sense for increasing time, but with the crucial distinction that the motion is now unstable, i.e. the spirals diverge to infinity. Fig. 5.13 depicts the members of these families which pass through the origin. The previous method of constructing the time-optimal switching curve may again be adopted to show that Γ is composed of the origin and the spiral arcs A_1O, B_1O leading to the origin, together with the sequences of arcs $\{A_{i+1}A_i\}_{i=1}^{\infty}$ and $\{B_{i+1}B_i\}_{i=1}^{\infty}$ extending end-to end along the positive and negative x_1 state axis. However, in this case, successive arcs of the latter sequences are *contracting* (in contrast to the previous study of the damped oscillator). In particular, the sequences of points $\{A_i\}$, $\{B_i\}$ on the x_1 axis again satisfy (5.59), but now with $-1 < \zeta < 0$, so that the x_1 coordinates of the sequences form geometric progressions which accumulate at the points A_∞, B_∞ given by

$$A_\infty : \begin{cases} x_1 = \lim\limits_{i \to \infty} \left(\exp(i\zeta\pi/\alpha) + 1 + 2 \sum\limits_{j=1}^{i-1} \exp(j\zeta\pi/\alpha) \right) = \dfrac{1 + \exp(\zeta\pi/\alpha)}{1 - \exp(\zeta\pi/\alpha)} \\ x_2 = 0 \end{cases} \tag{5.60a}$$

$$B_\infty : \begin{cases} x_1 = -\dfrac{1 + \exp(\zeta\pi/\alpha)}{1 - \exp(\zeta\pi/\alpha)} \\ x_2 = 0 \end{cases} \tag{5.60b}$$

The time-optimal switching curve is depicted in Fig. 5.14 for a parameter value of $\zeta = -\frac{1}{4}$. Now consider the p-path starting from A_∞ at time $t = 0$. It is readily

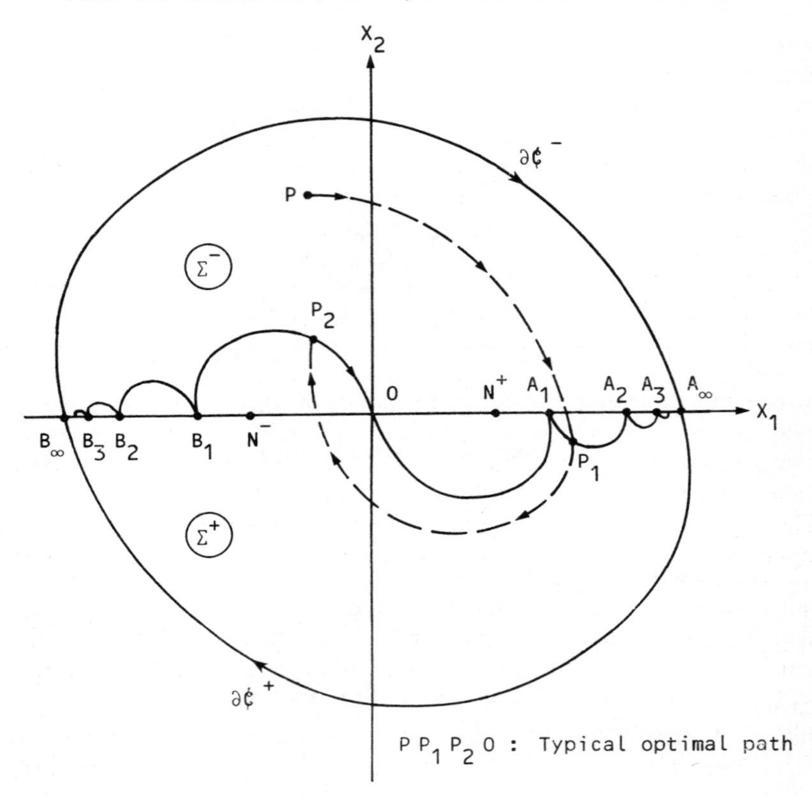

Fig. 5.14 *Unstable oscillator*
Domain of null controllability and time-optimal switching curve

verified from (5.54) that this p-path satisfies

$$x_1(t) = \exp(-\zeta t)\left[\frac{2\exp(\zeta\pi/\alpha)}{1-\exp(\zeta\pi/\alpha)}\right]\cos\alpha t + 1$$

$$x_2(t) = -\exp(-\zeta t)\left[\frac{2\exp(\zeta\pi/\alpha)}{1-\exp(\zeta\pi/\alpha)}\right]\sin\alpha t$$

Specifically, at time $t = \pi/\alpha$,

$$x_1(\pi/\alpha) = \exp(-\zeta\pi/\alpha)\left[\frac{-2\exp(\zeta\pi/\alpha)}{1-\exp(\zeta\pi/\alpha)}\right] + 1$$

$$= -\frac{1+\exp(\zeta\pi/\alpha)}{1-\exp(\zeta\pi/\alpha)}$$

$$x_2(\pi/\alpha) = 0$$

In other words, the p-path ($\partial\mathscr{C}^+$ of Fig. 5.14) from A_∞ leads to B_∞ and, by symmetry, the n-path ($\partial\mathscr{C}^-$ of Fig. 5.14) from B_∞ leads to A_∞, thereby forming the closed curve $\partial\mathscr{C} = \partial\mathscr{C}^+ \cup \partial\mathscr{C}^-$. From starting points in the region exterior to $\partial\mathscr{C}$ time-optimal paths to the origin do not exist, while minimum-time paths to the origin exist from all points in the region interior to $\partial\mathscr{C}$. Hence, $\partial\mathscr{C}$ defines the boundary of the domain of null controllability for the system.

5.2.3 Sensitivity of time-optimal systems to parameter variations

Practical implementation of the time-optimal feedback controls, derived for a variety of second-order single-input systems in Sections 5.2.1 and 5.2.2, all require the realisation of a nonlinear switching function $\Xi : \mathscr{C} \subseteq \mathbb{R}^2 \to \mathbb{R}$ which depends on some or all of the plant parameters; precise realisation of the control requires precise knowledge of these parameters. In practice, there will always be some parameter uncertainty (e.g. time-varying parameters) with consequent inaccuracy in the control realisation giving rise to deterioration in system performance (see e.g. Kreindler 1972, Becker 1980).

The effects of parameter uncertainty on the performance of nominally time-optimal feedback systems is briefly studied here. In all cases of parameter variation which lead to imprecise control realisation, the actual switching curve realised clearly will not coincide with the true time-optimal switching curve Γ so that the ability to employ the optimal subsystem control on Γ is lost. In other words, the actual (suboptimal) control input to the plant is almost everywhere determined by the *primary switching function* ξ; the *secondary* or *subsystem* switching function Ξ^s plays no essential rôle in generating the input. Hence, the control input is (almost everywhere) determined by a feedback law of the form

$$u(x) = -\operatorname{sgn}\,(\xi_p(x))$$

where the dependence of the nominally optimal primary switching function ξ_p: $\mathscr{C} \subseteq \mathbb{R}^2 \to \mathbb{R}$ on some plant parameter p has been indicated explicitly. In particular, the sensitivity to parameter variations in the following cases will be investigated:

(i) the plant gain parameter a for the double integrator system of Subsection 5.2.1(i);

(ii) the plant eigenvalue ratio α for the system of Subsection 5.2.1(v) with negative real and distinct eigenvalues.

5.2.3 (i) Double integrator system : sensitivity to gain parameter variation

This and related sensitivity problems for the nominally time-optimal double integrator system were initially studied by Sugiura (1966), Kreindler (1969) and Trieu and Pierre (1970); Zinober and Fuller (1973) gave the definitive study which is closely followed here.

The system under consideration is the following

$$\dot{x}_1 = x_2; \qquad \dot{x}_2 = \frac{u}{a}; \qquad |u| \le 1; \qquad a = \text{constant} > 0 \tag{5.61}$$

for which an *estimate* $\hat{a}$ of the true gain parameter value a is available, i.e. the (nominally optimal) feedback control is given by

$$u(x) = -\operatorname{sgn}(\xi_{\hat{a}}(x)); \qquad \hat{a} \neq a \tag{5.62a}$$

where, from (5.15), the switching function, realised on the basis of the parameter estimate $\hat{a}$, is

$$\xi_{\hat{a}}(x) = x_1 + \tfrac{1}{2}\hat{a}x_2|x_2| \tag{5.62b}$$

Let Γ_a and $\Gamma_{\hat{a}}$ denote the true time-optimal and approximate switching curves, respectively, i.e.

$$\Gamma_a = \{x: \xi_a(x) = 0\} \qquad \Gamma_{\hat{a}} = \{x: \xi_{\hat{a}}(x) = 0\} \tag{5.63}$$

Case 1: $\hat{a} > a$: In this case the control is realised on the basis of a parameter estimate which *exceeds* the true value. It will be shown that, under this control, on reaching $\Gamma_{\hat{a}}$ the state point follows a sliding path in $\Gamma_{\hat{a}}$ to the origin generated by a *constant* sliding control $u \equiv u_s \in (-1, 1)$. Specifically, applying the results of Sections 3.2 and 3.3 to the problem at hand, yields $\nabla\xi_{\hat{a}} = (1, \hat{a}|x_2|)'$ with $\langle\nabla\xi_{\hat{a}}, Ax\rangle = x_2$ and $\langle\nabla\xi_{\hat{a}}, b\rangle = (\hat{a}/a)|x_2|$. Noting that $\hat{a}/a > 1$, the conditions (3.15) for sliding on $\Gamma_{\hat{a}}$ (at points other than the origin) are clearly satisfied, viz.

(*a*) $|\langle\nabla\xi_{\hat{a}}, Ax\rangle| < |\langle\nabla\xi_{\hat{a}}, b\rangle|$

(*b*) $\langle\nabla\xi_{\hat{a}}, b\rangle > 0 \ \forall\ x \neq 0$

Moreover, from (3.18), the effective value of control at a point $x \in \Gamma_{\hat{a}}$ $(x \neq 0)$ is given by

$$u_s = -\frac{\langle\nabla\xi_{\hat{a}}, Ax\rangle}{\langle\nabla\xi_{\hat{a}}, b\rangle} = -\frac{a}{\hat{a}}\operatorname{sgn}(x_2) \in (-1, 1) \tag{5.64}$$

i.e. the sliding path in $\Gamma_{\hat{a}}$ to the origin is generated by a *constant* intermediate-value control $u \equiv u_s \in (-1, 1)$.

For a parameter estimate $\hat{a} = 2a$ (i.e. twice the true value), Fig. 5.15 depicts the resulting state portrait of the feedback system. With the exception of the symmetric pair of switchless trajectories AO and A′O which go directly to the origin under the constant controls $u \equiv \pm 1$ (and hence are time-optimal), all trajectories terminate with a sliding path in $\Gamma_{\hat{a}}$ to the origin. Consider, for example, a trajectory emanating from a point $x = (x_1, 0)'$, $x_1 \neq 0$, on the x_1 state axis. From (5.16), the minimum time to the origin is given by

$$t_2 = 2\sqrt{a|x_1|}$$

On the other hand, a straightforward calculation yields the result that the time from $x = (x_1, 0)'$ to the origin under the approximate control (5.62) (with $\hat{a} > a$), i.e. the time on the normal arc to $\Gamma_{\hat{a}}$ (generated by $u \equiv -\operatorname{sgn}(x_1)$) plus the time

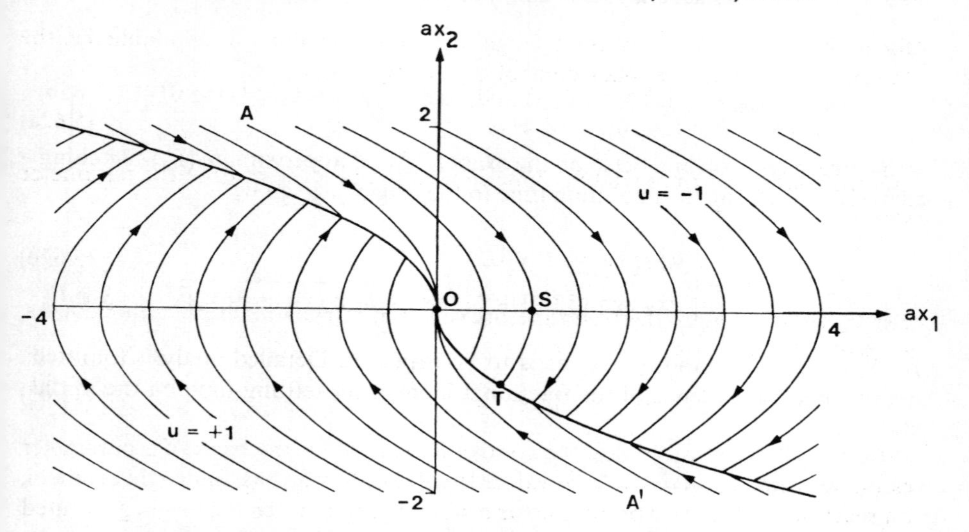

Fig. 5.15 *State portrait of feedback system with parameter estimate $\hat{a} = 2a$*

on the sliding path in $\Gamma_{\hat{a}}$ (generated by $u \equiv a/\hat{a}\ \mathrm{sgn}\ (x_1)$), is given by

$$\hat{t}_2 = \sqrt{2\left(1 + \frac{\hat{a}}{a}\right) a|x_1|}$$

Defining the fractional increase in settling time $\Delta \geq 0$ as

$$\Delta = \frac{\hat{t}_2 - t_2}{t_2}$$

then, for all non-zero initial states on the x_1 state axis, the same value of Δ is incurred, viz.

$$\Delta = \sqrt{\frac{1}{2}\left(1 + \frac{\hat{a}}{a}\right)} - 1$$

Clearly, $\Delta \to 0$ as $\hat{a} \to a$ as is to be expected.

For more general starting points $x = (x_1, x_2)' \notin \Gamma_{\hat{a}}$, the time to the origin under the approximate control (5.62) may be calculated as

$$\hat{t}_2 = \hat{T}_2(x) = \begin{cases} a|x_2| + \sqrt{2\left(\frac{\hat{a}}{a} - 1\right)[a|x_1| - \frac{1}{2}(ax_2)^2]}; & x \in \Sigma_1 \\ ax_2\ \mathrm{sgn}\ (\xi_{\hat{a}}(x)) \\ \quad + \sqrt{2\left(\frac{\hat{a}}{a} + 1\right)[ax_1\ \mathrm{sgn}\ (\xi_{\hat{a}}(x)) + \frac{1}{2}(ax_2)^2]}; \\ & x \notin \Sigma_1 \end{cases} \qquad (5.65a)$$

where

$$\Sigma_1 = \{x: \xi_{\hat{a}}(x) \geq 0;\ \xi_a(x) \leq 0\} \cup \{x: \xi_{\hat{a}}(x) \leq 0;\ \xi_a(x) \geq 0\} \tag{5.65b}$$

is the region lying on and between the exact (Γ_a) and approximate ($\Gamma_{\hat{a}}$) switching curves. From (5.16), the minimum time to the origin is given by

$$t_2 = T_2(x) = \begin{cases} a|x_2|; & x \in \Gamma_a \\ ax_2 \operatorname{sgn}(\xi_a(x)) + 2\sqrt{ax_1 \operatorname{sgn}(\xi_a(x)) + \frac{1}{2}(ax_2)^2}; & x \notin \Gamma_a \end{cases}$$

As $\hat{a} \to a$, $\Sigma_1 \to \Gamma_a$ and $\hat{t}_2 \to t_2$ as is to be expected. Detailed analysis (omitted here) of the dependence of the fractional increase in settling time on the initial state, i.e.

$$\Delta = \Delta(x) = \frac{\hat{T}_2(x) - T_2(x)}{T_2(x)}$$

reveals the following *worst case* condition (Zinober and Fuller, 1973): for a given $\hat{a} > a$, Δ takes its maximum value for starting points $x = (x_1, x_2)'$ on the curve $\Gamma_\gamma = \{x: x_1 + \frac{1}{2}\gamma a x_2 |x_2| = 0\}$ where

$$\gamma = \gamma\left(\frac{\hat{a}}{a}\right) = \left[7\left(\frac{\hat{a}}{a}\right)^2 + 2\left(\frac{\hat{a}}{a}\right) - 8\left(\frac{\hat{a}}{a} - 1\right)\sqrt{6\left(\frac{\hat{a}}{a}\right) - 2} - 5\right] \times \left(3 - \frac{\hat{a}}{a}\right)^{-2}$$

Fig. 5.16 depicts the fractional increase in settling time Δ as a function of the ratio $\hat{a}/a$ of estimated and true parameter values for initial states on the x_1 state axis and for initial states on the worst case curve Γ_γ. For example, if $\hat{a} = 2a$, then $\gamma = 27 - 8\sqrt{10}$ and evaluating Δ for $x \in \Gamma_\gamma$ gives the maximum fractional increase in settling time as

$$\Delta = \tfrac{1}{3}(\sqrt{10} - 2) \simeq 0{\cdot}3874$$

Hence, if the estimated parameter value exceeds the true value by a factor of 2, then the time to the origin from any initial state under the estimated control (5.62) is always within 38·74% of the true minimum time.

Case 2: $0 < \hat{a} < a$: If the control is realised on the basis of a gain parameter value $\hat{a}$ which is less than the true value a, then on the approximate switching curve $\Gamma_{\hat{a}} = \{x: \xi_{\hat{a}}(x) = 0\}$ the following relations hold:

$$|\langle \nabla\xi_{\hat{a}}(x), Ax\rangle| = |x_2| > \frac{\hat{a}}{a}|x_2| = |\langle \nabla\xi_{\hat{a}}(x), b\rangle| \ \forall\ x \neq 0$$

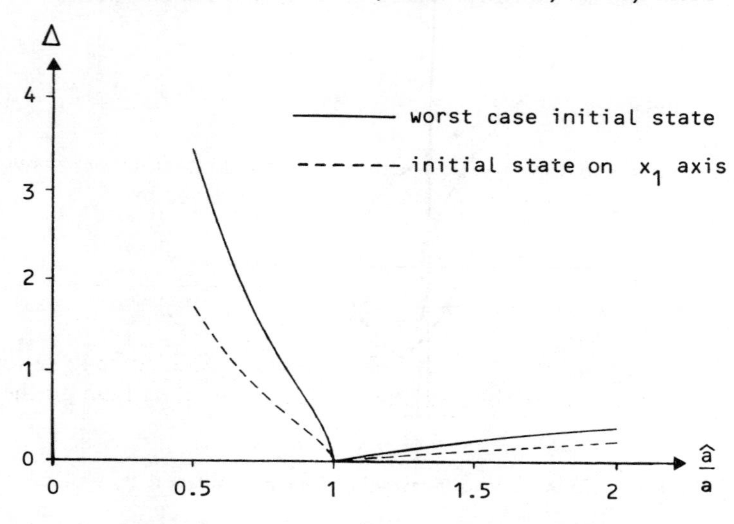

Fig. 5.16 *Fractional increase in time to the origin*

Consequently, in view of (3.14), it may be concluded that only regular switching behaviour can occur on $\Gamma_{\hat{a}}$, i.e. sliding trajectories or separatrices do not exist. Suppose that $x^0 \in \Gamma_{\hat{a}}$ is one such regular switch point, i.e.

$$x^0 \in \Gamma_{\hat{a}} \Leftrightarrow x_1^0 = -\tfrac{1}{2}\hat{a}x_2^0 |x_2^0|$$

then the coordinates of the next regular switch point x^1 of the feedback system may be calculated by integrating the state equations from $x(0) = x^0$ under the constant control $u \equiv -\text{sgn}\,(x_2^0)$ until the switching curve is again reached at time $t_1 > 0$ (see Fig. 5.17), viz.

$$x_1^1 = x_1(t_1) = x_1^0 + x_2^0 t_1 - \frac{\text{sgn}\,(x_2^0)}{2a} t_1^2$$

$$x_2^1 = x_2(t_1) = x_2^0 - \frac{\text{sgn}\,(x_2^0)}{a} t_1$$

$$x^1 \in \Gamma_{\hat{a}} \Leftrightarrow x_1^1 = -\tfrac{1}{2}\hat{a}x_2^1 |x_2^1|$$

Combining the above relations yields the result

$$t_1 = a(1 + \rho)|x_2^0|$$

$$x_2^1 = x_2(t_1) = -\rho x_2^0$$

$$x_1^1 = x_1(t_1) = -\rho^2 x_1^0$$

where

$$\rho = \sqrt{\left(1 - \frac{\hat{a}}{a}\right)\left(1 + \frac{\hat{a}}{a}\right)^{-1}} \in (0, 1); \qquad 0 < \hat{a} < a$$

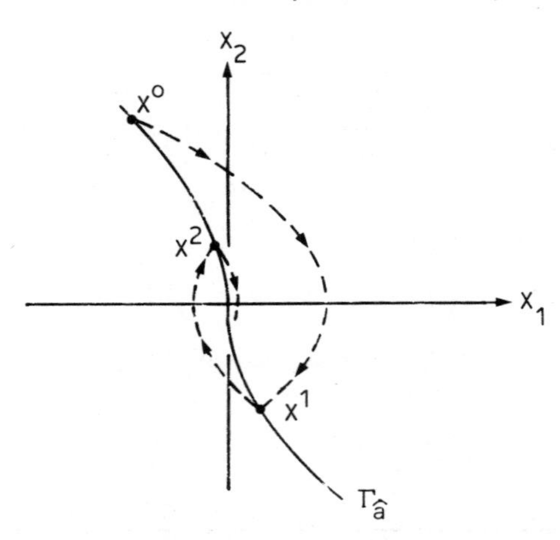

Fig. 5.17 *Constant-ratio trajectory of feedback system with* $0 < \hat{a} < a$

Hence, if $x = (x_1, x_2)' \in \Gamma_{\hat{a}}$ are the coordinates of *any* regular switch point of the feedback system, then $(-\rho^2 x_1, -\rho x_2)'$ are the state coordinates at the *next* switch, which occurs after a finite time $t_1 = (1 + \rho)a|x_2|$; it immediately follows that $(\rho^4 x_1, \rho^2 x_2)'$ are the coordinates of the subsequent switch, occurring after a further time interval of duration $t_2 = \rho t_1 = \rho(1 + \rho)a|x_2|$. Continuing this process, a countably infinite sequence of regular switch points is generated,

$$((-\rho^2)^i x_1, (-\rho)^i x_2)', \qquad i = 0, 1, 2, \ldots$$

which accumulate at the origin, i.e. $(-\rho^2)^i x_1 \to 0$ and $(-\rho)^i x_2 \to 0$ as $i \to \infty$ (since $0 < \rho < 1$), successive switches being separated in time by finite intervals of duration $t_i = p^{i-1} t_1$, $i = 1, 2, \ldots$. Such a trajectory is termed a *constant-ratio trajectory* with parameter $\rho \in (0, 1)$ (see also Fuller 1960c). Constant-ratio trajectories have the characteristic feature that, although the number of control switches is infinite, the state origin is attained in *finite* time due to the fact that the intervals t_i of control constancy form a convergent geometric progression, so that the total time t_f to the origin from $x = (x_1, x_2)' \in \Gamma_{\hat{a}}$ is given by

$$\begin{aligned} t_f &= \sum_{i=1}^{\infty} t_i = t_1(1 + \rho + \rho^2 + \cdots) \\ &= (1 - \rho)^{-1} t_1 \\ &= (1 + \rho)(1 - \rho)^{-1} a|x_2|; \qquad x = (x_1, x_2)' \in \Gamma_{\hat{a}} \end{aligned}$$

For more general starting points x, the total time to the origin is the sum of the time to reach $\Gamma_{\hat{a}}$ initially and the duration of the subsequent constant-ratio trajectory to the origin and is easily calculated as

$$t_f = T_f(x) = \begin{cases} ax_2 \operatorname{sgn}(\xi_{\hat{a}}(x)) + 2(1-\rho)^{-1} \\ \qquad \sqrt{(1+\rho^2)[ax_1 \operatorname{sgn}(\xi_{\hat{a}}(x)) + \frac{1}{2}(ax_2)^2]}; & x \notin \Gamma_{\hat{a}} \\ (1+\rho)(1-\rho)^{-1}a|x_2|; \quad x \in \Gamma_{\hat{a}} \end{cases} \tag{5.66a}$$

where

$$\rho = \sqrt{\left(1 - \frac{\hat{a}}{a}\right)\left(1 + \frac{\hat{a}}{a}\right)^{-1}} \in (0, 1); \quad 0 < \hat{a} < a \tag{5.66b}$$

For example, with an initial state $x = (x_1, 0)'$ on the x_1 state axis, the time to the origin is given by

$$t_f = 2(1-\rho)^{-1}\sqrt{(1+\rho^2)a|x_1|}$$

while the minimum time to the origin is $t_f^* = 2\sqrt{a|x_1|}$, yielding the fractional increase in settling time as

$$\Delta = (1-\rho)^{-1}\sqrt{1+\rho^2} - 1; \quad x = (x_1, 0)'$$

Detailed analysis of the dependence of Δ on the initial state for $0 < \hat{a} < a$, i.e.

$$\Delta = \Delta(x) = \frac{T_f(x) - T_2(x)}{T_2(x)} \quad \text{(with } T_2(x) \text{ defined as before)}$$

yields the following worst case condition (Zinober and Fuller, 1973): for a given $\hat{a} < a$, Δ takes its maximum value for starting points on the exact time-optimal switching curve Γ_a, in which case the worst-case fractional increase in settling time becomes

$$\Delta = 2[(1-\rho)^{-1}\sqrt{1+\rho^2} - 1]; \quad x \in \Gamma_a$$

i.e. *twice* the value incurred for initial states on the x_1 state axis; Fig. 5.16 depicts the dependence of Δ on the ratio $\hat{a}/a$ of estimated to true parameter values. For example, if $\hat{a} = \frac{1}{2}a$, then a worst-case calculation yields the value

$$\Delta = 2\sqrt{3}; \quad \hat{a} = \tfrac{1}{2}a$$

Hence, if the parameter estimate $\hat{a}$ is too small by a factor of 2, then times to the origin which exceed the corresponding minimum times by a factor of up to $2\sqrt{3}$ are to be expected. This is in stark contrast to the earlier example of $\hat{a} = 2a$ for which settling times which exceed the minimum times by 38·74%, at most, can occur. In other words, the sensitivity of the nominally optimal feedback system is significantly dependent on the 'direction' of parameter variation and is markedly more sensitive to 'inferior' parameter estimates, i.e. $\hat{a} < a$. Finally, note that for all parameter estimates $\hat{a} > 0$ the feedback system remains stable and the origin is attained in finite time from all initial states $x \in \mathbb{R}^2$.

5.2.3(ii) System with real and distinct eigenvalues : sensitivity to eigenvalue ratio variations in the control law

The system to be considered is that of Section 5.2.1(iv), viz.

$$\left.\begin{aligned} &\dot{x}_1 = \lambda_1 x_1 + x_2; \qquad \dot{x}_2 = \lambda_2 x_2 + \frac{u}{a}; \qquad |u| \leq 1; \\ &a = \text{constant} > 0 \\ &\lambda_1 < \lambda_2 < 0 \qquad \text{with} \qquad \alpha = \frac{\lambda_1}{\lambda_2} > 1 \end{aligned}\right\} \tag{5.67}$$

for which the primary time-optimal switching function is

$$\xi_\alpha(x) = x_1 + \frac{x_2}{(\lambda_1 - \lambda_2)} - \frac{\operatorname{sgn}(x_2)}{a\lambda_1(\lambda_1 - \lambda_2)}[1 - (1 - a\lambda_2 |x_2|)^\alpha] \tag{5.68}$$

where the dependence on the eigenvalue ratio α has been indicated explicitly. As in Subsection 5.2.1(iv), under the linear transformation

$$w = \begin{bmatrix} w_1 \\ w_2 \end{bmatrix} = \begin{bmatrix} a\lambda_1(\lambda_1 - \lambda_2) & a\lambda_1 \\ 0 & a\lambda_2 \end{bmatrix} \begin{bmatrix} x_1 \\ x_2 \end{bmatrix} = Px \tag{5.69}$$

the system equations may be expressed in the more convenient diagonalised form

$$\dot{w} = \Lambda w + \lambda u; \qquad |u| \leq 1 \tag{5.70a}$$

where

$$\Lambda = \begin{bmatrix} \lambda_1 & 0 \\ 0 & \lambda_2 \end{bmatrix} \quad \text{and} \quad \lambda = \begin{bmatrix} \lambda_1 \\ \lambda_2 \end{bmatrix} \tag{5.70b}$$

with the associated primary time-optimal switching function

$$\tilde{\xi}_\alpha(w) = w_1 + \operatorname{sgn}(w_2)[1 - (1 + |w_2|)^\alpha] \tag{5.71}$$

In practice it is to be expected that α will take values which render (5.71) difficult to realise exactly; some inaccuracy will frequently be introduced in the realisation. In other cases, if a true but 'difficult to realise' parameter value α were to be replaced by a neighbouring 'reasonable' value $\hat{\alpha} > 1$ (e.g. integer value), then the resulting perturbed feedback control

$$u_{\hat{\alpha}}(w) = -\operatorname{sgn}(\tilde{\xi}_{\hat{\alpha}}(w)) \tag{5.72}$$

would be considerably easier to implement. However, the following lemma (Ryan 1980b) shows that such (implicit or explicit) parameter variations completely destroy the time-optimality of the feedback system.

Lemma 5.2.1 : With the sole exception of a symmetric pair of switchless trajectories which go directly to the origin (when $\hat{\alpha} > \alpha$), for all $\hat{\alpha} \neq \alpha$ (with $\hat{\alpha} > 1$) the state origin cannot be attained in finite time on a (non-trivial) state trajectory generated by the perturbed feedback control (5.72).

Proof
As a rigorous proof is straightforward but lengthy, only a heuristic outline is given here. From (5.71), the following holds for the perturbed switching function

$$\tilde{\xi}_{\hat{\alpha}}(w) = w_1 - \hat{\alpha} w_2 + o(\|w\|); \qquad \|w\| \to 0$$

and, under transformation (5.69), the corresponding perturbed switching function $\xi_{\hat{\alpha}}(x)$ of the original x state vector is found to satisfy

$$\xi_{\hat{\alpha}}(x) = x_1 - \left[\frac{\hat{\alpha} - \alpha}{\lambda_1(\alpha - 1)}\right] x_2 + o(\|x\|); \qquad \|x\| \to 0 \tag{5.73a}$$

where, in terms of the x state vector, the perturbed feedback control is

$$u_{\hat{\alpha}}(x) = -\mathrm{sgn}\,(\xi_{\hat{\alpha}}(x)) \tag{5.73b}$$

Hence, for all parameter perturbations, a non-zero linear x_2 term is introduced in (5.73), the adverse effects of which will now be demonstrated. The exact time-optimal switching curve and the perturbed switching curve (in the x or w state plane) will be denoted by Γ_α and $\Gamma_{\hat{\alpha}}$, respectively, partitioning the state plane into the respective p and n control regions $\Sigma_\alpha^+, \Sigma_\alpha^-$ and $\Sigma_{\hat{\alpha}}^+, \Sigma_{\hat{\alpha}}^-$.

Case 1: $\hat{\alpha} > \alpha$: From (3.15), the conditions for sliding behaviour on the perturbed switching curve $\Gamma_{\hat{\alpha}}$ are

$$|\langle \nabla \tilde{\xi}_{\hat{\alpha}}(w), \Lambda w \rangle| < |\langle \nabla \tilde{\xi}_{\hat{\alpha}}(w), \lambda \rangle| \tag{5.74a}$$

$$\langle \nabla \tilde{\xi}_{\hat{\alpha}}(w), \lambda \rangle > 0 \tag{5.74b}$$

and, under these conditions, the state will follow a sliding path in $\Gamma_{\hat{\alpha}}$ under an effective sliding control input

$$u_s = -\frac{\langle \nabla \tilde{\xi}_{\hat{\alpha}}(w), \Lambda w \rangle}{\langle \nabla \tilde{\xi}_{\hat{\alpha}}(w), \lambda \rangle} \tag{5.74c}$$

A straightforward calculation reveals that, for $\hat{\alpha} > \alpha$, the above sliding conditions hold on a bounded sliding segment $\Gamma_{\hat{\alpha}}^s \subset \Gamma_{\hat{\alpha}}$ defined as

$$\Gamma_{\hat{\alpha}}^s = \{w\colon \tilde{\xi}_{\hat{\alpha}}(w) = 0;\ |w_2| < K^s\} \tag{5.75a}$$

where $K = K^s$ is the unique positive real solution of the equation

$$2\alpha - 2\hat{\alpha}(1 + K)^{\hat{\alpha}-1} + (\hat{\alpha} - \alpha)(1 + K)^{\hat{\alpha}} = 0 \tag{5.75b}$$

(Note that $K^s \to 0$ as $\hat{\alpha} \to \alpha$ as is to be expected). Furthermore, it may be shown that, under the perturbed feedback control (with $\hat{\alpha} > \alpha$), every non-trivial trajectory reaches $\Gamma_{\hat{\alpha}}^s$ after at most one regular switch in control† and subsequently follows $\Gamma_{\hat{\alpha}}^s$ to the origin. Now, in the x-state plane, the perturbed switching curve

† With the sole exception of the symmetric pair of switchless trajectories (comprising the exact time-optimal curve Γ_α) which go directly to the origin under the constant control $u \equiv \pm 1$.

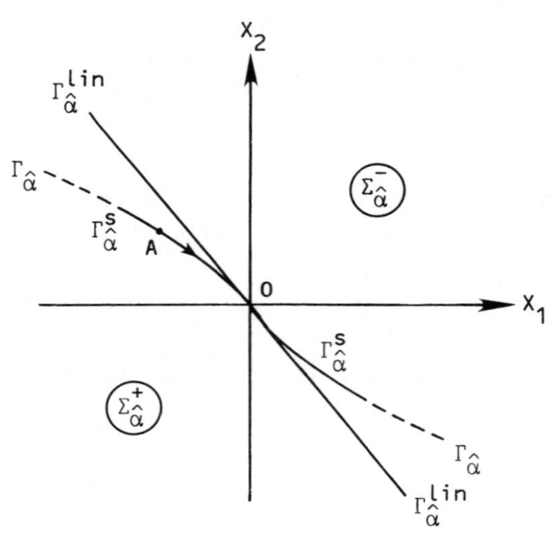

Fig. 5.18

$\Gamma_{\hat{\alpha}}$ lies in the second and fourth quadrants and is bounded above by its linearisation $\Gamma_{\hat{\alpha}}^{\text{lin}}$, viz.

$$\Gamma_{\hat{\alpha}}^{\text{lin}} = \left\{x \colon x_1 = \left[\frac{\hat{\alpha} - \alpha}{\lambda_1(\alpha - 1)}\right] x_2\right\} \tag{5.76}$$

in the second quadrant and bounded below by $\Gamma_{\hat{\alpha}}^{\text{lin}}$ in the fourth quadrant, as depicted in Fig. 5.18. It immediately follows that every sliding trajectory in $\Gamma_{\hat{\alpha}}^{s}$ approaches the origin asymptotically but is bounded *from below* by an exponential decay with time constant $\tau = -(\hat{\alpha} - \alpha)/[\lambda_1(\hat{\alpha} - 1)] > 0$, viz.

$$\|x(t)\| \geq M \exp\left[\lambda_1(\hat{\alpha} - 1)(\hat{\alpha} - \alpha)^{-1}t\right]; \qquad M \geq 0 \tag{5.77}$$

on a sliding trajectory $x(\cdot)$ in $\Gamma_{\hat{\alpha}}^{s}$. For example, on sliding path AO of Fig. 5.18, starting at $x(0) = x^{\text{A}} \in \Gamma_{\hat{\alpha}}^{s}$, (5.77) clearly holds with $M = |x_1^{\text{A}}|$ (or $M = \|x^{\text{A}}\|$). In summary, for $\hat{\alpha} > \alpha$, every trajectory of the feedback system (with $x(0) \notin \Gamma_{\hat{\alpha}}^{s}$) terminates in an asymptotically stable sliding path in $\Gamma_{\hat{\alpha}}^{s}$ which is exponentially bounded in norm from below so that the origin cannot be attained in finite time.

Case 2: $1 < \hat{\alpha} < \alpha$: In this case, the coefficient of the linear x_2 term in (5.73a) is negative, the effect of which is to introduce lobes on the perturbed switching curve $\Gamma_{\hat{\alpha}}$ which lie in the first and third quadrants of the x-state plane, as depicted in Fig. 5.19. As a direct result, the origin† becomes a *separatrix* of the feedback system, i.e. receded from by a p-path ($u = +1$) into the p-region $\Sigma_{\hat{\alpha}}^{+}$ and by an n-path ($u = -1$) into the n-region $\Sigma_{\hat{\alpha}}^{-}$, so that no regular switching

† In fact, a bounded segment of $\Gamma_{\hat{\alpha}}$ about the origin exists, each point of which is a separatrix (cf. Section 3.3) of the feedback system.

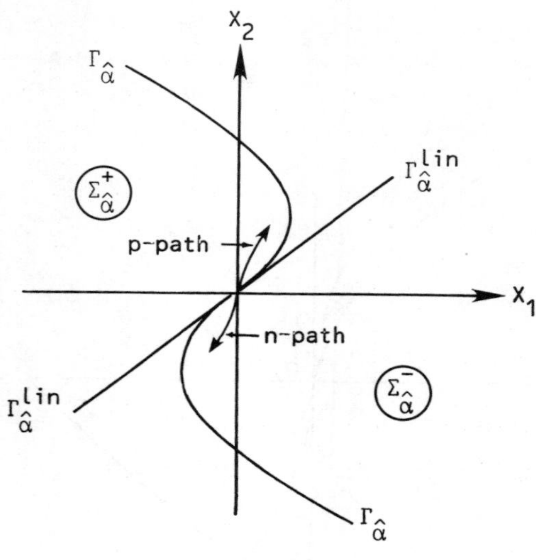

Fig. 5.19

trajectories to the state origin can exist. Moreover, it is easily verified that, for $1 < \hat{\alpha} < \alpha$, the sliding conditions (5.74) are never satisfied on $\Gamma_{\hat{\alpha}}$. In summary, under the perturbed feedback control (5.73) with $1 < \hat{\alpha} < \alpha$, a (non-trivial) state trajectory can never arrive at the origin, i.e. the origin is an unstable equilibrium point of the feedback system. A more detailed analysis reveals the following result.

Corollary 5.2.1: For each $1 < \hat{\alpha} < \alpha$, a limit cycle exists† towards which all (non-trivial) state trajectories converge under the perturbed feedback control.

Example 1: $\hat{\alpha} > \alpha$ *:* Consider system (5.67) with $\lambda_1 = -1$ and $\lambda_2 = -0{\cdot}375$ so that $\alpha = \frac{8}{3}$. Adopting the neighbouring value $\hat{\alpha} = 3$, then the perturbed switching function involves easily synthesised quadratic and cubic nonlinearities, viz.

$$\tilde{\xi}_{\hat{\alpha}}(w) = w_1 - 3w_2 - 3w_2|w_2| - w_2^3$$

The exact Γ_α and perturbed $\Gamma_{\hat{\alpha}}$ switch curves are depicted in the state plane in Fig. 5.20*a*. For an initial state $x(0) = x^A = (1, 0)'$, AB′O and ABO of Fig. 5.20*a* correspond to the trajectories under the optimal and perturbed feedback controls which, in turn, are shown (as functions of t) in Fig. 5.20*b*. Under the perturbed control, from A the state follows an n-path to $B \in \Gamma_{\hat{\alpha}}^s$ in the sliding segment $\Gamma_{\hat{\alpha}}^s$ from where it follows the sliding path BO, bounded below by an exponential decay with time constant $\tau = \frac{1}{6}$.

† Similar limit cycling behaviour was noted by O'Donnell (1964).

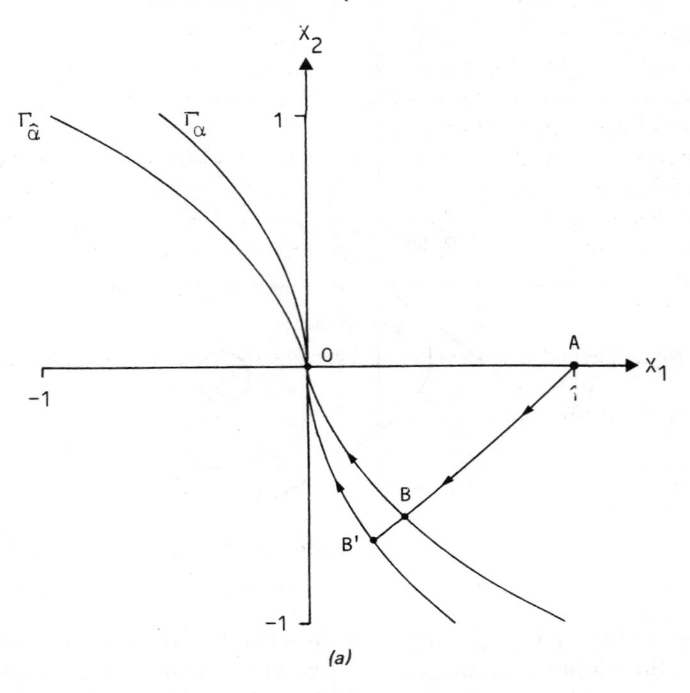

(a)

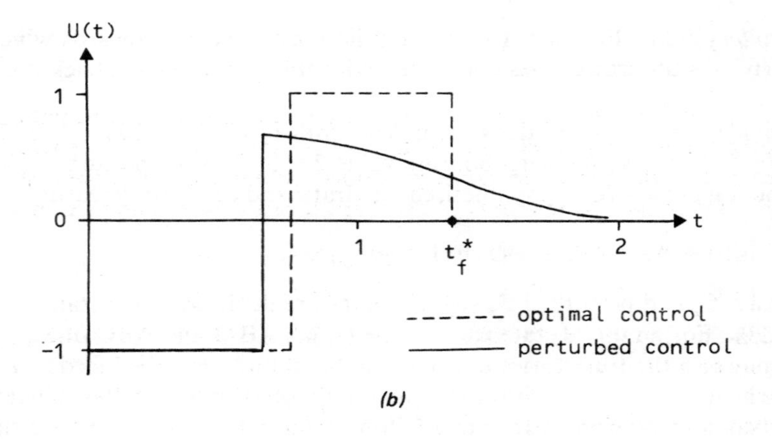

(b)

Fig. 5.20 *Example 1*: $\hat{\alpha} > \alpha$
(a) Optimal (Γ_{α}) and perturbed ($\Gamma_{\hat{\alpha}}$) switch curves
(b) Optimal and perturbed controls

Example 2: $1 < \hat{\alpha} < \alpha$: Consider system (5.67) with $\lambda_1 = -1$ and $\lambda_2 = -0{\cdot}3$ so that $\alpha = \frac{10}{3}$. Again adopting the neighbouring value $\hat{\alpha} = 3$ yields the readily synthesised switching function of Example 1. The exact Γ_α and perturbed $\Gamma_{\hat{\alpha}}$ switch curves are depicted in Fig. 5.21*a*. For the initial state $x(0) = x^A = (1, 0)'$, AB′O corresponds to the minimum-time trajectory generated by the optimal control of Fig. 5.21*b*. In agreement with the lemma and corollary, under the perturbed feedback control the state trajectory ABCD ... of Fig. 5.21*a* converges rapidly to a limit cycle; the associated control input is shown in Fig. 5.21*b*.

5.3 Two-input, second-order systems with negative, real and distinct eigenvalues

Proper, linear, autonomous, two-input systems of the general form

$$\dot{x} = Ax + Bu; \qquad x(t) \in \mathbb{R}^2; \qquad u(t) = \begin{bmatrix} u_1(t) \\ u_2(t) \end{bmatrix};$$

$$|u_i(t)| \leq 1, \qquad i = 1, 2 \tag{5.78}$$

will now be investigated. It is assumed that Rank $(B) = 2$, as otherwise the control problem can be reduced to the single-input case of the previous section. As before, the eigenvalues of the system A matrix will be denoted by λ_1 and λ_2. A two-input system with complex eigenvalues was studied in Example 2.6.2. Throughout this section, attention is restricted to systems with real eigenvalues. In addition, the eigenvalues are assumed to be negative and distinct; consequently, no loss in generality is incurred in adopting the diagonalised form for the matrix A, viz.

$$A = \begin{bmatrix} \lambda_1 & 0 \\ 0 & \lambda_2 \end{bmatrix} \quad \text{with} \quad B = \begin{bmatrix} b_1 & b_3 \\ b_2 & b_4 \end{bmatrix} \tag{5.79}$$

Since $\lambda_1, \lambda_2 < 0$, it follows from theorem 4.3.1 that a minimum-time path to the state origin exists from every initial state $x \in \mathbb{R}^2$, i.e. $\mathscr{C} = \mathbb{R}^2$. Moreover, from theorem 2.6.2, the piecewise constant, time-optimal control inputs $u_i(\cdot)$ are uniquely determined by relations (2.70) which, for the case at hand, become

$$\left.\begin{aligned} u_1(t) &= \operatorname{sgn}\,[\eta_1^0 b_1 \exp(-\lambda_1 t) + \eta_2^0 b_2 \exp(-\lambda_2 t)] \\ u_2(t) &= \operatorname{sgn}\,[\eta_1^0 b_3 \exp(-\lambda_1 t) + \eta_2^0 b_4 \exp(-\lambda_2 t)] \end{aligned}\right\} 0 \leq t \leq t_f \tag{5.80}$$

provided that the system satisfies the normality condition (2.68) which, in this case, may be written as

$$\det \begin{bmatrix} b_1 & \lambda_1 b_1 \\ b_2 & \lambda_2 b_2 \end{bmatrix} \neq 0; \qquad \det \begin{bmatrix} b_3 & \lambda_1 b_3 \\ b_4 & \lambda_2 b_4 \end{bmatrix} \neq 0 \tag{5.81a}$$

or, equivalently,

$$b_1 b_2(\lambda_1 - \lambda_2) \neq 0; \qquad b_3 b_4(\lambda_1 - \lambda_2) \neq 0 \tag{5.81b}$$

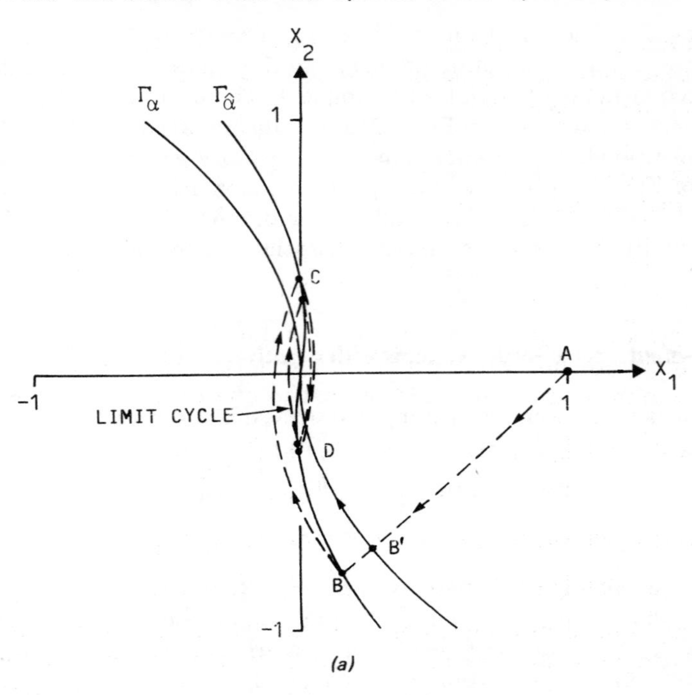

(a)

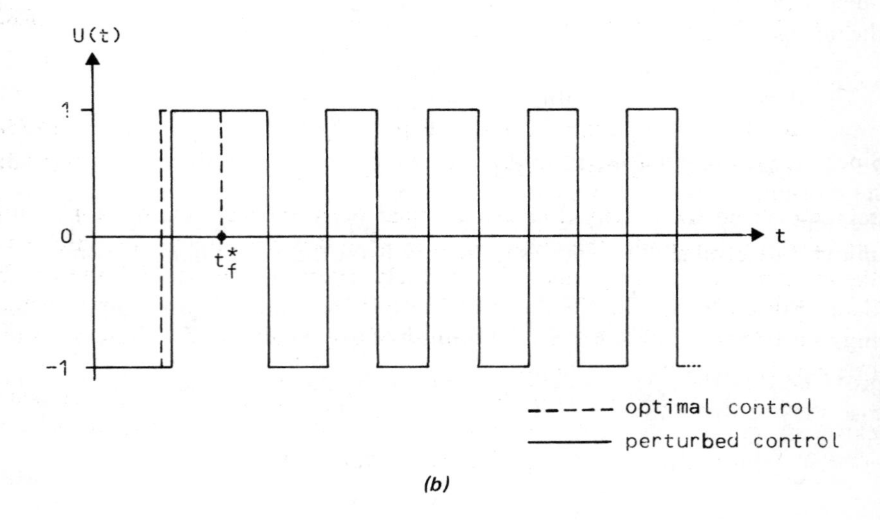

(b)

Fig. 5.21 Example 2: $1 < \hat{\alpha} < \alpha$
(*a*) Optimal (Γ_α) and perturbed ($\Gamma_{\hat{\alpha}}$) switch curves
(*b*) Optimal and perturbed controls

The non-uniqueness of optimal control, which results when (5.81) is not satisfied, is nicely illustrated by the following example of a non-normal system given by Hermes and LaSalle (1969).

5.3.1 Non-normal system: non-uniqueness of optimal control

Suppose that the system matrices are

$$A = \begin{bmatrix} -2 & 0 \\ 0 & -1 \end{bmatrix} \quad \text{and} \quad B = \begin{bmatrix} 1 & 1 \\ 1 & 0 \end{bmatrix}$$

Clearly, $b_1 b_2(\lambda_1 - \lambda_2) \neq 0$ but $b_3 b_4(\lambda_1 - \lambda_2) = 0$ so that the system is *not* normal. In accordance with the maximum principle (theorem 2.6.1) every state trajectory to the origin which is generated by an extremal control of the form

$$\begin{aligned} u_1(t) &= \text{sgn}\,[\eta_1^0 b_1 \exp(-\lambda_1 t) + \eta_2^0 b_2 \exp(-\lambda_2 t)] \\ &= \text{sgn}\,[\eta_1^0 \exp(2t) + \eta_2^0 \exp(t)] \\ u_2(t) &= \text{sgn}\,[\eta_1^0 b_3 \exp(-\lambda_1 t) + \eta_2^0 b_4 \exp(-\lambda_2 t)] \\ &= \text{sgn}\,[\eta_1^0 \exp(2t)] \end{aligned}$$

is optimal. Equivalently, integrating the state equations *in reverse time* $\tau = -t$ from the origin, under any pair of (reverse-time) extremal controls, yields a reverse-traced optimal path. In other words, the solution of the (reverse-time) equations

$$\frac{dx_1}{d\tau} = 2x_1 - u_1 - u_2; \qquad \frac{dx_2}{d\tau} = x_2 - u_1; \qquad x(0) = 0 \tag{5.82}$$

with

$$\left.\begin{aligned} u_1(\tau) &= \text{sgn}\,[\eta_1^0 \exp(-2\tau) + \eta_2^0 \exp(-\tau)] \quad &(5.83\text{a}) \\ u_2(\tau) &= \text{sgn}\,[\eta_1^0 \exp(-2\tau)] \quad &(5.83\text{b}) \end{aligned}\right\} \quad \tau \geq 0$$

regresses from the origin along an optimal path for each choice of $\eta^0 = (\eta_1^0, \eta_2^0)' \neq 0$. For example, the choice $\eta_1^0 = 0$, $\eta_2^0 \neq 0$ is permissible, in which case the extremal function u_2 fails to be uniquely determined by (5.83b) and can be defined in an infinity of different ways, taking *any* values between (and including) the values -1 and $+1$. Of the infinity of associated optimal state portraits, one may be constructed as follows.

Consider a (reverse-time) path from the origin generated by the constant extremal controls $u_1(\tau) = -1$, $u_2(\tau) = 1$, $\tau \geq 0$ (corresponding to $\eta_1^0 = 0$, $\eta_2^0 < 0$). Integrating (5.82), this path may be expressed as

$$\left.\begin{aligned} x_1(\tau) &= 0 \\ x_2(\tau) &= \exp(\tau) - 1 \end{aligned}\right\} \quad \tau \geq 0$$

which clearly coincides with the non-negative x_2 state semi-axis and is depicted

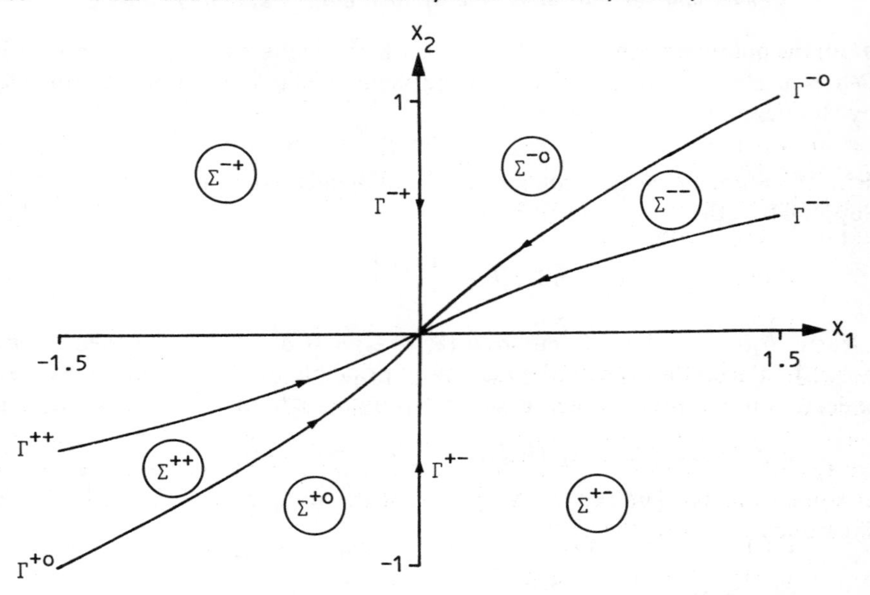

Fig. 5.22 *Non-normal system*
Time-optimal switching curves (non-unique)

(in forwards time) by Γ^{-+} of Fig. 5.22. By symmetry, the extremal controls $u_1(\tau) = 1$, $u_2(\tau) = -1$, $\tau \geq 0$ (corresponding to $\eta_1^0 = 0$, $\eta_2^0 > 0$) will generate the non-positive x_2 semi-axis, depicted (in forwards time) by Γ^{+-} of Fig. 5.22. Now suppose that u_2 is not held constant for all $\tau \geq 0$, but instead switches to the value 0 at some (reverse) time $\tau_1 \geq 0$. The state at this switch is given by

$$x(\tau_1) = (x_1(\tau_1), x_2(\tau_1))' = \begin{cases} (0, [\exp(\tau_1) - 1])'; & \eta_2^0 < 0 \ (\eta_1^0 = 0) \\ (0, -[\exp(\tau_1) - 1])'; & \eta_2^0 > 0 \ (\eta_1^0 = 0) \end{cases}$$

at which point the (reverse-time) state path will leave Γ^{-+} (if $\eta_2^0 < 0$) or Γ^{+-} (if $\eta_2^0 > 0$) and will subsequently follow the path

$$\left.\begin{aligned} x_1(\tau) &= \tfrac{1}{2}[\exp(2(\tau - \tau_1)) - 1] \\ x_2(\tau) &= \exp(\tau) - 1 \end{aligned}\right\} \quad \tau \geq \tau_1$$

if $\eta_2^0 < 0$, or the path

$$\left.\begin{aligned} x_1(\tau) &= -\tfrac{1}{2}[\exp(2(\tau - \tau_1)) - 1] \\ x_2(\tau) &= -[\exp(\tau) - 1] \end{aligned}\right\} \quad \tau \geq \tau_1$$

if $\eta_2^0 > 0$. These paths correspond to (i) the parabolic arc

$$(x_1 + \tfrac{1}{2}) = \tfrac{1}{2}\exp(-2\tau_1)(x_2 + 1)^2, \quad x_1, x_2 \geq 0; \qquad \text{if } \eta_2^0 < 0 \tag{5.84a}$$

or (ii) the parabolic arc

$$(x_1 - \tfrac{1}{2}) = -\tfrac{1}{2}\exp(-2\tau_1)(x_2 - 1)^2, \quad x_1, x_2 \leq 0; \qquad \text{if } \eta_2^0 > 0 \quad (5.84b)$$

As the switching time τ_1 increases from zero, parabolic arcs (5.84) fill the regions Σ^{-0} and Σ^{+0} of Fig. 5.22, lying between the curves Γ^{-0}, Γ^{-+} and Γ^{+0}, Γ^{+-}, respectively, where Γ^{-0} is the parabolic arc (5.84a) with $\tau_1 = 0$ and Γ^{+0} is the parabolic arc (5.84b) with $\tau_1 = 0$, i.e.

$$\Gamma^{-0} = \{x: x_1, x_2 \geq 0; x_1 + \tfrac{1}{2} = \tfrac{1}{2}(x_2 + 1)^2\} \quad (5.85a)$$

$$\Gamma^{+0} = \{x: x_1, x_2 \leq 0; x_1 - \tfrac{1}{2} = -\tfrac{1}{2}(x_2 - 1)^2\} \quad (5.85b)$$

In other words, Γ^{-0} (Γ^{+0}) is the (reverse-time) parabolic path from the origin generated by the constant extremal controls $u_1(\tau) = -1$ ($u_1(\tau) = +1$) and $u_2(\tau) = 0$, $\tau \geq 0$. Suppose now that u_2 is not held at the value zero along the entire length of path Γ^{-0} (or Γ^{+0}) but instead switches to the value -1 (or $+1$) at some (reverse) time $\tau_2 \geq 0$. At this point the state leaves Γ^{-0} (or Γ^{+0}) and subsequently follows the path

$$\left.\begin{aligned} x_1(\tau) &= \tfrac{1}{2}[\exp(2\tau) + \exp(2(\tau - \tau_2))] - 1 \\ x_2(\tau) &= \exp(\tau) - 1 \end{aligned}\right\} \quad \tau \geq \tau_2$$

if the switch at τ_2 occurs on Γ^{-0}, or the path

$$\left.\begin{aligned} x_1(\tau) &= -\tfrac{1}{2}[\exp(2\tau) + \exp(2(\tau - \tau_2))] + 1 \\ x_2(\tau) &= -[\exp(\tau) - 1] \end{aligned}\right\} \quad \tau \geq \tau_2$$

if the switch at τ_2 occurs on Γ^{+0}. These paths, respectively, correspond to the parabolic arcs

$$(x_1 + 1) = \tfrac{1}{2}[1 + \exp(-2\tau_2)](x_2 + 1)^2; \qquad x_1, x_2 \geq 0 \quad (5.86a)$$

and

$$(x_1 - 1) = -\tfrac{1}{2}[1 + \exp(-2\tau_2)](x_2 - 1)^2; \qquad x_1, x_2 \leq 0 \quad (5.86b)$$

As the switching time τ_2 increases from zero, parabolic arcs (5.86) fill the regions Σ^{--} and Σ^{++} of Fig. 5.22, lying between the curves Γ^{--}, Γ^{-0} and Γ^{++}, Γ^{+0}, respectively, where Γ^{--} is the parabolic arc (5.86a) with $\tau_2 = 0$ and Γ^{++} is the parabolic arc (5.86b) with $\tau_2 = 0$, i.e.

$$\Gamma^{--} = \{x: x_1, x_2 \geq 0; x_1 + 1 = (x_2 + 1)^2\} \quad (5.87a)$$

$$\Gamma^{++} = \{x: x_1, x_2 \leq 0; x_1 - 1 = -(x_2 - 1)^2\} \quad (5.87b)$$

Finally, by selecting η^0 such that $\eta_1^0 < 0$ and $\eta_1^0 + \eta_2^0 < 0$ (or $\eta_1^0 > 0$ and $\eta_1^0 + \eta_2^0 > 0$), it is easily verified from (5.83) that a switch in the u_1 control component can be generated at any (reverse) time $\tau_3 > 0$. In this case, the (reverse-time) trajectory will initially follow the curve Γ^{--} (or Γ^{++}) out from the origin until the u_1 switch occurs at the (reverse) time τ_3 and the state will subsequently follow the (reverse-time) path

$$\begin{aligned} x_1(\tau) &= \exp(2\tau)[1 - \exp(-2\tau_3)] \\ x_2(\tau) &= \exp(\tau)[1 - 2\exp(-\tau_3)] + 1 \end{aligned} \qquad \tau \geq \tau_3$$

if the switch at τ_3 occurs on Γ^{--}, or the path

$$\begin{aligned} x_1(\tau) &= -\exp(2\tau)[1 - \exp(-2\tau_3)] \\ x_2(\tau) &= -\exp(\tau)[1 - 2\exp(-\tau_3)] - 1 \end{aligned} \qquad \tau \geq \tau_3$$

if the switch at τ_3 occurs on Γ^{++}. The above paths, respectively, correspond to the parabolic arcs

$$x_1 = \frac{1 - \exp(-2\tau_3)}{[1 - 2\exp(-\tau_3)]^2}(x_2 - 1)^2; \qquad x_1, x_2 \geq 0 \tag{5.88a}$$

$$x_1 = -\frac{1 - \exp(-2\tau_3)}{[1 - 2\exp(-\tau_3)]^2}(x_2 + 1)^2; \qquad x_1, x_2 \leq 0 \tag{5.88b}$$

As the switching time τ_3 increases from zero, parabolic arcs (5.88) fill the remaining regions Σ^{+-} and Σ^{-+} of Fig. 5.22, lying between the curves Γ^{+-}, Γ^{--} and Γ^{-+}, Γ^{++}, respectively. Note that, for $\tau_3 = 0$, the lines Γ^{+-} and Γ^{-+} (i.e. the x_2 state axis) are recovered.

In summary, the above construction has yielded *an* optimal feedback control of the form

$$u(x) = \begin{bmatrix} u_1(x) \\ u_2(x) \end{bmatrix} = \begin{cases} \begin{bmatrix} +1 \\ +1 \end{bmatrix}; & x \in \Sigma^{++} \cup \Gamma^{++} \\ \begin{bmatrix} +1 \\ 0 \end{bmatrix}; & x \in \Sigma^{+0} \cup \Gamma^{+0} \\ \begin{bmatrix} +1 \\ -1 \end{bmatrix}; & x \in \Sigma^{+-} \cup \Gamma^{+-} \\ \begin{bmatrix} -1 \\ -1 \end{bmatrix}; & x \in \Sigma^{--} \cup \Gamma^{--} \\ \begin{bmatrix} -1 \\ 0 \end{bmatrix}; & x \in \Sigma^{-0} \cup \Gamma^{-0} \\ \begin{bmatrix} -1 \\ +1 \end{bmatrix}; & x \in \Sigma^{-+} \cup \Gamma^{-+} \end{cases}$$

However, as the normality condition does *not* hold in this case, it is stressed that the above synthesis is only one of infinitely many possible optimal syntheses, i.e. uniqueness of optimal control is lost.

5.3.2 Normal systems

Attention is now restricted to normal systems, i.e. (5.81) is assumed to hold (which carries the implication that all entries of the matrix B are non-zero) so that the optimal control is uniquely determined by (5.80). In this case, the piecewise constant controls take the values ± 1 only (each component changing sign at most once) so that the optimal state portrait is comprised of an appropriate composition of pp-paths ($u_1 = +1 = u_2$), pn-paths ($u_1 = +1$, $u_2 = -1$), np-paths ($u_1 = -1$, $u_2 = +1$) and nn-paths ($u_1 = -1 = u_2$). In the state plane, it is easily verified that the individual families of pp, pn, np, nn-paths have nodes at

$$\begin{aligned} N^{++} &= (-\lambda_1^{-1}(b_1 + b_3), -\lambda_2^{-1}(b_2 + b_4)) \\ N^{+-} &= (-\lambda_1^{-1}(b_1 - b_3), -\lambda_2^{-1}(b_2 - b_4)) \\ N^{-+} &= (\lambda_1^{-1}(b_1 - b_3), \lambda_2^{-1}(b_2 - b_4)) \\ N^{--} &= (\lambda_1^{-1}(b_1 + b_3), \lambda_2^{-1}(b_2 + b_4)) \end{aligned} \tag{5.89}$$

As in the example of Section 5.3.1, the optimal control synthesis will be obtained by characterising the family of reverse-time state paths from the origin generated by extremal controls, i.e. the family of solutions of the (reverse-time) state equations

$$\left.\begin{aligned} \frac{dx_1}{d\tau} &= -\lambda_1 x_1 - b_1 u_1 - b_3 u_2 \\ \frac{dx_2}{d\tau} &= -\lambda_2 x_2 - b_2 u_1 - b_4 u_2 \end{aligned}\right\} \quad x(0) = 0 \tag{5.90}$$

generated by (reverse-time) extremal controls

$$u_1(\tau) = \text{sgn}\ [\eta_1^0 b_1 \exp(\lambda_1 \tau) + \eta_2^0 b_2 \exp(\lambda_2 \tau)] \tag{5.91a}$$

$$u_2(\tau) = \text{sgn}\ [\eta_1^0 b_3 \exp(\lambda_1 \tau) + \eta_2^0 b_4 \exp(\lambda_2 \tau)] \tag{5.91b}$$

with $\eta^0 = (\eta_1^0, \eta_2^0)' = \text{constant} \neq 0$.

For convenience, write

$$\bar{u}_1 = b_1 u_1 + b_3 u_2; \qquad \bar{u}_2 = b_2 u_1 + b_4 u_2 \tag{5.92}$$

then (5.90) may be expressed as

$$\frac{dx_1}{d\tau} = -\lambda_1 x_1 - \bar{u}_1; \qquad \frac{dx_2}{d\tau} = -\lambda_2 x_2 - \bar{u}_2 \tag{5.93}$$

where the vector function $\bar{u} = (\bar{u}_1, \bar{u}_2)'$ takes the values

$$\begin{aligned} &(b_1 + b_3, b_2 + b_4)'; \qquad (b_1 - b_3, b_2 - b_4)'; \\ &\qquad (-b_1 + b_3, -b_2 + b_4)'; \qquad (-b_1 - b_3, -b_2 - b_4)' \end{aligned} \tag{5.94}$$

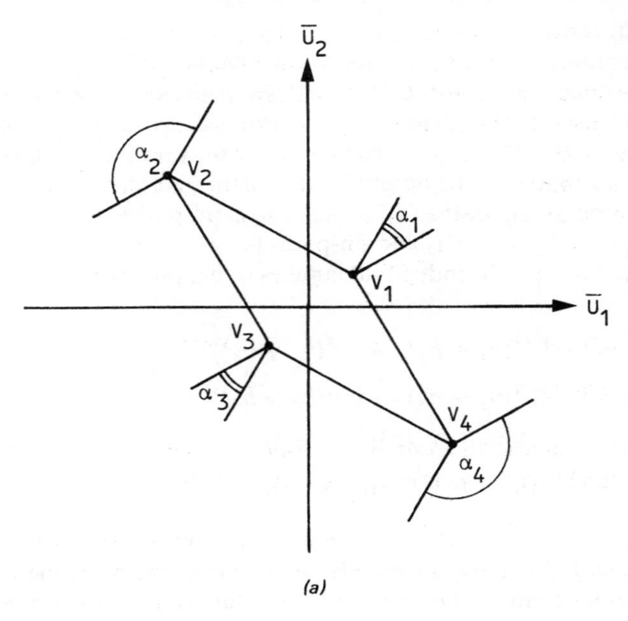

$\bar{U}_2$
$\bar{U}_1$
α_2
V_2
α_1
V_1
V_3
α_3
V_4
α_4

(a)

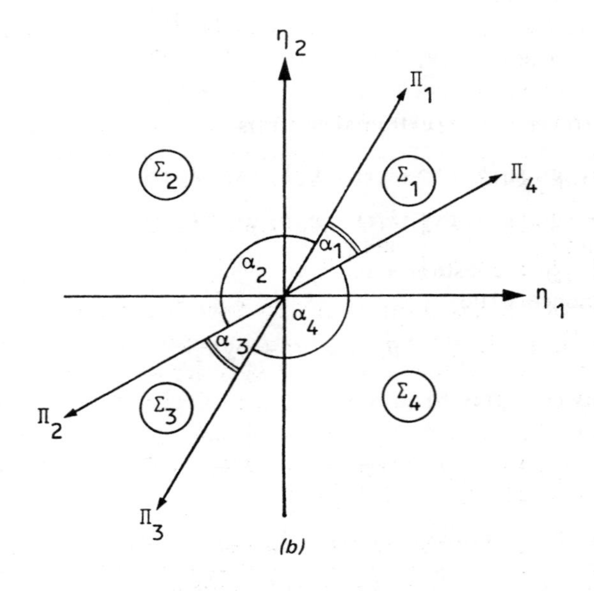

η_2
η_1
Π_1
Π_4
Π_2
Π_3
Σ_1
Σ_2
Σ_3
Σ_4
α_1
α_2
α_3
α_4

(b)

Fig. 5.23

only, corresponding to the vertices of a parallelogram in the $\bar{u}$-plane. Denote these vertices by v_1, v_2, v_3, v_4, numbered in an anticlockwise sense as illustrated in Fig. 5.23*a*, and denote, by $\Pi_1, \Pi_2, \Pi_3, \Pi_4$, the rays from the origin having directions normal to the edges v_1v_2, v_2v_3, v_3v_4, v_4v_1 of the parallelogram, as shown in Fig. 5.23*b*. These rays partition the plane into four regions $\Sigma_1, \Sigma_2, \Sigma_3, \Sigma_4$ labelled so that the angles α_i of Figs. 5.23*a* and 5.23*b* coincide. Now, in accordance with the maximum principle, an extremal control

$$\bar{u} = \begin{bmatrix} \bar{u}_1 \\ \bar{u}_2 \end{bmatrix} = \begin{bmatrix} b_1 u_1 + b_3 u_2 \\ b_2 u_1 + b_4 u_2 \end{bmatrix} = Bu$$

takes values only on the set of vertices $\{v_i\}$ such that

$$\langle \eta(\tau), Bu(\tau) \rangle = \langle \eta(\tau), \bar{u}(\tau) \rangle = \max_{v \in \{v_i\}} \langle \eta(\tau), v \rangle$$

Suppose that, at time τ, the vector $\eta(\tau)$ lies in the region Σ_i, then (see Fig. 5.23) the scalar product $\langle \eta(\tau), v \rangle$ takes its maximum value at vertex† v_i, i.e.

$$\bar{u}(\tau) = v_i \qquad \text{for} \qquad \eta(\tau) \in \Sigma_i \tag{5.94}$$

Equipped with the above properties, the optimal synthesis may be obtained. The synthesis falls into one of two possible categories depending on the 'shape' of the parallelogram $v_1v_2v_3v_4$ or, equivalently, the orientation of the rays Π_1, Π_2, Π_3, Π_4. In particular, if the straight lines $\Pi_1 \cup \Pi_3$ and $\Pi_2 \cup \Pi_4$ lie in the same quadrants (as in Fig. 5.23*b*), then the optimal synthesis is a *four-region synthesis* in which the domain of null controllability $\mathscr{C}$ is partitioned into four mutually exclusive regions of specified control effort separated by switching curves. On the other hand, if $\Pi_1 \cup \Pi_3$ and $\Pi_2 \cup \Pi_4$ lie in different quadrants, then a *two-region synthesis* occurs in which a single switching curve partitions the state plane into two regions of specified control effort. It may be verified that the orientation of the rays Π_i, as outlined above, can be characterised in terms of the sign of the product

$$D = \det \begin{bmatrix} b_1 & \lambda_1 b_1 \\ b_2 & \lambda_2 b_2 \end{bmatrix} \cdot \det \begin{bmatrix} b_3 & \lambda_1 b_3 \\ b_4 & \lambda_2 b_4 \end{bmatrix}$$

which, in view of (5.81), is always non-zero. Specifically, if D is positive, then the straight lines $\Pi_1 \cup \Pi_3$ and $\Pi_2 \cup \Pi_4$ lie in the same pair of quadrants, giving rise to a four-region synthesis; while, if D is negative, then the straight lines $\Pi_1 \cup \Pi_3$ and $\Pi_2 \cup \Pi_4$ lie in different quadrants and a two-region synthesis results.‡ For clarity of presentation, specific examples will be selected to illustrate each of the two possible forms of optimal synthesis; in principle, the same techniques can be adopted in other cases.

† This observation is a special case of the general result that the scalar product $\langle \eta, v \rangle$ ($\eta \in \mathbb{R}^n$, $v \in P \subset \mathbb{R}^n$) attains its maximum value at a unique vertex v_i of a closed convex polyhedron P if, and only if, $\langle \eta, e^i \rangle < 0$ for every edge e^i of P emanating from vertex v_i.

‡ For extensions of these results to higher-order multi-input systems, see e.g. Meeker and Puri (1971), Meeker (1978, 1980), Olsder (1975, 1980).

Example 5.3.1 : Four-region synthesis : Consider the case of

$$A = \begin{bmatrix} -2 & 0 \\ 0 & -1 \end{bmatrix} \quad \text{and} \quad B = \begin{bmatrix} 1 & 1 \\ 1 & 2 \end{bmatrix} \tag{5.95}$$

In this case, the pp, pn, np and nn-paths in the state plane constitute families of parabolae, with vertices at

$$\begin{aligned} N^{++} &= (1,\ 3) \\ N^{+-} &= (0,\ -1) \\ N^{-+} &= (0,\ 1) \\ N^{--} &= (-1,\ -3) \end{aligned} \tag{5.96}$$

respectively. Writing

$$\bar{u} = \begin{bmatrix} \bar{u}_1 \\ \bar{u}_2 \end{bmatrix} = Bu = \begin{bmatrix} u_1 + u_2 \\ u_1 + 2u_2 \end{bmatrix}$$

the parallelogram, with vertices v_i, may be constructed together with the rays Π_i and regions Σ_i as shown in Figs. 5.24*a* and 5.24*b*.

The optimal synthesis will be obtained by characterising the family of (reverse-time) solutions of (5.90), viz.

$$\left.\begin{aligned} \frac{dx_1}{d\tau} &= 2x_1 - u_1 - u_2 \\ \frac{dx_2}{d\tau} &= x_2 - u_1 - 2u_2 \end{aligned}\right\}; \qquad x(0) = 0 \tag{5.97}$$

generated by (reverse-time) extremal controls (5.91), viz.

$$\begin{aligned} u_1(\tau) &= \text{sgn}\ [\eta_1^0 \exp(-2\tau) + \eta_2^0 \exp(-\tau)] \\ u_2(\tau) &= \text{sgn}\ [\eta_1^0 \exp(-2\tau) + 2\eta_2^0 \exp(-\tau)] \end{aligned} \qquad \tau \geq 0 \tag{5.98}$$

with $\eta^0 = (\eta_1^0, \eta_2^0)' \neq 0$.

For the case at hand, the regions $\Sigma_i \subset \mathbb{R}^2$ may be expressed as

$$\begin{aligned} \Sigma_1 &= \{\eta\colon \eta_1 + \eta_2 > 0;\ \eta_1 + 2\eta_2 > 0\} \\ \Sigma_2 &= \{\eta\colon \eta_1 + \eta_2 < 0;\ \eta_1 + 2\eta_2 > 0\} \\ \Sigma_3 &= \{\eta\colon \eta_1 + \eta_2 < 0;\ \eta_1 + 2\eta_2 < 0\} \\ \Sigma_4 &= \{\eta\colon \eta_1 + \eta_2 > 0;\ \eta_1 + 2\eta_2 < 0\} \end{aligned} \tag{5.99}$$

Suppose that the initial (reverse-time) adjoint vector lies in the region Σ_1, i.e. $\eta^0 \in \Sigma_1$, then the (reverse-time) control (5.98) initially takes the value $u_1 = +1 = u_2$ (or, equivalently, $\bar{u}$ takes its value at vertex v_1). In this case, the state

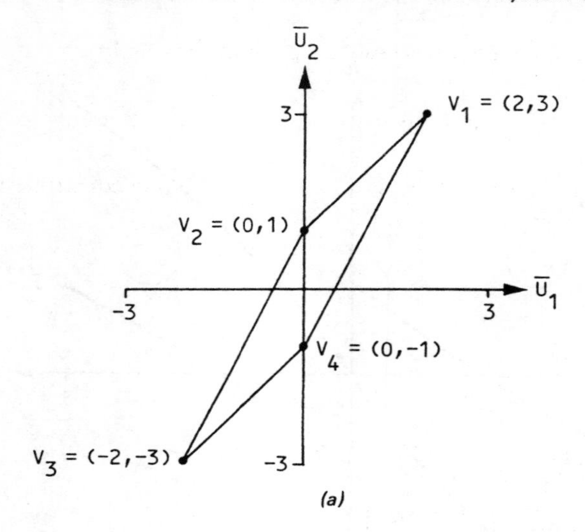

(a)

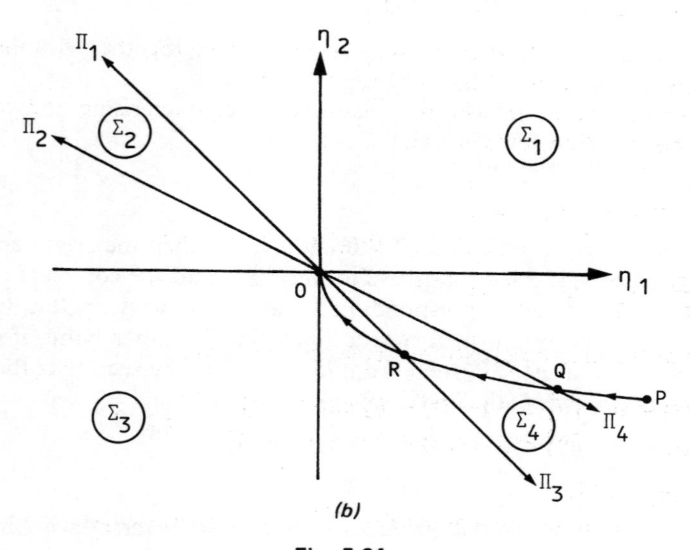

(b)

Fig. 5.24

initially backs out from the origin along the pp-path Γ^{++} of Fig. 5.25 given by

$$\left.\begin{aligned} x_1(\tau) &= 1 - \exp(2\tau) \\ x_2(\tau) &= 3(1 - \exp(\tau)) \end{aligned}\right\} \qquad \tau \geq 0$$

i.e. along the parabolic arc

$$\Gamma^{++} = \{x\colon x_1,\, x_2 \leq 0;\ 9x_1 - 6x_2 + x_2^2 = 0\} \tag{5.100}$$

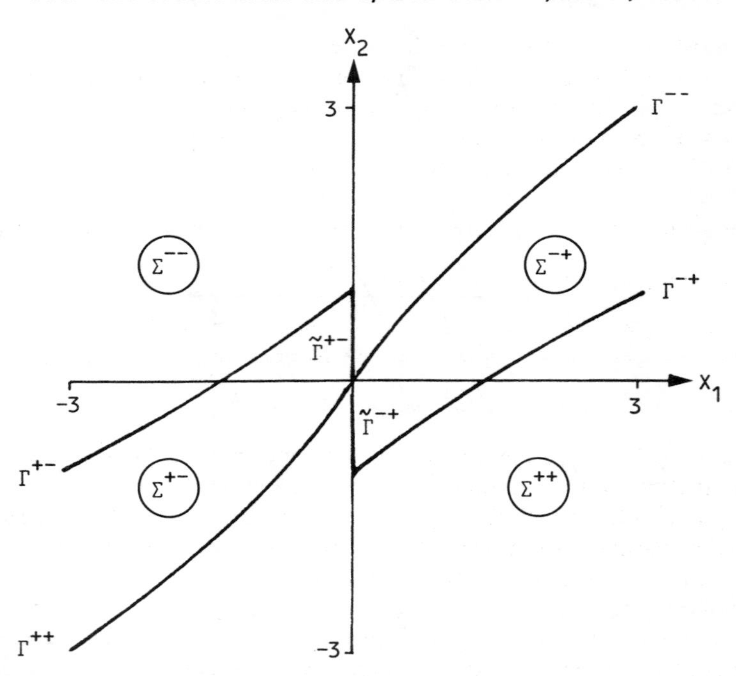

Fig. 5.25 *Example 5.3.1*
Four-region synthesis

If $\eta^0 = (\eta_1^0, \eta_2^0)' \in \Sigma_1$ is such that $\eta_1^0 > 0$ and $\eta_2^0 > 0$, then the arguments of the signum functions of (5.98) are positive for all $\tau \geq 0$ and the complete parabolic arc (5.100) is generated, corresponding to an optimal switchless trajectory ($u_1 \equiv 1 \equiv u_2$) to the origin in forwards time. On the other hand, if $\eta^0 = (\eta_1^0, \eta_2^0)' \in \Sigma_1$ is such that $\eta_2^0 < 0$, then, from (5.98), it may be seen that the control component u_2 switches at the first (reverse) switching time

$$\tau_1 = \ln\left(\frac{-\eta_1^0}{2\eta_2^0}\right) > 0$$

while the control component u_1 switches at the second (reverse) switching time

$$\tau_2 = \ln\left(\frac{-\eta_1^0}{\eta_2^0}\right)$$

Clearly, $\tau_2 - \tau_1 = \ln(2)$.

Hence, if $\eta^0 \in \Sigma_1$ and $\eta_2^0 < 0$, then the (reverse-time) optimal state path initially follows the pp-path Γ^{++} from the origin until the first switch (in component u_2) occurs at (reverse) time τ_1, the state then follows a pn-path, of duration $\tau_2 - \tau_1 = \ln(2)$, until the second switch (in component u_1) occurs at (reverse) time τ_2. Thus, to every optimal switch point $x^1 \in \Gamma^{++}$ there uniquely corresponds a switch point x^2 such that the (reverse) time of transition on the pn-path from x^1 to x^2 is a constant, viz. ln (2). Consequently, the curve Γ^{++} (on which u_2 changes sign) can be transformed to obtain the curve Γ^{+-} (on which u_1 changes sign). Specifically, let $(x_1^1, x_2^1)' = x^1 \in \Gamma^{++}$ be an arbitrary u_2 switch point on Γ^{++}, then the coordinates $(x_1^2, x_2^2)' = x^2 \in \Gamma^{+-}$ of the next (reverse-time) switch point (i.e. u_1 switch on Γ^{+-}) may be calculated by integrating the state equations along the pn-path from x^1 to x^2, of duration ln (2), specifically,

$$x_1^2 = 4x_1^1; \qquad x_2^2 = 2x_2^1 + 1$$

so that the curve Γ^{+-} of Fig. 5.25 can be constructed from Γ^{++} via the transformation $x_1 \mapsto \frac{1}{4}x_1, x_2 \mapsto \frac{1}{2}(x_2 - 1)$, giving, from (5.100), the characterisation

$$\Gamma^{+-} = \{x: x_1 \leq 0, x_2 \leq 1; 9x_1 - 14x_2 + x_2^2 + 13 = 0\} \tag{5.101}$$

For all $\tau > \tau_2$ (i.e. after the second reverse-time switch) the arguments of the signum functions in (5.98) remain negative and no further switches occur. Summarising, if $\eta^0 \in \Sigma_1$ is such that $\eta_2^0 < 0$, then the (reverse-time) adjoint solution follows the (parabolic) path PQRO of Fig. 5.24*b* with arc PQ in Σ_1 so that $\bar{u} = v_1$(or $u_1 = +1 = u_2$); at Q the adjoint solution enters the region Σ_4 so that $\bar{u}$ is transferred to vertex v_4, i.e. $\bar{u} = v_4$ (or $u_1 = +1, u_2 = -1$), corresponding to a (reverse-time) switch on Γ^{++}; at R the adjoint solution enters the region Σ_3 and the control vector is transferred to vertex v_3, i.e. $\bar{u} = v_3$ (or $u_1 = -1 = u_2$), corresponding to a (reverse-time) switch on Γ^{+-}; the (reverse-time) adjoint solution subsequently remains in Σ_3 and approaches the origin exponentially so that no further switch in control occurs and, in the state plane, the (reverse-time) state path reverses away from Γ^{+-}. The totality of such reverse-time trajectories, generated by (5.98) with $\eta^0 \in \Sigma_1$ and $\eta_2^0 < 0$, fills a region of the state plane lying above the curve Γ^{++}. A symmetrical construction, with $\eta^0 \in \Sigma_3$, yields the curves Γ^{--} and Γ^{-+} and the optimal (reverse-time) trajectories which fill a region of the state plane lying below Γ^{--}, where

$$\left.\begin{aligned} \Gamma^{--} &= \{x: x_1, x_2 \geq 0; 9x_1 - 6x_2 - x_2^2 = 0\} \\ \Gamma^{-+} &= \{x: x_1 \geq 0, x_2 \geq -1; 9x_1 - 14x_2 - x_2^2 - 13 = 0\} \end{aligned}\right\} \tag{5.102}$$

Finally, suppose $\eta^0 \in \Sigma_2$ (or $\eta^0 \in \Sigma_4$) then, in view of (5.97) and (5.98), the state will initially back out from the origin along the np-path (or pn-path)

$$\left.\begin{aligned} &x_1(\tau) = 0 \\ &x_2(\tau) = -[\exp(\tau) - 1] \qquad (\text{or} \quad x_2(\tau) = \exp(\tau) - 1) \end{aligned}\right\} \quad \tau \geq 0$$

which coincides with a segment of the nonpositive (or non-negative) x_2 state semi-axis. However, on such a path, it may be seen, from (5.98) and (5.99), that the u_1 control component must change sign at some reverse time not later than $\tau_3 = \ln(2)$, i.e. the np- (or pn-) path from the origin can be extended backwards along the negative (or positive) x_2 state axis only to the point with coordinates $(0, -1)$ (or $(0, +1)$), corresponding to the line segment $\tilde{\Gamma}^{-+}$ (or $\tilde{\Gamma}^{+-}$) of

Fig. 5.25. Hence, by a suitable choice of $\eta^0 \in \Sigma_2$ (or $\eta^0 \in \Sigma_4$), the control component u_1 can be made to switch at any point of $\tilde{\Gamma}^{-+}$ (or $\tilde{\Gamma}^{+-}$) from where the state will follow a (reverse-time) pp-path (or nn-path) with no further control switches. These paths fill the remaining regions of the state plane and complete the optimal synthesis.

Referring to Fig. 5.25, the optimal control for minimum-time transition to the state origin may now be expressed in feedback form as

$$u(x) = \begin{bmatrix} u_1(x) \\ u_2(x) \end{bmatrix} = \begin{cases} \begin{bmatrix} +1 \\ +1 \end{bmatrix}; & x \in \Sigma^{++} \cup \Gamma^{++} \\ \begin{bmatrix} +1 \\ -1 \end{bmatrix}; & x \in \Sigma^{+-} \cup \Gamma^{+-} \cup \tilde{\Gamma}^{+-} \\ \begin{bmatrix} -1 \\ +1 \end{bmatrix}; & x \in \Sigma^{-+} \cup \Gamma^{-+} \cup \tilde{\Gamma}^{-+} \\ \begin{bmatrix} -1 \\ -1 \end{bmatrix}; & x \in \Sigma^{--} \cup \Gamma^{--} \end{cases}$$

i.e. the optimal synthesis is comprised of a partition of the state plane into four regions of specified control effort. However, such a four-region synthesis does not always arise in the case of linear, normal, two-input second-order systems as the next example illustrates.

Example 5.3.2 : Two-region synthesis : Consider now the case of

$$A = \begin{bmatrix} -2 & 0 \\ 0 & -1 \end{bmatrix} \quad \text{and} \quad B = \begin{bmatrix} 1 & 1 \\ 1 & -2 \end{bmatrix} \tag{5.103}$$

As before, writing

$$\bar{u} = \begin{bmatrix} \bar{u}_1 \\ \bar{u}_2 \end{bmatrix} = Bu = \begin{bmatrix} u_1 + u_2 \\ u_1 - 2u_2 \end{bmatrix}$$

the parallelogram with vertices v_i may be constructed in the $\bar{u}$-plane, together with the rays Π_i and regions Σ_i as shown in Figs. 5.26*a* and 5.26*b*. This example differs from the previous case in that the straight lines $\Pi_1 \cup \Pi_3$ and $\Pi_2 \cup \Pi_4$ now lie in different quadrants. The regions Σ_i may be explicitly characterised as

$$\begin{aligned} \Sigma_1 &= \{\eta : \eta_1 + \eta_2 > 0; \eta_1 - 2\eta_2 < 0\} \\ \Sigma_2 &= \{\eta : \eta_1 + \eta_2 < 0; \eta_1 - 2\eta_2 < 0\} \\ \Sigma_3 &= \{\eta : \eta_1 + \eta_2 < 0; \eta_1 - 2\eta_2 > 0\} \\ \Sigma_4 &= \{\eta : \eta_1 + \eta_2 > 0; \eta_1 - 2\eta_2 > 0\} \end{aligned} \tag{5.104}$$

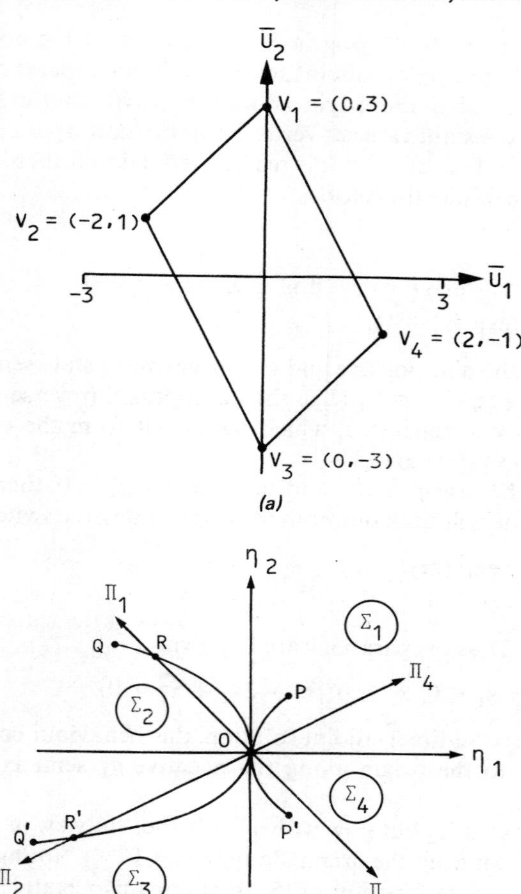

Fig. 5.26

As in the previous example, the optimal synthesis will be obtained by characterising the family of (reverse-time) solutions of (5.90), viz.

$$\left.\begin{aligned} \frac{dx_1}{d\tau} &= 2x_1 - u_1 - u_2 \\ \frac{dx_2}{d\tau} &= x_2 - u_1 + 2u_2 \end{aligned}\right\}; \qquad x(0) = 0 \tag{5.105}$$

generated by (reverse-time) extremal controls (5.91), viz.

$$u_1(\tau) = \text{sgn}\,[\eta_1^0 \exp(-2\tau) + \eta_2^0 \exp(-\tau)] \tag{5.106a}$$

$$u_2(\tau) = \text{sgn}\,[\eta_1^0 \exp(-2\tau) - 2\eta_2^0 \exp(-\tau)] \qquad \tau \geq 0 \tag{5.106b}$$

Consider initially a vector η^0 lying in Σ_1 (or Σ_3), e.g. at the point P (or P') of Fig. 5.26*b*. From P (or P'), the adjoint solution follows a (parabolic) path which exponentially approaches the origin and never leaves Σ_1 (or Σ_3). Hence, the control $\bar{u}$ takes its constant value at vertex v_1 (or v_3), corresponding to $u_1 \equiv +1$, $u_2 \equiv -1$ (or $u_1 \equiv -1, u_2 \equiv +1$). It is readily verified that these controls generate the (reverse-time) state trajectories

$$\left.\begin{aligned} x_1(\tau) &= 0 \\ x_2(\tau) &= \begin{cases} 3(1 - \exp(\tau)); & \text{if } \eta^0 \in \Sigma_1 \\ 3(\exp(\tau) - 1); & \text{if } \eta^0 \in \Sigma_3 \end{cases} \end{aligned}\right\} \quad \tau \geq 0$$

corresponding to the non-positive and non-negative x_2 state semi-axes, respectively. In other words, if $\eta^0 \in \Sigma_1$ (Σ_3), then an optimal (reverse-time) switchless pn- (np-) trajectory is generated, which backs out from the origin along the negative (positive) x_2 state axis.

Now consider the case $\eta^0 \in \Sigma_2$, and suppose that $\eta_2^0 = 0$, then it is clear from (5.106) that the state will back out from the origin along the switchless nn-path

$$\left.\begin{aligned} x_1(\tau) &= 1 - \exp(2\tau) \\ x_2(\tau) &= \exp(\tau) - 1 \end{aligned}\right\}; \quad \tau \geq 0$$

depicted in Fig. 5.27 as the parabolic arc Γ^{--}, viz.

$$\Gamma^{--} = \{x: x_1 \leq 0, x_2 \geq 0; x_1 + 2x_2 + x_2^2 = 0\} \tag{5.107}$$

In terms of the (reverse-time) adjoint solution, this behaviour corresponds to an exponential path to the origin along the negative η_1 semi-axis of Fig. 5.26*b*, lying entirely in Σ_2.

Again consider $\eta^0 \in \Sigma_2$ but now with $\eta_2^0 > 0$, then, in view of (5.106), the state will initially back out from the origin along the arc Γ^{--}. Noting that the sign of the argument of signum function of (5.106b) remains negative for all $\tau \geq 0$, it immediately follows that no switch in component u_2 can occur in this case. On the other hand, component u_1 will switch exactly once at the (reverse) time

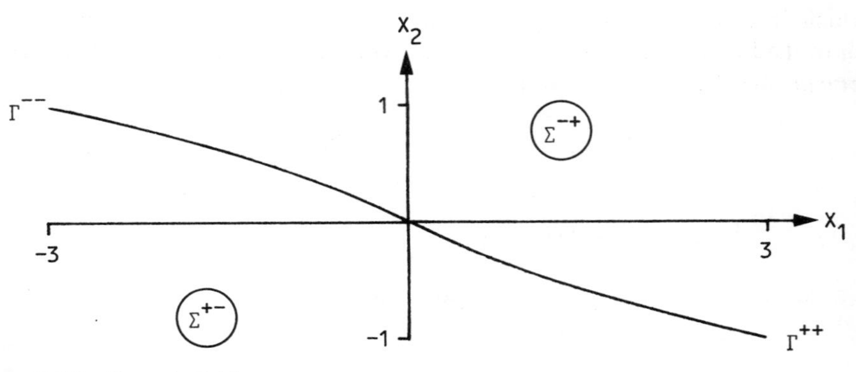

Fig. 5.27 *Example 5.3.2*
Two-region synthesis

$$\tau_1 = \ln\left(-\frac{\eta_1^0}{\eta_2^0}\right) > 0$$

Consequently, the state will initially back out from the origin along the curve Γ^{--} until the switch in u_1 occurs at (reverse) time τ_1; the state then leaves the curve Γ^{--} and subsequently follows a pn-path in the region of the left-half state plane lying below Γ^{--}. This behaviour corresponds to the typical adjoint solution QRO of Fig. 5.26*b* which starts at point Q in Σ_2 ($\Rightarrow \bar{u} = v_2$ or, equivalently, $u_1 = -1 = u_2$); after a finite time $\tau_1 = \ln(-\eta_1^0/\eta_2^0)$ the point R $\in \Pi_1$ is reached and the solution subsequently lies in Σ_1 ($\Rightarrow \bar{u} = v_1$ or, equivalently, $u_1 = +1$, $u_2 = -1$) and no further control switches occur.

Suppose now that $\eta_2^0 < 0$, with $\eta^0 \in \Sigma_2$ as before, so that no switch in the u_1 control component can occur; however, the u_2 component will switch exactly once at the (reverse) time

$$\tau_2 = \ln\left(-\frac{\eta_1^0}{2\eta_2^0}\right) > 0$$

In this case, the state again backs out from the origin along Γ^{--} until the u_2 switch occurs at (reverse) time τ_2; the state subsequently follows a (reverse-time) np-path in the region lying between Γ^{--} and the positive x_2 semi-axis. This behaviour corresponds to the typical (reverse-time) adjoint solution Q′R′O of Fig. 5.26*b*.

Combining the above cases of $\eta^0 \in \Sigma_2$, with $\eta_2^0 = 0$, $\eta_2^0 > 0$ and $\eta_2^0 < 0$, it may be seen that optimal paths which fill the left-half state plane have been constructed. By symmetry, for $\eta^0 \in \Sigma_4$, the switching curve Γ^{++} and optimal paths which fill the right-half state plane may be constructed, thereby completing the optimal synthesis. In particular, by symmetry and (5.107),

$$\Gamma^{++} = \{x\colon x_1 \geq 0,\ x_2 \leq 0;\ x_1 + 2x_2 - x_2^2 = 0\} \tag{5.108}$$

which, when combined with (5.107), yields the composite switching curve

$$\Gamma = \Gamma^{--} \cup \Gamma^{++} = \{x\colon x_1 + 2x_2 + x_2|x_2| = 0\} \tag{5.109}$$

which, in turn, partitions the state plane into two regions Σ^{-+} and Σ^{+-} lying above and below Γ, respectively, as shown in Fig. 5.27. The optimal control is then given by

$$u = \begin{bmatrix} u_1 \\ u_2 \end{bmatrix} = \begin{cases} \begin{bmatrix} +1 \\ +1 \end{bmatrix}; & x \in \Gamma^{++} \\[2ex] \begin{bmatrix} +1 \\ -1 \end{bmatrix}; & x \in \Sigma^{+-} \\[2ex] \begin{bmatrix} -1 \\ +1 \end{bmatrix}; & x \in \Sigma^{-+} \\[2ex] \begin{bmatrix} -1 \\ -1 \end{bmatrix}; & x \in \Gamma^{--} \end{cases}$$

which can be written in explicit feedback form as

$$u = \begin{bmatrix} u_1(x) \\ u_2(x) \end{bmatrix} = \begin{bmatrix} -\text{sgn}\,(\Xi_1(x)) \\ -\text{sgn}\,(\Xi_2(x)) \end{bmatrix} \tag{5.110a}$$

where

$$\Xi_1(x) = \begin{cases} \xi(x); & \xi(x) \neq 0 \\ x_2; & \xi(x) = 0 \end{cases} \tag{5.110b}$$

$$\Xi_2(x) = \begin{cases} -\xi(x); & \xi(x) \neq 0 \\ x_2; & \xi(x) = 0 \end{cases} \tag{5.110c}$$

and

$$\xi(x) = x_1 + 2x_2 + x_2|x_2| \tag{5.110d}$$

5.4 Second-order nonlinear systems

Finally, the time-optimal regulator problem for single-input nonlinear systems will be briefly studied but not in fully generality. In particular, the system equations are assumed to be of the form

$$\dot{x}_1 = x_2; \qquad \dot{x}_2 = f(x_2, u); \qquad |u| \leq 1 \tag{5.111a}$$

where the function f is continuously differentiable in both arguments and satisfies the relation

$$\frac{\partial f}{\partial u}(x_2, u) > 0 \qquad \text{for all } x_2, u \tag{5.111b}$$

The Hamiltonian function associated with the time-optimal regulator problem for this system is

$$H(x, u, \eta) = -1 + \langle \eta, \dot{x} \rangle = -1 + \eta_1 x_2 + \eta_2 f(x_2, u) \tag{5.112}$$

In accordance with the maximum principle (theorem 2.4.1 and corollary 2.4.1), if $u(t)$, $0 \leq t \leq t_f$, is an optimal control generating a minimum-time trajectory $x(t)$, $0 \leq t \leq t_f$, then there exists a solution $\eta(t)$, $0 \leq t \leq t_f$, of the adjoint equation

$$\dot{\eta} = -\frac{\partial H}{\partial x} \Leftrightarrow \dot{\eta}_1 = 0;\ \dot{\eta}_2 = -\eta_1 - \left(\frac{\partial f}{\partial x_2}\right)\eta_2 \tag{5.113}$$

such that

$$H(x(t), u(t), \eta(t)) = \max_{|u| \leq 1} H(x(t), u, \eta(t)) = 0 \tag{5.114}$$

almost everywhere on $[0, t_f]$.

From (5.112) and (5.114) it may be seen that the optimal control must maximise $f(x_2, u)$ whenever $\eta_2 > 0$ and must minimise $f(x_2, u)$ whenever $\eta_2 < 0$. Now, condition (5.111b) implies that $f(x_2, u)$ is maximised with respect to u by the choice $u = +1$ and is minimised by the choice $u = -1$. Hence, the optimal control must satisfy the relation

$$u(t) = \text{sgn}\,(\eta_2(t)); \qquad \eta_2(t) \neq 0 \tag{5.115}$$

Note that the optimal control is not determined when $\eta_2(t) = 0$. However, it is easily seen, from (5.113), that if the latter singular condition holds on some finite interval I (i.e. $\eta_2(t) = 0 \ \forall\ t \in I$) then η_1 must also vanish on the interior of this interval, with the result that $H(x(t), u(t), \eta(t)) = -1 \ \forall\ t \in \text{int}\,(I)$ which clearly contradicts (5.114). In other words, $\eta_2(t) = 0$ only at isolated instants (a set of measure zero) and the control is determined almost everywhere by (5.115). It now remains to determine the form of the function $\eta_2(\cdot)$ which, on solving (5.113), may be expressed as

$$\eta_2(t) = \exp\left\{-\int_0^t \left[\frac{\partial f}{\partial x_2}(x_2(s), u(s))\right] ds\right\}$$
$$\times \left[\eta_2^0 - \eta_1^0 \int_0^t \exp\left\{\int_0^s \left[\frac{\partial f}{\partial x_2} x_2(\sigma), u(\sigma))\right] d\sigma\right\} ds\right] \tag{5.116}$$

Now, as the real exponential function is positive-valued, it may be seen, from (5.116), that the function $\eta_2(\cdot)$ can have *at most* one zero on $[0, t_f]$ corresponding to at most one switch in control. Hence, any piecewise constant control $u(\cdot)$ with $|u(t)| = 1$ and *at most* one discontinuity satisfies the maximum principle; a time-optimal state portrait (when it exists) must consist of state paths to the origin generated by such controls.

It is remarked that the above analysis of a *nonlinear* system is based on theorem 2.4.1 which only provides a *necessary* condition for optimality (whereas the previous studies of *linear* systems were based on the *necessary* and *sufficient* conditions of theorems 2.5.2, 2.5.3 and 2.6.1). Consequently, it cannot be immediately concluded that a synthesis derived on the above basis is, in fact, optimal. However, optimality can be concluded in many cases (as in the following example) if a candidate synthesis derived on the basis of the maximum principle is regular in the sense of Boltyanskii (1966).

Example 5.4.1 : Consider, for example, system (5.111) with

$$f(x_2, u) = -x_2|x_2| + u \tag{5.117}$$

which models the controlled motion of a mass subject to a nonlinear drag or friction force (see also Athans and Falb 1966). The p- and n-paths for this system are depicted in Fig. 5.28. In view of the above results, the time-optimal synthesis can be constructed as in Fig. 5.29 where the switching curve $\Gamma = \Gamma^+ \cup \Gamma^-$ consists of the p-path (Γ^+) and n-path (Γ^-) leading to the origin and

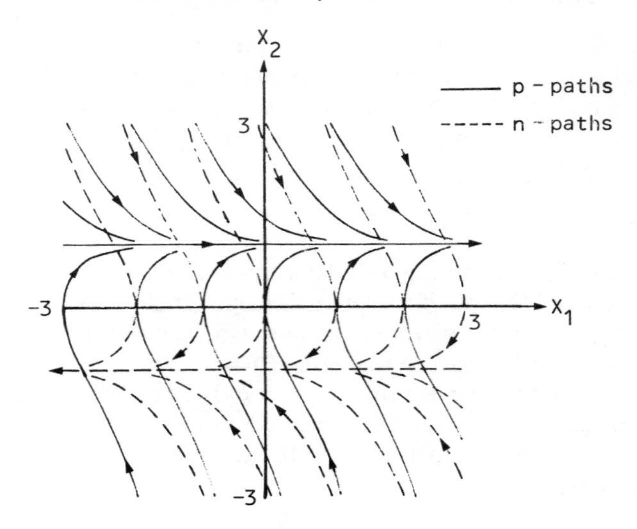

Fig. 5.28 *Example 5.4.1*

partitions the state plane into the regions Σ^+ and Σ^- wherein the optimal control takes the values $+1$ and -1, respectively.

The above example of a time-optimal nonlinear system is seen to exhibit the structural features of a second-order linear system with real eigenvalues in the sense that the optimal control is piecewise constant with *at most* one discontinuity. Other nonlinear systems, such as the controlled Van der Pol oscillator, are

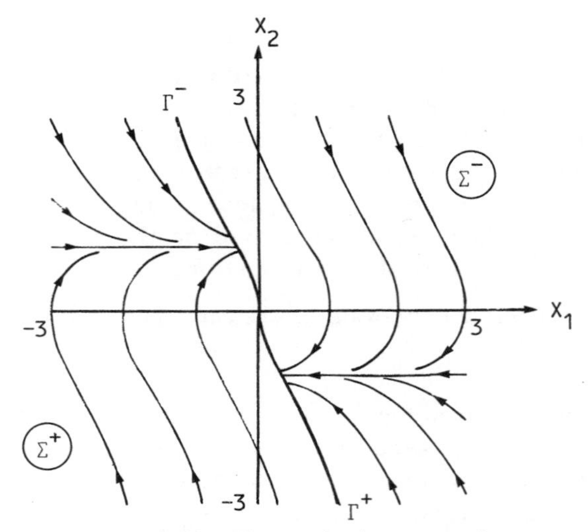

Fig. 5.29 *Time-optimal state portrait*

structurally akin to the linear case with complex eigenvalues for which the number of control switches is unbounded. For time-optimal control studies of these and other nonlinear systems, the reader is referred to Almuzara and Flügge-Lotz (1968), Atanackovic (1978), Athans and Falb (1966), Davies (1970, 1971, 1972), Flügge-Lotz (1968), Glen (1973) and Lee and Markus (1961, 1967).

Chapter 6

Third-order time-optimal control system synthesis

6.1 Introduction

Synthesis of time-optimal control for third-order systems will be considered in this chapter. While some *partial* constructions of the optimal switching surfaces have been obtained for systems with complex eigenvalues (Flügge-Lotz and Mih Yin, 1961; Flügge-Lotz and Titus, 1962, 1963; Sakawa and Hayashi, 1963), complete characterisations of the feedback controls and associated switching surfaces are currently available only for a limited number of single-input systems with real eigenvalues. Attention is exclusively restricted to the latter class of systems in the present chapter and the synthesis problem is solved (as in Ryan, 1974) in a number of specific cases by exploiting the time-optimality of the strategy of Gulko *et al.*, as in the second-order investigations of Section 5.2.1.

The systems considered are assumed to be of the general form (4.23), shown in Fig. 6.1, viz.

$$\dot{x}_1 = \lambda_1 x_1 + x_2; \qquad \dot{x}_2 = \lambda_2 x_2 + x_3; \qquad \dot{x}_3 = \lambda_3 x_3 + \frac{u}{a};$$

$$|u| \le 1; \qquad a > 0 \tag{6.1a}$$

or

$$\dot{x} = Ax + bu; \qquad |u| \le 1 \qquad \text{where}$$

$$A = \begin{bmatrix} \lambda_1 & 1 & 0 \\ 0 & \lambda_2 & 1 \\ 0 & 0 & \lambda_3 \end{bmatrix} \text{ and } B = \begin{bmatrix} 0 \\ 0 \\ 1/a \end{bmatrix} \tag{6.1b}$$

Recalling the synthesis technique outlined in Section 4.5.2, the derivation of the time-optimal feedback regulator for (6.1) presupposes knowledge of the sub-

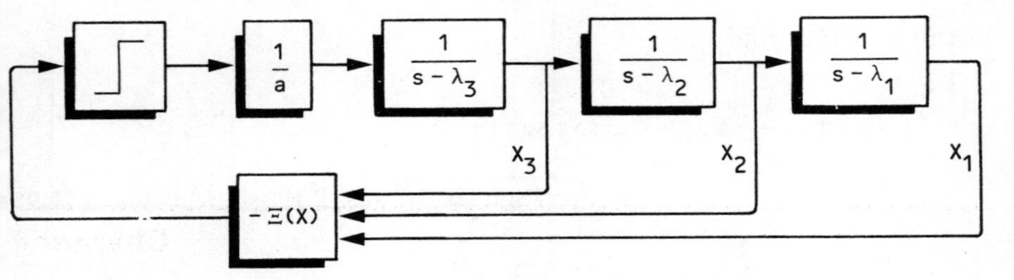

Fig. 6.1

system optimal switching function Ξ^s and the dependence of the single subsystem switching time $t_1 = T_1^s(x^s)$ and final time $t_2 = T_2^s(x^s)$ on the initial state of the second-order subsystem

$$\dot{x}_2 = \lambda_2 x_2 + x_3; \quad \dot{x}_3 = \lambda_3 x_3 + \frac{u}{a}; \quad |u| \leq 1; \quad x^s = (x_2, x_3)' \in \mathbb{R}^2 \tag{6.2}$$

Second-order systems of the form (6.2) were studied in the previous chapter and the results thereof which are relevant to the third-order problem are summarised below.

6.2 Second-order subsystems

6.2.1 Subsystem I : double integrator ; $\lambda_2 = \lambda_3 = 0$

Using the results of Section 5.2.1(i), the subsystem functions Ξ^s: $\mathbb{R}^2 \to \mathbb{R}$, T_i^s: $\mathbb{R}^2 \to [0, \infty)$ are given by

$$\Xi^s(x^s) = \begin{cases} \xi^s(x^s) = x_2 + \tfrac{1}{2}ax_3|x_3|; & \xi^s(x^s) \neq 0 \\ x_3; & \xi^s(x^s) = 0 \end{cases} \tag{6.3}$$

$$T_1^s(x^s) = ax_3 \operatorname{sgn}(\Xi^s) + \sqrt{ax_2 \operatorname{sgn}(\Xi^s) + \tfrac{1}{2}(ax_3)^2} \tag{6.4a}$$

$$T_2^s(x^s) = ax_3 \operatorname{sgn}(\Xi^s) + 2\sqrt{ax_2 \operatorname{sgn}(\Xi^s) + \tfrac{1}{2}(ax_3)^2} \tag{6.4b}$$

6.2.2 Subsystem II : eigenvalues in simple ratio ; $\lambda_2 = 2\lambda$, $\lambda_1 = \lambda$, $\lambda \neq 0$

Interpreting the results of Sections 5.2.1(v) and 5.2.1(vii), the subsystem functions Ξ^s: $\mathscr{C}^s \to \mathbb{R}$, T_i^s: $\mathscr{C}^s \to [0, \infty)$ in this case are given by

$$\Xi^s(x^s) = \begin{cases} \xi^s(x^s) = x_2 + \tfrac{1}{2}ax_3|x_3|; & \xi^s(x^s) \neq 0 \\ x_3; & \xi^s(x^s) = 0 \end{cases} \tag{6.5}$$

identical to (6.3), and

$$T_1^s(x^s) = -\frac{1}{\lambda} \ln \{1 - \lambda[ax_3 \operatorname{sgn}(\Xi^s) + \sqrt{ax_2 \operatorname{sgn}(\Xi^s) + \tfrac{1}{2}(ax_3)^2}]\}; \qquad x^s \in \mathscr{C}^s \tag{6.6a}$$

$$T_2^s(x^s) = -\frac{1}{\lambda} \ln \{1 - \lambda[ax_3 \operatorname{sgn}(\Xi^s) + 2\sqrt{ax_2 \operatorname{sgn}(\Xi^s) + \tfrac{1}{2}(ax_3)^2}]\}; \qquad x^s \in \mathscr{C}^s \tag{6.6b}$$

where $\mathscr{C}^s$ is the subsystem domain of null controllability given by

$$\mathscr{C}^s = \begin{cases} \mathbb{R}^2; & \lambda < 0 \\ \left\{ x^s \in \mathbb{R}^2 : ax_3 \operatorname{sgn}(\Xi^s) + 2\sqrt{ax_2 \operatorname{sgn}(\Xi^s) + \tfrac{1}{2}(ax_3)^2} < \dfrac{1}{\lambda} \right\}; & \lambda > 0 \end{cases} \tag{6.7}$$

6.2.3 *Subsystem III : integrator-plus-lag; $\lambda_2 = 0$; $\lambda_3 < 0$*

Interpreting the results of Section 5.2.1(ii), the subsystem functions $\Xi^s: \mathbb{R}^2 \to \mathbb{R}$ and $T_i^s: \mathbb{R}^2 \to [0, \infty)$ in this case are

$$\Xi^s(x^s) = \begin{cases} \xi^s(x^s) = x_2 - \dfrac{x_3}{\lambda_3} - \dfrac{\operatorname{sgn} x_3}{a\lambda_3^2} \ln(1 - a\lambda_3|x_3|); & \xi^s(x^s) \neq 0 \\ x_3; & \xi^s(x^s) = 0 \end{cases} \tag{6.8}$$

and

$$T_1^s(x^s) = -\frac{1}{\lambda_3} \ln \times \{1 + \sqrt{1 - [1 - a\lambda_3 x_3 \operatorname{sgn}(\Xi^s)] \exp[-a\lambda_3(\lambda_3 x_2 - x_3) \operatorname{sgn}(\Xi^s)]}\} - a(\lambda_3 x_2 - x_3) \operatorname{sgn}(\Xi^s) \tag{6.9a}$$

$$T_2^s(x^s) = -\frac{2}{\lambda_3} \ln \times \{1 + \sqrt{1 - [1 - a\lambda_3 x_3 \operatorname{sgn}(\Xi^s)] \exp[-a\lambda_3(\lambda_3 x_2 - x_3) \operatorname{sgn}(\Xi^s)]}\} - a(\lambda_3 x_2 - x_3) \operatorname{sgn}(\Xi^s) \tag{6.9b}$$

6.2.4 Subsystem IV : $\lambda_2 = 0; \lambda_3 > 0$

Interpreting the results of Section 5.2.1(iii) for this unstable version of subsystem III, the functions Ξ^s: $\mathscr{C}^s \to \mathbb{R}$ and T_i^s: $\mathscr{C}^s \to [0, \infty)$ are

$$\Xi^s(x^s) = \begin{cases} \xi^s(x^s) = x_2 - \dfrac{x_3}{\lambda_3} - \dfrac{\text{sgn}\,(x_3)}{a\lambda_3^2} \ln\,(1 - a\lambda_3|x_3|); & \xi^s(x^s) \neq 0 \\ x_3; \quad \xi^s(x^s) = 0 & x^s \in \mathscr{C}^s \end{cases} \tag{6.10}$$

i.e. (6.8) restricted to $\mathscr{C}^s \subset \mathbb{R}^2$, and

$$T_1^s(x^s) = \frac{1}{\lambda_3} \ln \times \left[1 - \sqrt{1 - [1 - a\lambda_3 x_3\,\text{sgn}\,(\Xi^s)]\exp\,[-a\lambda_3(\lambda_3 x_2 - x_3)\,\text{sgn}\,(\Xi^s)]}\right]^{-1} - a(\lambda_3 x_2 - x_3)\,\text{sgn}\,(\Xi^s); \qquad x^s \in \mathscr{C}^s \tag{6.11a}$$

$$T_2^s(x^s) = \frac{2}{\lambda_3} \ln \times \left[1 - \sqrt{1 - [1 - a\lambda_3 x_3\,\text{sgn}\,(\Xi^s)]\exp\,[-a\lambda_3(\lambda_3 x_2 - x_3)\,\text{sgn}\,(\Xi^s)]}\right]^{-1} - a(\lambda_3 x_2 - x_3)\,\text{sgn}\,(\Xi^s); \qquad x^s \in \mathscr{C}^s \tag{6.11b}$$

where

$$\mathscr{C}^s = \left\{x^s \in \mathbb{R}^2 : |x_3| < \frac{1}{a\lambda_3}\right\} \tag{6.12}$$

is the subsystem domain of null controllability.

6.3 Third-order time-optimal synthesis

Recalling the general approach of Section 4.5.2, the time-optimal feedback control for the third-order system (6.1) may be explicitly obtained in those cases where λ_2 and λ_3 are such that they define one of the above four second-order subsystems. Specifically, from (4.49), the time-optimal switching function Ξ: $\mathscr{C} \subseteq \mathbb{R}^3 \to \mathbb{R}$ is given by

$$\Xi(x) = \begin{cases} \xi(x); & \xi(x) \neq 0 \\ \Xi^s(x^s); & \xi(x) = 0 \end{cases} \tag{6.13}$$

where, from (4.48), $\xi(x) = \exp\{-\lambda_1 T_2^s(x^s)\}\beta(x)$ and

$$\begin{bmatrix} \beta(x) \\ 0 \\ 0 \end{bmatrix} = \exp[AT_2^s(x^s)]\Big\{x - \operatorname{sgn}(\Xi^s(x^s))$$

$$\times \left[\left[\int_0^{T_1^s(x^s)} \exp(-As)b\, ds - \int_{T_1^s(x^s)}^{T_2^s(x^s)} \exp(-As)b\, ds\right]\right\} \qquad (6.14)$$

with

$$A = \begin{bmatrix} \lambda_1 & 1 & 0 \\ 0 & \lambda_2 & 1 \\ 0 & 0 & \lambda_3 \end{bmatrix} \quad \text{and} \quad B = \begin{bmatrix} 0 \\ 0 \\ \frac{1}{a} \end{bmatrix}$$

and Ξ^s, T_1^s, T_2^s defined as for Subsystems I, II, III or IV as appropriate.

On this basis, the time-optimal switching function Ξ will be determined for the following systems (where $G(s)$ denotes the transfer function from Laplace transformed input u to output x_1):

(*a*) Triple integrator: $G(s) = \dfrac{1}{as^3}$

(*b*) Eigenvalues in simple ratio:

$$G(s) = \frac{1}{a(s - 3\lambda)(s - 2\lambda)(s - \lambda)}; \qquad \lambda \neq 0$$

(*c*) Subsystem I-plus-lag: $G(s) = \dfrac{1}{as^2(s - \lambda)}$; $\quad \lambda < 0$

(*d*) Subsystem II-plus-lag:

$$G(s) = \frac{1}{a(s - \lambda_1)(s - 2\lambda_3)(s - \lambda_3)}; \qquad \lambda_1 < 0, \lambda_3 \neq 0$$

(*e*) Subsystem III-plus-lag:

$$G(s) = \frac{1}{as(s - \lambda_1)(s - \lambda_3)}; \qquad \lambda_1 < 0; \lambda_3 < 0$$

(*f*) Subsystem IV-plus-lag:

$$G(s) = \frac{1}{as(s - \lambda_1)(s - \lambda_3)}; \qquad \lambda_1 < 0; \lambda_3 > 0$$

In the case of systems (*a*) and (*b*), the minimum-time isochronal surfaces will be expressed in explicit algebraic form. Furthermore, for systems (*b*), (*d*) and (*f*), when one or more of the eigenvalues are positive, the domain of null controllability $\mathscr{C}$ will be characterised (in other cases $\mathscr{C} = \mathbb{R}^3$).

System (*a*) was studied initially by Feldbaum (1955) and subsequently considered by many authors including Oldenburger and Thompson (1963), Chaudhuri and Chaudhury (1964a), Windall (1970) and Fuller (1971, 1974b). These results were extended to system (*b*), with eigenvalues in simple ratio, by Fuller (1973d). System (*c*) has been investigated by Oldenburger and Thompson (1963), Athans and Falb (1966) and Pavlov (1966) who has also considered system (*e*); system (*f*) has been investigated by Malek-Zavarei (1980). Systems (*c*) and (*e*) have been treated by Chaudhuri and Chaudhury (1964a, b); however, the expressions presented therein for the associated time-optimal switching functions appear to be incorrect (see Ryan, 1974). Finally, related third-order investigations are contained in Moroz (1969a, b).

6.4 System (a): triple integrator: $\lambda_1 = \lambda_2 = \lambda_3 = 0$

From theorem 4.3.1, it may be concluded that all states are null controllable, i.e. $\mathscr{C} = \mathbb{R}^3$. Now

$$\exp(At) = \begin{bmatrix} 1 & t & \dfrac{t^2}{2} \\ 0 & 1 & t \\ 0 & 0 & 1 \end{bmatrix}$$

and the appropriate subsystem is Subsystem I with switching function Ξ^s and times T_i^s given by (6.3) and (6.4), respectively. Hence (6.14) can be readily evaluated to give

$$\beta(x) = \xi(x) = x_1 + x_2 T_2^s(x^s) + \tfrac{1}{2}x_3[T_2^s(x^s)]^2 - \frac{\operatorname{sgn}(\Xi^s(x^s))}{6a}\{[T_2^s(x^s)]^3 - 2[T_2^s(x^s) - T_1^s(x^s)]^3\}$$

which, combined with (6.4), gives (after some algebraic manipulation)

$$\xi(x) = x_1 + ax_2x_3 \operatorname{sgn}(\Xi^s) + \frac{a^2x_3^3}{3} + a^{-1}[ax_2 \operatorname{sgn}(\Xi^s) + \tfrac{1}{2}(ax_3)^2]^{3/2} \operatorname{sgn}(\Xi^s) \qquad (6.15a)$$

where

$$\Xi^s = \Xi^s(x^s) = \begin{cases} \xi^s(x^s) = x_2 + \tfrac{1}{2}ax_3|x_3|; & \xi^s(x^s) \neq 0 \\ x_3; & \xi^s(x^s) = 0 \end{cases} \qquad (6.15b)$$

The time-optimal feedback control may now be explicitly written as

$$u = u(x) = -\operatorname{sgn}(\Xi(x)) \tag{6.16a}$$

where the optimal switching function Ξ is given by

$$\Xi(x) = \begin{cases} \xi(x); & \xi(x) \neq 0 \\ \Xi^s(x^s); & \xi(x) = 0 \end{cases} \tag{6.16b}$$

The switching function (6.15a) was first obtained by Feldbaum (1955). Fuller (1974b) has shown that the fractional power appearing in (6.15a) and the consequent difficulties in realizing the optimal control can be avoided by defining the function

$$\tilde{\xi}(x) = A\,|A| + B^3 \tag{6.17a}$$

where

$$A = A(x) = x_1 + ax_2x_3 \operatorname{sgn}(\Xi^s) + \tfrac{1}{3}a^2x_3^3 \tag{6.17b}$$

and

$$B = B(x^s) = x_2 + \tfrac{1}{2}ax_3^2 \operatorname{sgn}(\Xi^s) \tag{6.17c}$$

giving an 'equivalent' optimal switching function

$$\tilde{\Xi}(x) = \begin{cases} \tilde{\xi}(x); & \tilde{\xi}(x) \neq 0 \\ \Xi^s(x^s); & \tilde{\xi}(x) = 0 \end{cases} \tag{6.18}$$

equivalent in the sense that, if the function Ξ in (6.16a) is replaced by $\tilde{\Xi}$, then the resulting feedback control is also optimal. The set

$$\Gamma = \{x\colon \xi(x) = 0\} = \{x\colon \tilde{\xi}(x) = 0\} \tag{6.19}$$

constitutes the time-optimal switching surface and is composed of the set of all state trajectories which go the origin under a piecewise constant control taking the values ± 1 only and with at most *one* discontinuity. Γ partitions the state space into two mutually exclusive regions Σ^+ and Σ^- of positive and negative control ($u = +1$ if $x \in \Sigma^+$ and $u = -1$ if $x \in \Sigma^-$) where

$$\Sigma^+ = \{x\colon \xi(x) < 0\} = \{x\colon \tilde{\xi}(x) < 0\}$$

$$\text{and} \quad \Sigma^- = \{x\colon \xi(x) > 0\} = \{x\colon \tilde{\xi}(x) > 0\} \tag{6.20}$$

Recalling that the time-optimal triple integrator system exhibits the property of special invariance (see Section 3.4.3) which, in turn, enables the system motion to be depicted in the two-dimensional (Θ, Φ)-space as in Section 3.5.4, then referring to Fig. 3.11, the optimal switching surface Γ in state space maps to the curve $\Gamma = \Gamma^+ \cup C^+ \cup \Gamma^- \cup C^-$ in (Θ, Φ)-space, the region Σ^+ maps to the region lying below Γ and the region Σ^- maps to the region lying above Γ in (Θ, Φ) space. For initial states $x \in \Gamma$, the optimal control is identical to the optimal subsystem control, with at most two intervals of constancy and with the switching and final times given by (6.4). For all other initial states $x \notin \Gamma$, the optimal control contains precisely three intervals of constant control with discontinuities at times $t_1, t_2 \in (0, t_3)$, the origin being attained in minimum time t_3.

The dependence of the times t_i on the initial state x will now be determined algebraically (as in Ryan 1977a), i.e. explicit representations of the functions T_i: $\mathbb{R}^3 \to [0, \infty)$ will be obtained such that $t_i = T_i(x)$, $i = 1, 2, 3$. The set $\Upsilon_{t_3} = \{x: T_3(x) = t_3\}$ constitutes the t_3-optimal isochrone for the problem.

6.5 Isochronal surfaces for the triple integrator system

For an initial state $x(0) = x \notin \Gamma(\Rightarrow \xi(x) \neq 0)$, integrating the state equations along the optimal two-switch trajectory from $x(0) = x$ to the origin $x(t_3) = 0$ gives

$$0 = x_1(t_3) = x_1 + x_2 t_3 + \tfrac{1}{2}x_3 t_3^2 - \frac{\operatorname{sgn}(\xi(x))}{6a} \times [t_3^3 - 2(t_3 - t_1)^3 + 2(t_3 - t_2)^3] \tag{6.21a}$$

$$0 = x_2(t_3) = x_2 + x_3 t_3 - \frac{\operatorname{sgn}(\xi(x))}{2a} \times [t_3^2 - 2(t_3 - t_1)^2 + 2(t_3 - t_2)^2] \tag{6.21b}$$

$$0 = x_3(t_3) = x_3 - \frac{\operatorname{sgn}(\xi(x))}{a}[t_3 - 2(t_3 - t_1) + 2(t_3 - t_2)] \tag{6.21c}$$

where t_1 and t_2 denote the control switching times. To avoid excessive notation, the term $\operatorname{sgn}(\xi(x))$ will henceforth be written as $\operatorname{sgn}(\xi)$. Solving (6.21) for t_1, t_2, t_3 in terms of the initial state $x = (x_1, x_2, x_3)'$ will yield the required results. With this objective, from (6.21c), it follows that

$$(t_3 - t_1) = (t_3 - t_2) + \tfrac{1}{2}[t_3 - ax_3 \operatorname{sgn}(\xi)] \tag{6.22}$$

and substituting in (6.21b) gives

$$(t_3 - t_2) = \tfrac{1}{4}[t_3 - ax_3 \operatorname{sgn}(\xi)] - [ax_2 \operatorname{sgn}(\xi) + \tfrac{1}{2}(ax_3)^2][t_3 - ax_3 \operatorname{sgn}(\xi)]^{-1} \tag{6.23}$$

Substituting (6.22) and (6.23) in (6.21a) yields a quartic equation in t_3, viz.

$$t_3^4 + c_1 t_3^3 + c_2 t_3^2 + c_3 t_3 + c_4 = 0 \tag{6.24a}$$

where

$$\left.\begin{aligned} c_1 &= -4ax_3 \operatorname{sgn}(\xi) \\ c_2 &= -2[8ax_2 \operatorname{sgn}(\xi) + (ax_3)^2] \\ c_3 &= -4[8ax_1 - \tfrac{1}{3}(ax_3)^3] \operatorname{sgn}(\xi) \\ c_4 &= 32a^2 x_1 x_3 - 16(ax_2)^2 - \tfrac{1}{3}(ax_3)^4 \end{aligned}\right\} \tag{6.24b}$$

On writing†

$$s = t_3 - ax_3 \text{ sgn } (\xi) \tag{6.25}$$

the cubic term is eliminated from (6.24), which becomes

$$s^4 + \alpha_1 s^2 + \alpha_2 s + \alpha_3 = 0 \tag{6.26a}$$

where

$$\left.\begin{aligned} \alpha_1 &= c_2 - \tfrac{3}{8}c_1^2 = -16\gamma_2 \\ \alpha_2 &= \tfrac{1}{8}c_1^3 - \tfrac{1}{2}c_1c_2 + c_3 = -32\gamma_1 \\ \alpha_3 &= c_4 - \tfrac{1}{4}c_1c_3 + \frac{c_1^2c_2}{16} - \frac{3c_1^4}{256} = -16\gamma_2^2 \end{aligned}\right\} \tag{6.26b}$$

with

$$\left.\begin{aligned} \gamma_1 &= ax_1 \text{ sgn } (\xi) + a^2x_2x_3 + \tfrac{1}{3}(ax_3)^3 \text{ sgn } (\xi) \\ \gamma_2 &= ax_2 \text{ sgn } (\xi) + \tfrac{1}{2}(ax_3)^2 \end{aligned}\right\} \tag{6.26c}$$

The following Subsections 6.5.1 and 6.5.2 are concerned with (i) characterising all four roots of the quartic equation (6.26) and (ii) selecting from these roots the *unique admissible root*, i.e. the single root which corresponds, via (6.22), (6.23) and (6.25), to values t_i satisfying $0 < t_1 < t_2 < t_3$. The analysis of Subsections 6.5.1 and 6.5.2 is standard but of a rather involved algebraic nature; the analytical detail can be by-passed without affecting the ensuing treatment, in which case the reader may, at this point, progress directly to the required expression (6.66) for the unique admissible root s_1 of (6.26).

6.5.1 Roots of the quartic equation

Adopting the standard method (see e.g. Uspensky 1948), (6.26) is first written as $s^4 = -\alpha_1 s^2 - \alpha_2 s - \alpha_3$ and adding the quantity $(ys^2 + \frac{1}{4}y^2)$ to both sides gives

$$(s^2 + \tfrac{1}{2}y)^2 = (y - \alpha_1)s^2 - \alpha_2 s + (\tfrac{1}{4}y^2 - \alpha_3) \tag{6.27}$$

y is now determined so that the right-hand side of (6.27) becomes the square of a linear expression in s, i.e. so that

$$(y - \alpha_1)s^2 - \alpha_2 s + (\tfrac{1}{4}y^2 - \alpha_3) = (vs + \mu)^2 \tag{6.28}$$

implying that

$$\alpha_2^2 = 4(y - \alpha_1)(\tfrac{1}{4}y^2 - \alpha_3)$$

or, in expanded form,

$$y^3 - \alpha_1 y^2 - 4\alpha_3 y + (4\alpha_1\alpha_3 - \alpha_2^2) = 0 \tag{6.29}$$

† Note that $s > 0$ as otherwise $t_3 \leq |ax_3|$ which implies that the time to the origin is less than or equal to the minimum time to the subspace origin for the first-order subsystem $\dot{x}_3 = u/a$, $|u| \leq 1$, which is clearly impossible for $x \notin \Gamma$.

Now, in view of (6.26b),

$$\alpha_3 = -\frac{\alpha_1^2}{16}$$

and, hence, (6.29) simplifies to

$$y(y - \tfrac{1}{2}\alpha_1)^2 - (\tfrac{1}{4}\alpha_1^3 + \alpha_2^2) = 0 \tag{6.30}$$

It suffices to select any root $y = \tilde{y}$ of the cubic equation (6.30), called the *cubic resolvent* of the quartic equation (6.26), in order to satisfy (6.28); then (6.27) may be written in the form

$$(s^2 + \tfrac{1}{2}\tilde{y})^2 = (vs + \mu)^2 \tag{6.31}$$

where, from (6.26b) and (6.28),

$$\left.\begin{aligned} v^2 &= (\tilde{y} - \alpha_1) = \tilde{y} + 16\gamma_2 \\ 2v\mu &= -\alpha_2 = 32\gamma_1 \\ \mu^2 &= \tfrac{1}{4}\tilde{y}^2 - \alpha_3 = \tfrac{1}{4}\tilde{y}^2 + 16\gamma_2^2 \end{aligned}\right\} \tag{6.32}$$

Defining μ as the *positive* square root of $(\frac{1}{4}\tilde{y}^2 + 16\gamma_2^2)$, i.e.

$$\mu = \sqrt{\tfrac{1}{4}\tilde{y}^2 + 16\gamma_2^2} \tag{6.33}$$

then, from (6.32), v may be written as†

$$v = \sqrt{\tilde{y} + 16\gamma_2}\ \mathrm{sgn}\,(\gamma_1) \tag{6.34}$$

Now, (6.31) splits into two quadratic equations, viz.

$$s^2 + \tfrac{1}{2}\tilde{y} = vs + \mu$$

$$s^2 + \tfrac{1}{2}\tilde{y} = -(vs + \mu)$$

which are readily solved to yield the four roots s_i $(i = 1, 2, 3, 4)$ of the quartic equation (6.26), viz.

$$\left.\begin{aligned} s_1 &= \tfrac{1}{2}v + \sqrt{\tfrac{1}{4}v^2 - \tfrac{1}{2}\tilde{y} + \mu} \\ s_2 &= \tfrac{1}{2}v - \sqrt{\tfrac{1}{4}v^2 - \tfrac{1}{2}\tilde{y} + \mu} \\ s_3 &= -\tfrac{1}{2}v + \sqrt{\tfrac{1}{4}v^2 - \tfrac{1}{2}\tilde{y} - \mu} \\ s_4 &= -\tfrac{1}{2}v - \sqrt{\tfrac{1}{4}v^2 - \tfrac{1}{2}\tilde{y} - \mu} \end{aligned}\right\} \tag{6.35}$$

where $\tilde{y}$ is a root of the cubic resolvent (6.30) and the parameters v, μ are determined in terms of $\tilde{y}$ and the coefficients α_i of (6.26) via (6.33) and (6.34). The existence and uniqueness of the time-optimal control imply that one, and only one, of the roots (6.35), i.e. *the unique admissible root*, can correspond, via (6.22),

† Owing to the special form of (6.30), it is easily verified that $\tilde{y} + 16\gamma_2 \geq 0$ so that the square root term in (6.34) is a well defined non-negative real number.

(6.23) and (6.25), to the required values of t_i satisfying the inequalities $0 < t_1 < t_2 < t_3$. The selection of the unique admissible positive real root from (6.35) will be discussed in due course. First, the root $\tilde{y}$ of the cubic resolvent must be determined.

From (6.26) and (6.30), the cubic resolvent may be written as

$$y(y + 8\gamma_2)^2 - 32^2(\gamma_1^2 - \gamma_2^3) = 0 \tag{6.36}$$

Setting

$$y = z - \tfrac{16}{3}\gamma_2 \tag{6.37}$$

and substituting in (6.36) eliminates the quadratic term to give

$$z^3 + k_1 z + k_2 = 0 \tag{6.38a}$$

where

$$\left.\begin{aligned} k_1 &= -\tfrac{64}{3}\gamma_2^2 \\ k_2 &= -32^2(\gamma_1^2 - \tfrac{26}{27}\gamma_2^3) \end{aligned}\right\} \tag{6.38b}$$

The cubic equation of the special form (6.38) can be solved by the following standard method (see Uspensky 1948):

First introduce two unknowns v and w by setting

$$z = v + w \tag{6.39}$$

then, substituting in (6.38) yields the equation

$$v^3 + w^3 + (k_1 + 3vw)(v + w) + k_2 = 0 \tag{6.40}$$

This problem is indeterminate unless a second relation between v and w is specified; taking as this relation

$$k_1 + 3vw = 0 \Leftrightarrow vw = -\tfrac{1}{3}k_1 \tag{6.41a}$$

then

$$v^3 + w^3 = -k_2 \tag{6.41b}$$

so that the solution of the cubic (6.38) may be obtained, via (6.39), by solving the system of equations (6.41). In particular, from (6.41),

$$\begin{aligned} v^3 = \Lambda_1 &= -\tfrac{1}{2}k_2 + \sqrt{\tfrac{1}{4}k_2^2 + \tfrac{1}{27}k_1^3} \\ &= \frac{32^2}{2}\left[(\gamma_1^2 - \tfrac{26}{27}\gamma_2^3) + \sqrt{(\gamma_1^2 - \gamma_2^3)(\gamma_1^2 - \tfrac{25}{27}\gamma_2^3)}\right] \end{aligned} \tag{6.42a}$$

and

$$\begin{aligned} w^3 = \Lambda_2 &= -\tfrac{1}{2}k_2 - \sqrt{\tfrac{1}{4}k_2^2 + \tfrac{1}{27}k_1^3} \\ &= \frac{32^2}{2}\left[(\gamma_1^2 - \tfrac{26}{27}\gamma_2^3) - \sqrt{(\gamma_1^2 - \gamma_2^3)(\gamma_1^2 - \tfrac{25}{27}\gamma_2^3)}\right] \end{aligned} \tag{6.42b}$$

Hence, the three possible values of v and w are

$$v_1 = (\Lambda_1)^{1/3}; \qquad v_2 = \psi(\Lambda_1)^{1/3}; \qquad v_3 = \psi^2(\Lambda_1)^{1/3} \tag{6.43a}$$

$$w_1 = (\Lambda_2)^{1/3}; \qquad w_2 = \psi(\Lambda_2)^{1/3}; \qquad w_3 = \psi^2(\Lambda_2)^{1/3} \tag{6.43b}$$

where

$$\psi = \tfrac{1}{2}(-1 + i\sqrt{3}) \quad \text{and} \quad \psi^2 = \tfrac{1}{2}(-1 - i\sqrt{3}) = \bar{\psi} \tag{6.43c}$$

From (6.39), (6.41) and (6.43), it may be concluded that the roots z_i of the cubic equation (6.38) are

$$\left.\begin{aligned} z_1 &= v_1 + w_1 = (\Lambda_1)^{1/3} + (\Lambda_2)^{1/3} \\ z_2 &= v_2 + w_3 = \psi(\Lambda_1)^{1/3} + \psi^2(\Lambda_2)^{1/3} \\ z_3 &= v_3 + w_2 = \psi^2(\Lambda_1)^{1/3} + \psi(\Lambda_2)^{1/3} \end{aligned}\right\} \tag{6.44}$$

Now defining the region $\Sigma_1 \subset \mathbb{R}^3$ of the state space as

$$\Sigma_1 = \{x\colon \gamma_1^2 - \gamma_2^3 \leq 0; \gamma_1^2 - \tfrac{25}{27}\gamma_2^3 > 0\} \tag{6.45}$$

it follows that

$$x \notin \Sigma_1 \Rightarrow (\gamma_1^2 - \gamma_2^3)(\gamma_1^2 - \tfrac{25}{27}\gamma_2^3) \geq 0 \tag{6.46}$$

which, in view of (6.42), implies that, for each state x not contained in Σ_1, both Λ_1 and Λ_2 are real-valued; furthermore, Λ_1 and Λ_2 are distinct, with the sole exception of the specific case of $x \notin \Sigma_1$ in which

$$\gamma_1^2 - \tfrac{25}{27}\gamma_2^3 = 0 \Rightarrow \Lambda_1 = \Lambda_2 = -\frac{32^2}{54}\gamma_2^3 \tag{6.47}$$

Hence, it may be concluded from (6.43) and (6.44) that, for $x \notin \Sigma_1$, only the root z_1 in (6.44) is real; again, with the sole exception of the specific case (6.47) for which all three roots in (6.44) are real, viz. $z_1 = -\frac{16}{3}\gamma_2$, $z_2 = \frac{8}{3}\gamma_2 = z_3$. Select, as the root $\tilde{y}$ of the cubic resolvent, that root which corresponds, via (6.37), to the real root z_1, i.e.

$$\tilde{y} = z_1 - \tfrac{16}{3}\gamma_2 = (\Lambda_1)^{1/3} + (\Lambda_2)^{1/3} - \tfrac{16}{3}\gamma_2 \tag{6.48}$$

or, in view of (6.42), for $x \notin \Sigma_1$:

$$\begin{aligned} \tilde{y} = {} & 8\left[(\gamma_1^2 - \tfrac{26}{27}\gamma_2^3) + \sqrt{(\gamma_1^2 - \gamma_2^3)(\gamma_1^2 - \tfrac{25}{27}\gamma_2^3)}\right]^{1/3} \\ & + 8\left[(\gamma_1^2 - \tfrac{26}{27}\gamma_2^3) - \sqrt{(\gamma_1^2 - \gamma_2^3)(\gamma_1^2 - \tfrac{25}{27}\gamma_2^3)}\right]^{1/3} - \tfrac{16}{3}\gamma_2 \end{aligned} \tag{6.49a}$$

where,

$$\left.\begin{aligned} \gamma_1 &= ax_1 \operatorname{sgn}(\xi) + a^2 x_2 x_3 + \tfrac{1}{3}(ax_3)^3 \operatorname{sgn}(\xi) \\ \gamma_2 &= ax_2 \operatorname{sgn}(\xi) + \tfrac{1}{2}(ax_3)^2 \end{aligned}\right\} \xi = \xi(x) \neq 0 \tag{6.49b}$$

Consider now the case where x is contained in Σ_1, i.e.

$$x \in \Sigma_1 \Rightarrow (\gamma_1^2 - \gamma_2^3)(\gamma_1^2 - \tfrac{25}{27}\gamma_2^3) \leq 0 \tag{6.50}$$

With the single exception of the specific case of $x \in \Sigma_1$ for which

$$\gamma_1^2 - \gamma_2^3 = 0 \Rightarrow \Lambda_1 = \Lambda_2 = \frac{32^2}{54}\gamma_2^3 \tag{6.51}$$

it may be seen, from (6.42), that Λ_1 and Λ_2 form a complex conjugate pair. However, it is possible to determine the roots (6.44) by extracting the cube root of Λ_1 trignometrically, as follows. Expressing Λ_1 in complex form

$$\Lambda_1 = \frac{32^2}{2}\left[(\gamma_1^2 - \tfrac{26}{27}\gamma_2^3) + i\sqrt{-(\gamma_1^2 - \gamma_2^3)(\gamma_1^2 - \tfrac{25}{27}\gamma_2^3)}\right] \tag{6.52}$$

The modulus l of Λ_1 is easily shown to be†

$$l = \frac{32^2}{54}\gamma_2^3 \Rightarrow l^{1/3} = \tfrac{8}{3}\gamma_2 > 0 \tag{6.53}$$

and the argument ω of Λ_1 is

$$\omega = \tan^{-1}\left(\frac{\sqrt{|(\gamma_1^2 - \gamma_2^3)(\gamma_1^2 - \tfrac{25}{27}\gamma_2^3)|}}{(\gamma_1^2 - \tfrac{26}{27}\gamma_2^3)}\right) \tag{6.54}$$

where the range of the function $\tan^{-1}$ is interpreted such that $0 \le \omega < \pi$, i.e. ω is taken in the first or second quadrant according as $(\gamma_1^2 - \frac{26}{27}\gamma_2^3)$ is positive or negative. Hence, $\Lambda_1 = le^{i\omega}$ and

$$\Lambda_1^{1/3} = l^{1/3}e^{i\omega/3} \tag{6.55}$$

Moreover, since Λ_2 is conjugate to Λ_1,

$$\Lambda_2^{1/3} = l^{1/3}e^{-i\omega/3} \tag{6.56}$$

Also, from (6.43c)

$$\psi = e^{i2\pi/3}; \qquad \psi^2 = \bar{\psi} = e^{-i2\pi/3} \tag{6.57}$$

and consequently, the three roots (6.44) of the cubic (6.38) are real and are given by

$$\left.\begin{aligned} z_1 &= l^{1/3}(e^{i\omega/3} + e^{-i\omega/3}) = 2l^{1/3}\cos\left(\frac{\omega}{3}\right) \\ z_2 &= l^{1/3}(e^{i(\omega+2\pi)/3} + e^{-i(\omega+2\pi)/3}) = 2l^{1/3}\cos\left(\frac{\omega+2\pi}{3}\right) \\ z_3 &= l^{1/3}(e^{i(\omega-2\pi)/3} + e^{-i(\omega-2\pi)/3}) = 2l^{1/3}\cos\left(\frac{\omega-2\pi}{3}\right) \end{aligned}\right\} \tag{6.58}$$

† For each $x \in \Sigma_1$ it may be verified that

$$\gamma_2 = ax_2 \operatorname{sgn}(\xi) + \tfrac{1}{2}(ax_3)^2 > 0 \Rightarrow l > 0$$

In order to facilitate the later determination of the unique admissible root of the underlying quartic equation (6.26), the required root $\tilde{y}$ of the cubic resolvent is taken to be that root which corresponds, via (6.37), to root z_2 of (6.58), i.e. for $x \in \Sigma_1$,

$$\tilde{y} = z_2 - \tfrac{16}{3}\gamma_2 = \tfrac{16}{3}\gamma_2\left[\cos\left(\frac{2\pi + \omega}{3}\right) - 1\right] \tag{6.59}$$

where γ_2 and ω are defined by (6.26c) and (6.54).

Summarising, with $\Sigma_1 \subset \mathbb{R}^3$ defined by (6.45)

(*a*) if $x \notin \Sigma_1$, then the required root $\tilde{y}$ of the cubic resolvent is given by (6.49); while

(*b*) if $x \in \Sigma_1$, then $\tilde{y}$ is given by (6.59).

With $\tilde{y}$ defined as above and with v and μ determined by (6.33) and (6.34), the four roots of the quartic equation (6.26) are defined by (6.35); one, and only one, of these roots (the unique admissible root), correspond, via (6.22), (6.23) and (6.25), to t_i values satisfying $0 < t_1 < t_2 < t_3$ (by existence and uniqueness of the optimal control).

6.5.2 Unique admissible root of quartic equation

The four roots of the quartic (6.26) are given by

$$\begin{aligned}
s_1 &= \sqrt{\tfrac{1}{4}\tilde{y} + 4\gamma_2}\,\operatorname{sgn}(\gamma_1) + [4\gamma_2 - \tfrac{1}{4}\tilde{y} + \sqrt{\tfrac{1}{4}\tilde{y}^2 + 16\gamma_2^2}]^{1/2}\\
s_2 &= \sqrt{\tfrac{1}{4}\tilde{y} + 4\gamma_2}\,\operatorname{sgn}(\gamma_1) - [4\gamma_2 - \tfrac{1}{4}\tilde{y} + \sqrt{\tfrac{1}{4}\tilde{y}^2 + 16\gamma_2^2}]^{1/2}\\
s_3 &= -\sqrt{\tfrac{1}{4}\tilde{y} + 4\gamma_2}\,\operatorname{sgn}(\gamma_1) + [4\gamma_2 - \tfrac{1}{4}\tilde{y} - \sqrt{\tfrac{1}{4}\tilde{y}^2 + 16\gamma_2^2}]^{1/2}\\
s_4 &= -\sqrt{\tfrac{1}{4}\tilde{y} + 4\gamma_2}\,\operatorname{sgn}(\gamma_1) - [4\gamma_2 - \tfrac{1}{4}\tilde{y} - \sqrt{\tfrac{1}{4}\tilde{y}^2 + 16\gamma_2^2}]^{1/2}
\end{aligned} \tag{6.60}$$

It is evident that the unique admissible root must be the largest of the (one or more) positive real roots in (6.60); it will be established that s_1 fulfills this condition. It may be verified that s_1 is real and positive for all admissible $\tilde{y}$, γ_1 and γ_2; as $s_1 \geq s_2$, root s_2 may be disregarded. If root s_3 is real, then s_4 is also real with $s_3 \geq s_4$ so that root s_4 may also be disregarded. Of the two remaining candidates s_1 and s_3, it may be verified by inspection that, for all $\gamma_1 \geq 0$, root s_1 is the unique admissible root. Hence the case $\gamma_1 < 0$ only need be considered. In particular, for $\gamma_1 < 0$ the remaining candidates are

$$\begin{aligned}
s_1 &= -\sqrt{\tfrac{1}{4}\tilde{y} + 4\gamma_2} + [4\gamma_2 - \tfrac{1}{4}\tilde{y} + \sqrt{\tfrac{1}{4}\tilde{y}^2 + 16\gamma_2^2}]^{1/2}\\
s_3 &= \sqrt{\tfrac{1}{4}\tilde{y} + 4\gamma_2} + [4\gamma_2 - \tfrac{1}{4}\tilde{y} - \sqrt{\tfrac{1}{4}\tilde{y}^2 + 16\gamma_2^2}]^{1/2}
\end{aligned} \tag{6.61}$$

Now, as $\tfrac{1}{4}\tilde{y} + 4\gamma_2 \geq 0$ (see footnote on page 167), it follows that $4\gamma_2 - \tfrac{1}{4}\tilde{y} \leq 8\gamma_2$ and hence s_3 is complex if $\gamma_2 \leq 0$ so that s_1 is again the unique admissible root. Consequently, s_3 remains a candidate only for the case $\gamma_1 < 0$ and $\gamma_2 > 0$, in which case it may be verified that if $\tilde{y}$ belongs to the closed interval $[-\tfrac{32}{3}\gamma_2, 0]$ then s_3 is real and positive, whereas if $\tilde{y}$ lies outside this interval, then s_3 is complex (and s_1 is again the unique admissible root). Consider now the behaviour of the roots (6.61) as $\tilde{y}$ ranges over the interval $[-\frac{32}{3}\gamma_2, 0]$; it is straightforward to show that

$$\begin{aligned} s_1 &= s_3 \quad &\text{if} \quad &\tilde{y} = -\tfrac{8}{3}\gamma_2 \\ s_1 &< s_3 \quad &\text{if} \quad &\tilde{y} \in (-\tfrac{8}{3}\gamma_2, 0] \\ s_1 &> s_3 \quad &\text{if} \quad &\tilde{y} \in [-\tfrac{32}{3}\gamma_2, -\tfrac{8}{3}\gamma_2) \end{aligned} \tag{6.62}$$

Finally it will be shown that $\tilde{y} \notin (-\frac{8}{3}\gamma_2, 0]$ so that s_3 can never exceed s_1; this will establish s_1 as the unique admissible root. Consider again the cubic resolvent (6.36), viz.

$$f(y) = y(y + 8\gamma_2)^2 - 32^2(\gamma_1^2 - \gamma_2^3) = 0 \tag{6.63}$$

Differentiating,

$$\frac{df}{dy} = (3y + 8\gamma_2)(y + 8\gamma_2)$$

and, hence, for $y = -\frac{8}{3}\gamma_2$

$$\frac{df}{dy} = 0 \quad \text{and} \quad f(y) = -32^2(\gamma_1^2 - \tfrac{25}{27}\gamma_2^3)$$

while, for $y = -8\gamma_2$

$$\frac{df}{dy} = 0 \quad \text{and} \quad f(y) = -32^2(\gamma_1^2 - \gamma_2^3)$$

Recalling that the case $\gamma_1 < 0$, $\gamma_2 > 0$ only need be considered, (a) for $x \notin \Sigma_1$ ($\Rightarrow \gamma_1^2 - \gamma_2^3 > 0; \gamma_1^2 - \frac{25}{27}\gamma_2^3 \geq 0$) the general form of the graph of the cubic resolvent is shown in Fig. 6.2. With the exception of the special case for which $\gamma_1^2 - \frac{25}{27}\gamma_2^3 = 0$, only one real root $\tilde{y}$ exists; moreover, this root satisfies†

$$\tilde{y} \leq -\tfrac{32}{3}\gamma_2 \tag{6.64}$$

and is given by (6.49).

(b) For $x \in \Sigma_1$ ($\Rightarrow \gamma_1^2 - \gamma_2^3 \leq 0, \gamma_1^2 - \frac{25}{27}\gamma_2^3 > 0$), the general form of the graph of the cubic resolvent is shown in Fig. 6.3. In this case all three roots are real and negative. The root $\tilde{y}$ is defined by (6.59) and corresponds to the most negative of the three roots. Hence, in view of (6.64)

$$-\tfrac{32}{3}\gamma_2 < \tilde{y} \leq -8\gamma_2 \tag{6.65}$$

Combining (6.64) and (6.65), it follows that if $\gamma_1 < 0$, $\gamma_2 > 0$ then $\tilde{y} \notin (-8\gamma_2, 0]$ which, in view of (6.62), implies that $s_1 > s_3$. This establishes the unique admiss-

† If $\gamma_1^2 - \frac{25}{27}\gamma_2^3 = 0$, then all roots of (6.63) are real and negative, with $\tilde{y} = -\frac{32}{3}\gamma_2$ and the remaining two roots take the value $\frac{8}{3}\gamma_2$.

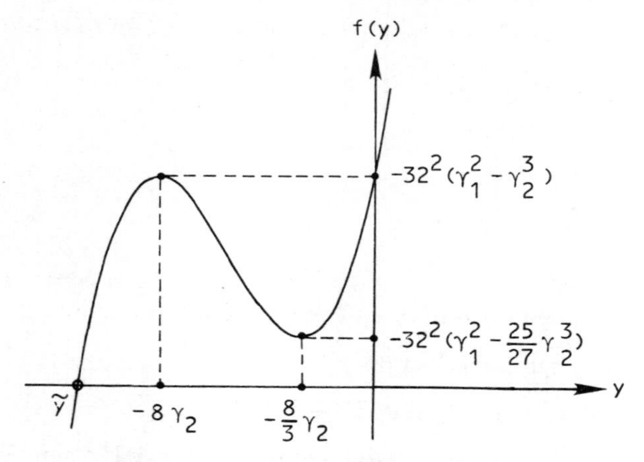

Fig. 6.2 *Graph of cubic resolvent for $x \notin \Sigma_1$*

ible root as

$$s_1 = S(x) = \sqrt{\tfrac{1}{4}\tilde{y} + 4\gamma_2}\ \mathrm{sgn}\ (\gamma_1) + [4\gamma_2 - \tfrac{1}{4}\tilde{y} + \sqrt{\tfrac{1}{4}\tilde{y}^2 + 16\gamma_2^2}]^{1/2} \tag{6.66a}$$

with

$$\tilde{y} = \tilde{y}(x) = \begin{cases} 8[f_1(x) + \sqrt{f_2(x)f_3(x)}]^{1/3} & \\ \quad + 8[f_1(x) - \sqrt{f_2(x)f_3(x)}]^{1/3} - \frac{16}{3}\gamma_2; & \text{if} \quad x \notin \Sigma_1 \\ \frac{16}{3}\gamma_2(x)\left[\cos\dfrac{2\pi + \omega}{3} - 1\right]; & \\ \omega = \omega(x) = \tan^{-1}\left(\dfrac{\sqrt{|f_2(x)f_3(x)|}}{f_1(x)}\right); & \text{if} \quad x \in \Sigma_1 \end{cases} \tag{6.66b}$$

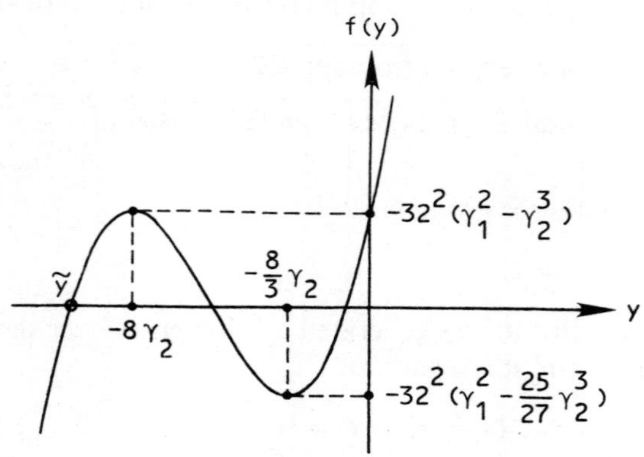

Fig. 6.3 *Graph of cubic resolvent for $x \in \Sigma_1$*

where the range of $\tan^{-1}$ is taken as $[0, \pi)$ with $\omega = \pi/2$ for $f_1(x) = 0$ and where

$$\Sigma_1 = \{x: f_2(x) \leq 0; f_3(x) > 0\} \tag{6.66c}$$

$$f_1(x) = \gamma_1(x)^2 - \tfrac{26}{27}\gamma_2(x)^3 \tag{6.66d}$$

$$f_2(x) = \gamma_1(x)^2 - \gamma_2(x)^3 \tag{6.66e}$$

$$f_3(x) = \gamma_1(x)^2 - \tfrac{25}{27}\gamma_2(x)^3 \tag{6.66f}$$

with

$$\gamma_1(x) = ax_1 \operatorname{sgn}(\xi) + a^2 x_2 x_3 + \tfrac{1}{3}(ax_3)^3 \operatorname{sgn}(\xi) \tag{6.66g}$$

$$\gamma_2(x) = ax_2 \operatorname{sgn}(\xi) + \tfrac{1}{2}(ax_3)^2 \tag{6.66h}$$

$$\xi = \xi(x) = x_1 + ax_2 x_3 \operatorname{sgn}(\Xi^s) + \tfrac{1}{3}a^2 x_3^3 + a^{-1}[ax_2 \operatorname{sgn}(\Xi^s) + \tfrac{1}{2}(ax_3)^2]^{3/2} \operatorname{sgn}(\Xi^s) \tag{6.66i}$$

$$\Xi^s = \begin{cases} \xi^s(x^s) = x_2 + \tfrac{1}{2}ax_3|x_3|; & \xi^s(x^s) \neq 0 \\ x_3; & \xi^s(x^s) = 0 \end{cases} \tag{6.66j}$$

where, at the outset, the condition $x \notin \Gamma \Rightarrow \xi(x) \neq 0$ was imposed. This characterisation enables the optimal switching times t_1, t_2 and minimum time t_3 to be expressed (using (6.22), (6.23) and (6.25)) as explicit functions $T_i: \mathbb{R}^3 \to [0, \infty)$ as follows (Ryan 1977a).

$$t_3 = T_3(x) = \begin{cases} S(x) + ax_3 \operatorname{sgn}(\xi); & \xi = \xi(x) \neq 0 \\ T_2^s(x^s); & \xi = \xi(x) = 0 \end{cases} \tag{6.67a}$$

$$t_2 = T_2(x) = \begin{cases} \tfrac{3}{4}S(x) + \gamma_2(x)S(x)^{-1} + ax_3 \operatorname{sgn}(\xi); & \xi = \xi(x) \neq 0 \\ T_1^s(x^s); & \xi = \xi(x) = 0 \end{cases} \tag{6.67b}$$

$$t_1 = T_1(x) = \begin{cases} \tfrac{1}{4}S(x) + \gamma_2(x)S(x)^{-1} + ax_3 \operatorname{sgn}(\xi); & \xi = \xi(x) \neq 0 \\ T_1^s(x^s); & \xi = \xi(x) = 0 \end{cases} \tag{6.67c}$$

where T_1^s, T_2^s: $\mathbb{R}^2 \to [0, \infty)$ are the optimal subsystem times (6.4), viz.

$$\begin{aligned} T_1^s(x^s) &= ax_3 \operatorname{sgn}(\Xi^s) + \sqrt{ax_2 \operatorname{sgn}(\Xi^s) + \tfrac{1}{2}(ax_3)^2} \\ T_2^s(x^s) &= ax_3 \operatorname{sgn}(\Xi^s) + 2\sqrt{ax_2 \operatorname{sgn}(\Xi^s) + \tfrac{1}{2}(ax_3)^2} \end{aligned} \tag{6.68}$$

The t-isochronal surface is given by

$$\Upsilon_t = \{x: T_3(x) = t\} \tag{6.69}$$

and the family of such surfaces generated as t increases over the interval $[0, \infty)$ expands to fill the entire state space, i.e.

$$\bigcup_{t \geq 0} \Upsilon_t = \bigcup_{t \geq 0} \{x: T_3(x) = t\} = \mathscr{C} = \mathbb{R}^3 \tag{6.70}$$

6.5.3 Illustrative examples

Consider an initial state on the x_1 state axis, i.e. $x = (x_1, 0, 0)'$. From (6.66g) and (6.66h),

$$\gamma_1(x) = a|x_1|; \qquad \gamma_2(x) = 0$$

and from (6.66b)

$$\tilde{y}(x) = (32a|x_1|)^{2/3}$$

Hence, from (6.67), the minimum time t to the origin is given by

$$t_3 = T_3(x) = S(x) = \sqrt{\tilde{y}} = (32a|x_1|)^{1/3} \tag{6.71a}$$

and the optimal switching times are

$$t_1 = T_1(x) = \tfrac{1}{4}S(x) = \tfrac{1}{4}(32a|x_1|)^{1/3} \tag{6.71b}$$

$$t_2 = T_2(x) = \tfrac{3}{4}S(x) = \tfrac{3}{4}(32a|x_1|)^{1/3} \tag{6.71c}$$

These special results were also obtained by Burmeister (1961) and Wolek (1971).

Now consider an initial state on the x_2 state axis, i.e. $x = (0, x_2, 0)'$. In this case

$$\gamma_1(x) = 0; \qquad \gamma_2(x) = a|x_2|$$

and, from (6.66b),

$$\tilde{y}(x) = \tfrac{8}{3}a|x_2|[(15\sqrt{3} - 26)^{1/3} - (15\sqrt{3} + 26)^{1/3} - 2]$$

or, noting that $15\sqrt{3} - 26 = (\sqrt{3} - 2)^3$ and $15\sqrt{3} + 26 = (\sqrt{3} + 2)^3$,

$$\tilde{y}(x) = -16a|x_2|$$

Hence, from (6.67a), the minimum time to the origin is given by

$$t_3 = T_3(x) = [4(2 + \sqrt{5})a|x_2|]^{1/2} \tag{6.72}$$

6.6 Time-optimal triple integrator system: sensitivity to parameter variation

In Section 5.2.3, the sensitivity of a nominally time-optimal double integrator system to variations in the gain parameter was discussed. Here, these investigations are extended to the triple integrator system of Section 6.4, again following the treatment of Zinober and Fuller (1973).

The system under consideration is the following:

$$\dot{x}_1 = x_2; \qquad \dot{x}_2 = x_3; \qquad \dot{x}_3 = \frac{u}{a}; \qquad |u| \leq 1; \qquad a = \text{constant} > 0$$

for which an estimate $\hat{a}$ of the gain parameter a only, is available. Hence, the nominally optimal feedback control is given by†

$$u = u(x) = -\operatorname{sgn}(\xi_{\hat{a}}(x)); \qquad \hat{a} \neq a \tag{6.73}$$

where the dependence of the switching function on the parameter estimate $\hat{a}$ has been explicitly indicated and, from (6.15),

$$\xi_{\hat{a}}(x) = x_1 + \hat{a}x_2 x_3 \operatorname{sgn}(\Xi^s_{\hat{a}}) + \frac{\hat{a}^2 x_3^3}{3} + \hat{a}^{-1}[\hat{a}x_2 \operatorname{sgn}(\Xi^s_{\hat{a}}) + \tfrac{1}{2}(\hat{a}x_3)^2]^{3/2} \operatorname{sgn}(\Xi^s_{\hat{a}}) \tag{6.74a}$$

with

$$\Xi^s_{\hat{a}} = \Xi^s_{\hat{a}}(x^s) = \begin{cases} \xi^s_{\hat{a}}(x) = x_2 + \tfrac{1}{2}\hat{a}x_3|x_3|; & \xi^s_{\hat{a}}(x^s) \neq 0 \\ x_3; & \xi^s_{\hat{a}}(x^s) = 0 \end{cases} \tag{6.74b}$$

6.6.1 Case 1 : $\hat{a} > a$

In this case the control is realised on the basis of a gain parameter estimate which exceeds the true value. It will be shown that, under this control, on reaching the estimated switching surface

$$\Gamma_{\hat{a}} = \{x\colon \xi_{\hat{a}}(x) = 0\} \tag{6.75}$$

the state point follows a sliding path on $\Gamma_{\hat{a}}$ to the origin generated by a piecewise constant control with $|u| \equiv u_s \in (0, 1)$.

Specifically, applying the results of Sections 3.2 and 3.3,

$$\nabla\xi_{\hat{a}} = \begin{bmatrix} 1 \\ \hat{a}x_3 \operatorname{sgn}(\Xi^s_{\hat{a}}) + \tfrac{3}{2}\sqrt{\hat{a}x_2 \operatorname{sgn}(\Xi^s_{\hat{a}}) + \tfrac{1}{2}(\hat{a}x_3)^2} \\ \hat{a}x_2 \operatorname{sgn}(\Xi^s_{\hat{a}}) + \hat{a}^2 x_3^2 + \tfrac{3}{2}\hat{a}x_3 \operatorname{sgn}(\Xi^s_{\hat{a}})\sqrt{\hat{a}x_2 \operatorname{sgn}(\Xi^s_{\hat{a}}) + \tfrac{1}{2}(\hat{a}x_3)^2} \end{bmatrix} \tag{6.76}$$

with

$$\langle\nabla\xi_{\hat{a}}, Ax\rangle = x_2 + \hat{a}x_3^2 \operatorname{sgn}(\Xi^s_{\hat{a}}) + \tfrac{3}{2}x_3\sqrt{\hat{a}x_2 \operatorname{sgn}(\Xi^s_{\hat{a}}) + \tfrac{1}{2}(\hat{a}x_3)^2} \tag{6.77a}$$

and

$$\langle\nabla\xi_{\hat{a}}, b\rangle = \frac{\hat{a}}{a}[x_2 \operatorname{sgn}(\Xi^s_{\hat{a}}) + \hat{a}x_3^2 + \tfrac{3}{2}x_3 \operatorname{sgn}(\Xi^s_{\hat{a}})\sqrt{\hat{a}x_2 \operatorname{sgn}(\Xi^s_{\hat{a}}) + \tfrac{1}{2}(\hat{a}x_3)^2}] \tag{6.77b}$$

† As discussed in Subsection 5.2.3, for $\hat{a} \neq a$, the actual control input is (almost everywhere) determined by the primary switching function $\xi_{\hat{a}}$.

Completing the square in (6.77b) yields

$$\langle \nabla\xi_{\hat{a}}, b\rangle = a^{-1}[(\tfrac{3}{4}\hat{a}x_3 \operatorname{sgn}(\Xi_{\hat{a}}^s) + \sqrt{\hat{a}x_2 \operatorname{sgn}(\Xi_{\hat{a}}^s) + \tfrac{1}{2}(\hat{a}x_3)^2})^2 - \tfrac{1}{16}(\hat{a}x_3)^2]$$
$$> 0 \qquad \text{for all } x \neq 0$$

so that sliding condition (3.15b) is satisfied for all $x \neq 0$. Clearly, from (6.77),

$$|\langle \nabla\xi_{\hat{a}}, Ax\rangle| = \frac{\hat{a}}{a} |\langle \nabla\xi_{\hat{a}}, b\rangle|$$
$$< |\langle \nabla\xi_{\hat{a}}, b\rangle| \qquad \text{for all } \hat{a} > a$$

i.e. sliding condition (3.15a) is also satisfied. Hence, each (non-zero) point of the estimated switching surface $\Gamma_{\hat{a}}$ is a sliding point† so that, on reaching the surface $\Gamma_{\hat{a}}$, the state point is subsequently constrained to remain on $\Gamma_{\hat{a}}$. But $\Gamma_{\hat{a}}$ is the time-optimal switching surface for a triple integrator system with gain parameter $\hat{a}$, or, equivalently, for the triple integrator system with gain parameter a and control constraint

$$|u| \leq \frac{a}{\hat{a}}$$

Hence, $\Gamma_{\hat{a}}$ is comprised of the set of paths which go to the origin under a piecewise constant control with $|u| = a/\hat{a}$ and at most two intervals of control constancy; for the original system it may be concluded that, on reaching $\Gamma_{\hat{a}}$, the state subsequently follows such a (sliding) path in $\Gamma_{\hat{a}}$ to the origin.

For example, if $\hat{a} = 2a$, i.e. if the feedback control realisation is based on a gain parameter estimate which exceeds the true value by a factor of two, then the (estimated) switching surface maps to the curve $\Gamma_{\hat{a}}$ of Fig. 6.4 in (Θ, Φ)-space. For a starting point $x^0 = (x_1^0, 0, 0)'$, $x_1^0 > 0$, on the positive x_1 state semi-axis, the state trajectory to the origin maps to the path QRS$^-$ of Fig. 6.4. From Q, the n-path QR ($u = -1$) to $\Gamma_{\hat{a}}$ is followed, the sliding path RS$^-$ is subsequently generated by the piecewise constant sliding control input $u_s(\cdot)$, with $u_s = +a/\hat{a} = \frac{1}{2}$ from R to S$^-$ and $u_s = -a/\hat{a} = -\frac{1}{2}$ at S$^-$; at S$^-$ the state remains stationary in (Θ, Φ)-space, while, in z-space, the ray through S$^-$ is followed to the state origin. From (6.71a), the minimum time t_3 to the origin from Q is given by

$$t_3 = T_3(x) = (32a|x_1^0|)^{1/3}$$

† Note that in the derivation (Sections 3.2 and 3.3) of conditions for sliding, the switching surface was assumed to be smooth in a neighbourhood of the point in question. For the case at hand, the switching surface Γ_a is smooth everywhere except at a corner corresponding to the one-dimensional curve $\Gamma_{\hat{a}}^s = \{x \in \Gamma_{\hat{a}}: \xi_{\hat{a}}^s(x^s) = x_2 + \frac{1}{2}\hat{a}x_3|x_3| = 0\}$ which, when projected into the (x_2, x_3) subspace, corresponds to the (nominally) time-optimal switching curve for the double integrator system so that the analysis of Subsection 5.2.3 is valid.

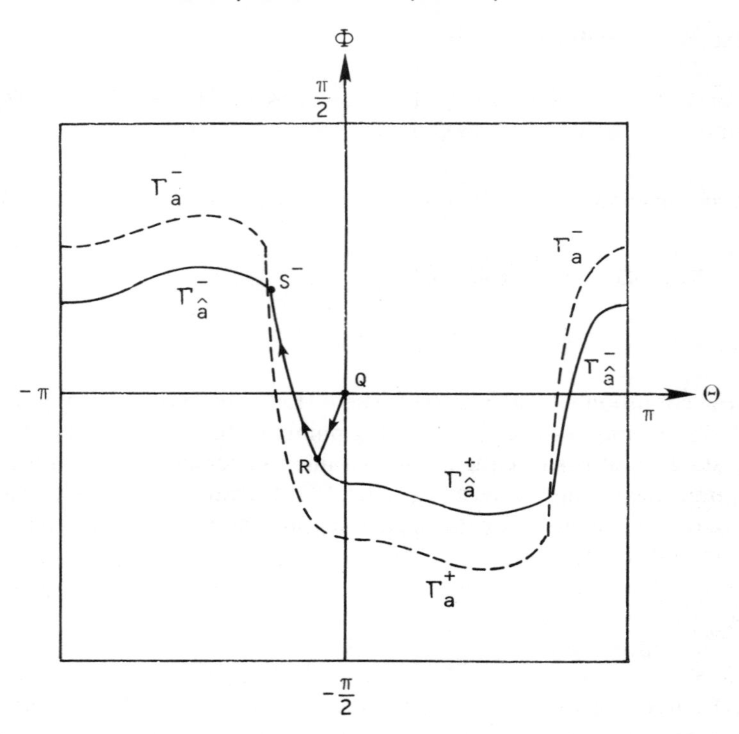

Fig. 6.4 $\hat{a} = 2a$
Time-optimal (Γ_a) and estimated ($\Gamma_{\hat{a}}$) switching surfaces in (Θ, Φ)-space

while a straightforward but lengthy calculation gives the time to the origin from Q under the estimated feedback control (6.73) as

$$t_f = \hat{T}_3(x) = \left[1 + \frac{\hat{a}}{a} + 2\sqrt{\frac{\hat{a}}{2a}\left(1 + \frac{\hat{a}}{a}\right)}\right] \times \left[\frac{1}{6} + \frac{\hat{a}}{2a} + \frac{1}{3}\left(\frac{\hat{a}}{a}\right)^2 + \frac{1}{2}\sqrt{\frac{\hat{a}}{2a}}\left(1 + \frac{\hat{a}}{a}\right)^{3/2}\right]^{-1/3} (a|x_1^0|)^{1/3}$$

so that the fractional increase in time to the origin for $\hat{a}/a \geq 1$ becomes

$$\Delta = \frac{t_f - t_3}{t_3} = F\left(\frac{\hat{a}}{a}\right) = \left[1 + \frac{\hat{a}}{a} + 2\sqrt{\frac{\hat{a}}{2a}\left(1 + \frac{\hat{a}}{a}\right)}\right] \times \left[\frac{16}{3} + 16\left(\frac{\hat{a}}{a}\right) + \frac{32}{3}\left(\frac{\hat{a}}{a}\right)^2 + 16\sqrt{\frac{\hat{a}}{2a}}\left(1 + \frac{\hat{a}}{a}\right)^{3/2}\right]^{-1/3} - 1$$

For example, the fractional increase in time to the origin incurred on path QRS^- of Fig. 6.4, with $\hat{a}/a = 2$, is $\Delta = F(2) \simeq 0{\cdot}183$, i.e. a parameter estimate error of 100% causes an increase in time to the origin over the theoretical minimum time only of the order of 18·3% for all initial states on the x_1 state axis.

In summary, for all $\hat{a}/a > 1$, the feedback system is stable and the origin is attained in finite time from all initial states $x(0) \in \mathbb{R}^3$. For a quantitative assessment of the nominally-optimal feedback system, the reader is referred to the work of Zinober and Fuller (1973) which contains computed values of the fractional increase in time to the origin incurred for a wide range of initial states and parameter ratios $\hat{a}/a$.

6.6.2 Case 2 : $0 < \hat{a} < a$

This case has also been studied by Zinober and Fuller (1973) and Fuller and Zinober (1977), whose results are briefly described here.

From (6.77), but now with $0 < \hat{a} < a$, it may be seen that the regular switching condition (3.14) holds at all non-zero points of the switching surface $\Gamma_{\hat{a}}$, i.e. all trajectories of the feedback system are regular switching trajectories; sliding motion or separatrices do not occur. Within the regular switching behaviour, the special class of *constant-ratio trajectories* can be isolated as in the case of the double integrator feedback system of Subsection 5.2.3(i). For the triple integrator system under consideration, a constant-ratio trajectory is defined as follows: suppose a control switch occurs at point $x^0 = (x_1^0, x_2^0, x_3^0)' \in \Gamma_{\hat{a}}$, then, on a constant-ratio trajectory, the next control switch occurs, after a finite time t_1, at the point

$$x^1 = (x_1^1, x_2^1, x_3^1)' = (-\rho^3 x_1^0, -\rho^2 x_2^0, -\rho x_3^0)' \in \Gamma_{\hat{a}}$$

where ρ is a non-negative constant, referred to as the constant-ratio parameter. After a further time interval of duration $t_2 = \rho t_1$, the latter switch is, in turn, followed by a switch at the point

$$x^2 = (x_1^2, x_2^2, x_3^2)' = (-\rho^3 x_1^1, -\rho^2 x_2^1, -\rho x_3^1)'$$
$$= (\rho^6 x_1^0, \rho^4 x_2^0, \rho^2 x_3^0)' \in \Gamma_{\hat{a}}$$

This process continues, thereby generating a sequence of switch points (in $\Gamma_{\hat{a}}$) for which the ith switch is given by

$$x^i = (x_1^i, x_2^i, x_3^i)' = ((-1)^i \rho^{3i} x_1^0, (-1)^i \rho^{2i} x_2^0, (-1)^i \rho^i x_3^0)' \tag{6.78a}$$

with the interval between successive switches (at x^{i-1} and x^i) given by

$$t_i = \rho^{i-1} t_1 \tag{6.78b}$$

where t_1 is the duration of the first constant control interval between switch points x^0 and x^1. Such constant-ratio trajectories are discussed in detail by Grensted and Fuller (1965) and Fuller (1971).

For definiteness, suppose that the initial switch point x^0 corresponds to an np control switch so that the trajectory from x^0 to x^1 is generated by the positive constant control $u(t) = +1, 0 \le t \le t_1$, yielding the result

$$\left.\begin{aligned} -\rho^3 x_1^0 &= x_1^1 = x_1(t_1) = x_1^0 + x_2^0 t_1 + x_3^0 \frac{t_1^2}{2} + \frac{t_1^3}{6a} \\ -\rho^2 x_2^0 &= x_2^1 = x_2(t_1) = x_2^0 + x_3^0 t_1 + \frac{t_1^2}{2a} \\ -\rho x_3^0 &= x_3^1 = x_3(t_1) = x_3^0 + \frac{t_1}{a} \end{aligned}\right\} \tag{6.79}$$

Solving (6.79) for x_1^0, x_2^0, x_3^0 gives

$$\left.\begin{aligned} x_1^0 &= -\frac{t_1^3}{6a}\frac{1 - 3\rho + \rho^2}{(1 + \rho^2)(1 + \rho^3)} \\ x_2^0 &= \frac{t_1^2}{2a}\frac{1 - \rho}{(1 + \rho)(1 + \rho^2)} \\ x_3^0 &= -\frac{t_1}{a}\frac{1}{(1 + \rho)} \end{aligned}\right\} \tag{6.80}$$

Transformation of the np-switch point (6.80) into (Θ, Φ)-space (via (3.40) and (3.44)) yields the np-switch point (Θ^0, Φ^0) which is *independent* of t_1 and a. As the constant-ratio parameter ρ increases from zero, the point (Θ^0, Φ^0) traces a locus in (Θ, Φ)-space which starts at the singular point C^+ (for $\rho = 0$) and approaches the singular point D^- as $\rho \to \infty$; by symmetry, the locus of pn-switch points may be obtained, starting at the singular point C^- and approaching the singular point D^+ as $\rho \to \infty$. These loci are depicted in Fig. 6.5 (see also Fuller 1971). For the feedback system to generate such a constant-ratio trajectory, the following conditions must hold:

(*a*) the constant-ratio np-switch point (6.80) corresponds to an np-regular switch point of the feedback system for some $\rho \ge 0$ or, equivalently, the constant-ratio locus of np- (pn-) switch points in (Θ, Φ)-space interests the switching curve $\Gamma_{\hat{a}}$ (in (Θ, Φ)-space) at an np- (pn-) regular switch point;
(*b*) the p-path (n-path) joining any pair of successive constant-ratio switch points does not intersect the switching surface $\Gamma_{\hat{a}}$ at any intermediate point.

Fuller and Zinober (1977) have established that the above conditions are satisfied for all $0 < \hat{a} < a$, so that a constant-ratio trajectory with parameter ρ (dependent on $\hat{a}/a$) exists in all such cases. It may be seen from (6.78) that $x^i \to 0$ as $i \to \infty$ for all values of ρ such that $0 \le \rho < 1$, i.e. the motion is stable for $\rho \in [0, 1)$; moreover, the origin is attained from the starting point x^0 (np constant-ratio switch) in *finite* time t_f, viz.

$$t_f = \sum_{i=1}^{\infty} t_i = t_1 \sum_{i=1}^{\infty} \rho^{i-1} = t_1(1 - \rho)^{-1}; \qquad 0 < \rho < 1 \tag{6.81}$$

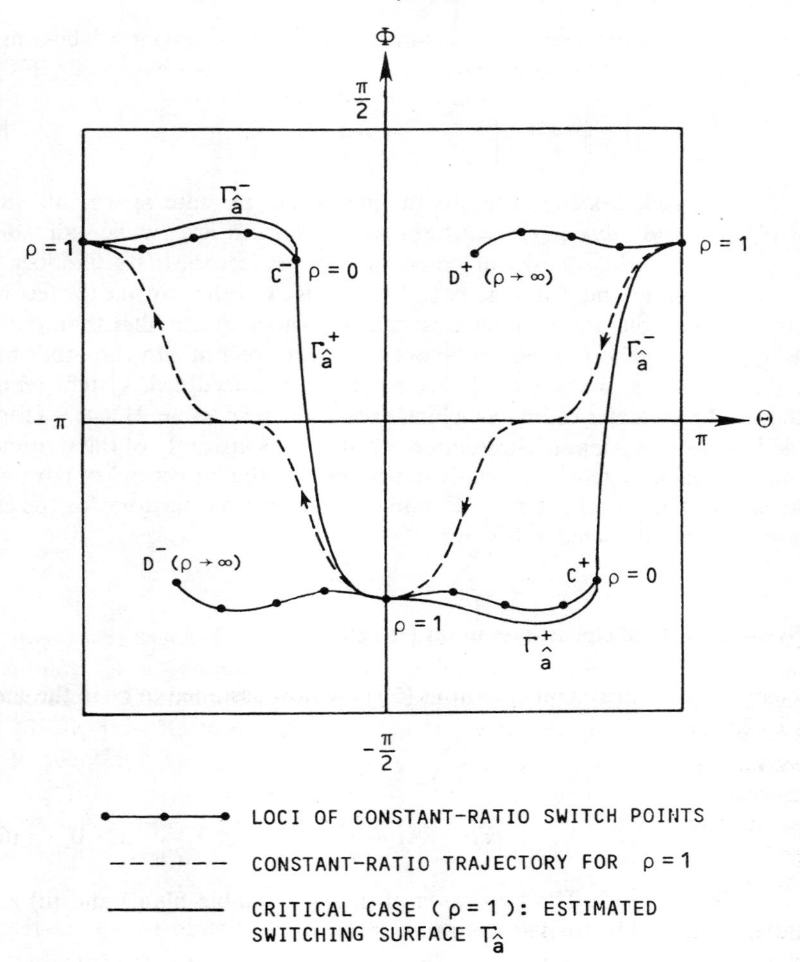

Fig. 6.5

On the other hand, all values of $\rho > 1$ yield divergent (unstable) constant-ratio trajectories. The critical case of $\rho = 1$ corresponds to a limit cycle in state space; from (6.80), it may be seen that an np-switch point of such a limit cycle is given by

$$x_1^0 = \frac{t_1^3}{6a}; \qquad x_2^0 = 0; \qquad x_3^0 = -\frac{t_1}{a}$$

Substituting the above values in (6.74) yields the result

$$\xi_{\hat{a}}(x^0) = \frac{t_1^3}{24}\left(1 - \left(\frac{\hat{a}}{a}\right)^2\left[1 + \frac{3}{2\sqrt{2}}\right]\right)$$

so that, if $\xi_{\hat{a}}(x^0) = 0$, then the constant-ratio np-switch point x^0 lies on the switching surface $\Gamma_{\hat{a}}$. Hence, if

$$\frac{\hat{a}}{a} = \left(\frac{\hat{a}}{a}\right)_c = \left[1 + \frac{3}{2\sqrt{2}}\right]^{-1/2} \simeq 0{\cdot}6964 \tag{6.82}$$

then the feedback system generates a limit cycle in state space; all values $\hat{a}/a < (\hat{a}/a)_c$ yield *divergent* constant-ratio trajectories, while all values $(\hat{a}/a)_c < \hat{a}/a < 1$ yield *convergent* constant-ratio trajectories (see Zinober and Fuller, 1973; Fuller and Zinober, 1977, for details). In other words, the feedback system can only tolerate parameter estimates $\hat{a}$ which are smaller than the true value by a factor of up to 0·6964 before instability sets in. On the other hand, from the preceding analysis of the case $\hat{a}/a > 1$, the feedback system remains stable for all parameter estimates which exceed the true value. Hence, as for the double integrator system of Subsection 5.2.3(i), the sensitivity of the nominally optimal feedback system is crucially dependent on the 'direction' of parameter perturbation. The switching curve $\Gamma_{\hat{a}}$ and constant-ratio trajectory for the critical case (6.82) are depicted in Fig. 6.5.

6.7 System (*b*): Real eigenvalues in simple ratio

The eigenvalues λ_i in system equations (6.1) are now assumed to be in the simple ratio 3 : 2 : 1, i.e.

$$\dot{x}_1 = 3\lambda x_1 + x_2; \qquad \dot{x}_2 = 2\lambda x_2 + x_3;$$

$$\dot{x}_3 = \lambda x_3 + \frac{u}{a}; \qquad |u| \leq 1; \qquad a > 0 \tag{6.83}$$

with $\lambda =$ constant $\neq 0$. The two cases of (i) $\lambda < 0$ (stable plant) and (ii) $\lambda > 0$ (unstable plant) will be treated separately.

6.7.1 *Case (i): negative real eigenvalues, $\lambda < 0$*

From theorem 4.3.1, it follows that $\mathscr{C} = \mathbb{R}^3$. Furthermore, the system equivalence property of Section 3.6 can be invoked to conclude that the time-optimal feedback control is independent of the parameter λ and is identical to that of the triple integrator system, i.e. the time-optimal feedback control is given by

$$u = u(x) = -\operatorname{sgn}(\Xi(x)) = -\operatorname{sgn}(\tilde{\Xi}(x)) \tag{6.84}$$

where the optimal switching function $\Xi: \mathbb{R}^3 \to \mathbb{R}$ (or $\tilde{\Xi}: \mathbb{R}^3 \to \mathbb{R}$) is given by (6.15) (or by (6.17) and (6.18)). Moreover, recalling the time transformation (3.52) of Subsection 3.6.1, it immediately follows, from the triple integrator results (6.67), that the optimal switching times t_1, t_2 and minimum time to the origin t_3, for the system under consideration, are given as explicit functions of the initial state by the relations

$$t_3 = T_3(x) = \begin{cases} -\frac{1}{\lambda} \ln [1 - \lambda[S(x) + ax_3 \operatorname{sgn}(\xi)]]; & \xi = \xi(x) \neq 0 \\ T_2^s(x^s); & \xi = \xi(x) = 0 \end{cases} \tag{6.85a}$$

$$t_2 = T_2(x) = \begin{cases} -\frac{1}{\lambda} \ln [1 - \lambda[\frac{3}{4}S(x) + \gamma_2(S(x))^{-1} + ax_3 \operatorname{sgn}(\xi)]]; & \xi = \xi(x) \neq 0 \\ T_1^s(x^s); & \xi = \xi(x) = 0 \end{cases} \tag{6.85b}$$

$$t_1 = T_1(x) = \begin{cases} -\frac{1}{\lambda} \ln [1 - \lambda[\frac{1}{4}S(x) + \gamma_2(S(x))^{-1} + ax_3 \operatorname{sgn}(\xi)]]; & \xi = \xi(x) \neq 0 \\ T_1^s(x^s); & \xi = \xi(x) = 0 \end{cases} \tag{6.85c}$$

where the functions S and γ_2 are defined as in (6.66) and T_i^s: $\mathbb{R}^2 \to [0, \infty)$, $i = 1, 2$, are the subsystem switching and minimum times given by (6.6).

In this case ($\lambda < 0$) the family of isochronal surfaces expand to fill the entire state space

$$\bigcup_{t_3 \geq 0} \Upsilon_{t_3} = \bigcup_{t_3 \geq 0} \{x: T_3(x) = t_3\} = \mathscr{C} = \mathbb{R}^3 \tag{6.86}$$

6.7.2 Case (ii) : positive real eigenvalues ($\lambda > 0$)

The system equations are again given as in (6.83) but now with the important distinction that the parameter λ is positive ($\lambda > 0$), i.e. the uncontrolled plant is unstable. In this case not all initial states can be driven to the origin but only those states $x \in \mathscr{C}$ in the domain of null controllability $\mathscr{C} \subset \mathbb{R}^3$, in other words, only sufficiently small disturbances away from the zero state can be nullified by the constrained control. Within $\mathscr{C}$ the system equivalence property can again be invoked to give the optimal feedback control (independent of λ) in the form (6.84) but now restricted to $\mathscr{C}$; moreover, for $x \in \mathscr{C}$, the switching and minimum times t_i are again given by (6.85). In particular, consider (6.85a), or equivalently,

$$t_3 = T_3(x) = \begin{cases} \frac{1}{\lambda} \ln \left[1 - \lambda[S(x) + ax_3 \operatorname{sgn}(\xi)]\right]^{-1}; \\ \qquad \xi = \xi(x) \neq 0, \quad x \in \mathscr{C} \\ T_2^s(x^s) = \frac{1}{\lambda} \ln \left[1 - \lambda[ax_3 \operatorname{sgn}(\Xi^s) \right. \\ \qquad \left. + 2\sqrt{ax_2 \operatorname{sgn}(\Xi^s) + \frac{1}{2}(ax_3)^2}]\right]^{-1}; \\ \qquad \xi = \xi(x) = 0, \quad x \in \mathscr{C} \end{cases} \tag{6.87}$$

It is evident from (6.87) that (as $\lambda > 0$)

$$t_3 \to \infty \quad \text{as} \quad S(x) + ax_3 \operatorname{sgn}(\xi) \to \frac{1}{\lambda} \quad \text{if } \xi = \xi(x) \neq 0$$

and

$$t_3 \to \infty \quad \text{as} \quad ax_3 \operatorname{sgn}(\Xi^s) + 2\sqrt{ax_2 \operatorname{sgn}(\Xi^s) + \tfrac{1}{2}(ax_3)^2} \to \frac{1}{\lambda} \quad \text{if } \xi = \xi(x) = 0$$

so that the domain of null controllability may be expressed as

$$\mathscr{C} = \{x \colon S(x) + ax_3 \operatorname{sgn}(\xi) < \frac{1}{\lambda};\ \xi = \xi(x) \neq 0\}$$
$$\cup \{x \colon ax_3 \operatorname{sgn}(\Xi^s) + 2\sqrt{ax_2 \operatorname{sgn}(\Xi^s) + \tfrac{1}{2}(ax_3)^2} < \frac{1}{\lambda};\ \xi(x) = 0\} \tag{6.88}$$

6.8 System (*c*): Double integrator-plus-lag

The system equations are

$$\dot{x}_1 = \lambda x_1 + x_2; \qquad \dot{x}_2 = x_3; \qquad \dot{x}_3 = \frac{u}{a};$$
$$|u| \leq 1; \qquad a > 0; \qquad \lambda < 0 \tag{6.89a}$$

or

$$\dot{x} = Ax + bu; \qquad |u| \leq 1 \tag{6.89b}$$

where

$$A = \begin{bmatrix} \lambda & 1 & 0 \\ 0 & 0 & 1 \\ 0 & 0 & 0 \end{bmatrix} \quad \text{and} \quad b = \begin{bmatrix} 0 \\ 0 \\ \frac{1}{a} \end{bmatrix} \tag{6.89c}$$

From theorem 4.3.1 all states are null controllable ($\mathscr{C} = \mathbb{R}^3$). The second-order subsystem is the double integrator, i.e. Subsystem I of Section 6.2, and the switching function $\xi(x)$ may be calculated via (6.14). To facilitate this calculation the following matrix is introduced

$$P = \begin{bmatrix} 1 & 1 & 0 \\ 0 & -\lambda & 1 \\ 0 & 0 & -\lambda \end{bmatrix} \quad \text{with} \quad P^{-1} = \begin{bmatrix} 1 & \frac{1}{\lambda} & \frac{1}{\lambda^2} \\ 0 & -\frac{1}{\lambda} & -\frac{1}{\lambda^2} \\ 0 & 0 & -\frac{1}{\lambda} \end{bmatrix} \tag{6.90}$$

with the properties that

$$P^{-1}\begin{bmatrix} \beta(x) \\ 0 \\ 0 \end{bmatrix} = \begin{bmatrix} \beta(x) \\ 0 \\ 0 \end{bmatrix} \tag{6.91}$$

and

$$P^{-1} \exp (At)P = \exp (P^{-1}APt) = \exp (\Lambda t) \tag{6.92a}$$

where

$$\Lambda = P^{-1}AP = \begin{bmatrix} \lambda & 0 & 0 \\ 0 & 0 & 1 \\ 0 & 0 & 0 \end{bmatrix} \quad \text{and} \quad \exp (\Lambda t) = \begin{bmatrix} \exp (\lambda t) & 0 & 0 \\ 0 & 1 & t \\ 0 & 0 & 1 \end{bmatrix} \tag{6.92b}$$

Combining (6.14), (6.91) and (6.92) yields the result

$$\begin{bmatrix} \beta(x) \\ 0 \\ 0 \end{bmatrix} = \exp (\Lambda T_2^s)\Bigg[P^{-1}x - \operatorname{sgn} (\Xi^s)$$

$$\times \left[\int_0^{T_1^s} \exp (-\Lambda s)P^{-1}b \, ds - \int_{T_1^s}^{T_2^s} \exp (-\Lambda s)P^{-1}b \, ds \right]\Bigg] \tag{6.93}$$

where, for notational convenience, the dependence of Ξ^s, T_1^s, T_2^s on the initial (subsystem) state x^s has been suppressed. Due to the special form of Λ, which decouples the first component of the vector equation (6.93), it immediately follows that

$$\beta(x) = \exp (\lambda T_2^s)\Bigg[x_1 + \frac{x_2}{\lambda} + \frac{x_3}{\lambda^2} - \frac{\operatorname{sgn} (\Xi^s)}{a\lambda^3}$$

$$\times [\exp (-\lambda T_2^s) - 2 \exp (-\lambda T_1^s) + 1]\Bigg] \tag{6.94}$$

giving

$$\xi(x) = \exp (-\lambda T_2^s)\beta(x)$$

as

$$\xi(x) = x_1 + \frac{x_2}{\lambda} + \frac{x_3}{\lambda^2} - \frac{\operatorname{sgn}(\Xi^s)}{a\lambda^3} \times [\exp(-\lambda T_2^s(x^s)) - 2\exp(-\lambda T_1^s(x^s)) + 1] \tag{6.95a}$$

where, from (6.3) and (6.4),

$$T_1^s(x^s) = ax_3 \operatorname{sgn}(\Xi^s) + \sqrt{ax_2 \operatorname{sgn}(\Xi^s) + \tfrac{1}{2}(ax_3)^2} \tag{6.95b}$$

$$T_2^s(x^s) = 2T_1^s(x^s) - ax_3 \operatorname{sgn}(\Xi^s) \tag{6.95c}$$

$$\Xi^s = \Xi^s(x^s) = \begin{cases} \xi^s(x^s) = x_2 + \tfrac{1}{2}ax_3|x_3|; & \xi^s(x^s) \neq 0 \\ x_3; & \xi^s(x^s) = 0 \end{cases} \tag{6.95d}$$

Hence, the time-optimal control is explicitly characterised by

$$u = u(x) = -\operatorname{sgn}[\Xi(x)]; \qquad \Xi(x) = \begin{cases} \xi(x); & \xi(x) \neq 0 \\ \Xi^s(x^s); & \xi(x) = 0 \end{cases} \tag{6.96}$$

6.9 System (*d*): Subsystem II-plus-lag: $\lambda_1 < 0; \lambda_2 = 2\lambda_3; \lambda_3 \neq 0$

The system A matrix is

$$A = \begin{bmatrix} \lambda_1 & 1 & 0 \\ 0 & 2\lambda_3 & 0 \\ 0 & 0 & \lambda_3 \end{bmatrix}: \qquad \lambda_1 < 0; \qquad \lambda_3 \neq 0 \tag{6.97}$$

and three cases arise which are treated separately below.

6.9.1 *Case 1: distinct eigenvalues: $\lambda_1 \neq 2\lambda_3$; $\lambda_1 \neq \lambda_3$*

(i) $\lambda_3 < 0$: Suppose initially that λ_3 is negative so that all states are null controllable ($\mathscr{C} = \mathbb{R}^3$). Proceeding as in the previous example, the following nonsingular matrix is introduced

$$P = \begin{bmatrix} 1 & 1 & 1 \\ 0 & -(\lambda_1 - 2\lambda_3) & -(\lambda_1 - \lambda_3) \\ 0 & 0 & \lambda_3(\lambda_1 - \lambda_3) \end{bmatrix} \tag{6.98}$$

such that

$$P^{-1}\begin{bmatrix} \beta(x) \\ 0 \\ 0 \end{bmatrix} = \begin{bmatrix} \beta(x) \\ 0 \\ 0 \end{bmatrix}$$

and $P^{-1}\exp(At)P = \exp(\Lambda t)$ where

$$\Lambda = \begin{bmatrix} \lambda_1 & 0 & 0 \\ 0 & 2\lambda_3 & 0 \\ 0 & 0 & \lambda_3 \end{bmatrix}; \qquad \exp(\Lambda t) = \begin{bmatrix} \exp(\lambda_1 t) & 0 & 0 \\ 0 & \exp(2\lambda_3 t) & 0 \\ 0 & 0 & \exp(\lambda_3 t) \end{bmatrix}$$

Consequently, (6.14) is equivalent to

$$\begin{bmatrix} \beta(x) \\ 0 \\ 0 \end{bmatrix} = \exp(\Lambda T_2^s)\Bigg[P^{-1}x - \operatorname{sgn}(\Xi^s)$$

$$\times \left[\int_0^{T_1^s} \exp(-\Lambda s)P^{-1}b\, ds - \int_{T_1^s}^{T_2^s} \exp(-\Lambda s)P^{-1}b\, ds\right]\Bigg] \tag{6.99}$$

which is easily calculated to give

$$\begin{aligned}\xi(x) &= \exp(-\lambda_1 T_2^s)\beta(x) \\ &= x_1 + \frac{x_2}{(\lambda_1 - 2\lambda_3)} + \frac{x_3}{(\lambda_1 - \lambda_3)(\lambda_1 - 2\lambda_3)} \\ &\quad - \frac{\operatorname{sgn}(\Xi^s)}{a\lambda_1(\lambda_1 - \lambda_3)(\lambda_1 - 2\lambda_3)} \\ &\quad \times [\exp(-\lambda_1 T_2^s) - 2\exp(-\lambda_1 T_1^s) + 1]\end{aligned} \tag{6.100}$$

and which, on substituting for $T_i^s = T_i^s(x^s)$ from (6.6), becomes

$$\begin{aligned}\xi(x) &= x_1 + \frac{x_2}{(\lambda_1 - 2\lambda_3)} + \frac{x_3}{(\lambda_1 - \lambda_3)(\lambda_1 - 2\lambda_3)} \\ &\quad - \frac{\operatorname{sgn}(\Xi^s)}{a\lambda_1(\lambda_1 - \lambda_3)(\lambda_1 - 2\lambda_3)} \\ &\quad \times ([\![1 - \lambda_3[ax_3 \operatorname{sgn}(\Xi^s) + 2R(x^s)]]\!]^{\lambda_1/\lambda_3} \\ &\quad - 2[\![1 - \lambda_3[ax_3 \operatorname{sgn}(\Xi^s) + R(x^s)]]\!]^{\lambda_1/\lambda_3} + 1)\end{aligned} \tag{6.101a}$$

where

$$R(x^s) = \sqrt{ax_2 \operatorname{sgn}(\Xi^s) + \tfrac{1}{2}(ax_3)^2} \tag{6.101b}$$

$$\Xi^s = \Xi^s(x^s) = \begin{cases} \xi^s(x^s) = x_2 + \frac{1}{2}ax_3|x_3|; & \xi^s(x^s) \neq 0 \\ x_3; & \xi^s(x^s) = 0 \end{cases} \tag{6.101c}$$

Hence, the time-optimal switching function $\Xi: \mathbb{R}^3 \to \mathbb{R}$ may be written as

$$\Xi(x) = \begin{cases} \xi(x); & \xi(x) \neq 0 \\ \Xi^s(x^s); & \xi(x) = 0 \end{cases} \tag{6.102}$$

(ii) $\lambda_3 > 0$: Suppose now that λ_3 is positive, i.e. the plant to be controlled consists of an *unstable* Subsystem II in cascade with a first-order lag. It is easily verified that, *when restricted to* $\mathscr{C} \subset \mathbb{R}^3$, (6.101) and (6.102) again define the time-optimal feedback solution, where the domain of null controllability $\mathscr{C}$ is given by

$$\mathscr{C} = \{x : x^s \in \mathscr{C}^s\} \tag{6.103a}$$

with, from (6.7),

$$\mathscr{C}^s = \left\{ x^s = (x_2, x_3)' : ax_3 \operatorname{sgn}(\Xi^s) + 2\sqrt{ax_2 \operatorname{sgn}(\Xi^s) + \tfrac{1}{2}(ax_3)^2} < \frac{1}{\lambda_3} \right\} \tag{6.103b}$$

In other words, all states

$$x = \begin{bmatrix} x_1 \\ x^s \end{bmatrix}$$

with null-controllable sub-state $x^s \in \mathscr{C}^s$ are null controllable.†

6.9.2 Case 2: repeated eigenvalues: $\lambda_1 = \lambda_2 = 2\lambda;\ \lambda_3 = \lambda \neq 0$

(i) $\lambda < 0$: Initially it is assumed that the plant to be controlled is stable, i.e. $\lambda < 0$, in which case $\mathscr{C} = \mathbb{R}^3$. Setting

$$P = \begin{bmatrix} 1 & 0 & 1 \\ 0 & 1 & -\lambda \\ 0 & 0 & \lambda^2 \end{bmatrix} \tag{6.104}$$

then $P^{-1} \exp(At)P = \exp(\Lambda t)$, where

$$\Lambda = \begin{bmatrix} 2\lambda & 1 & 0 \\ 0 & 2\lambda & 0 \\ 0 & 0 & \lambda \end{bmatrix} \quad \text{and} \quad \exp(\Lambda t) = \begin{bmatrix} \exp(2\lambda t) & t\exp(2\lambda t) & 0 \\ 0 & \exp(2\lambda t) & 0 \\ 0 & 0 & \exp(\lambda t) \end{bmatrix} \tag{6.105}$$

† To see this, note that all states $x \in \mathscr{N}$ in a sufficiently small neighbourhood $\mathscr{N} = \{x : \|x\| < \epsilon\}$ of the origin are null-controllable. With $\mathscr{C}$ defined by (6.103), an admissible control steering $x \in \mathscr{C}$ to the origin can be constructed as follows: first drive the substate $x^s \in \mathscr{C}^s$ to the subspace origin; then set the control to zero and allow the remaining non-zero x_1 component to decay exponentially ($\lambda_1 < 0$) until the neighbourhood $\mathscr{N}$ is attained, from where the state can be driven to the origin by means of a (non-zero) admissible control.

The above argument can be extended to the case of any plant consisting of an *unstable* subplant (with domain of null controllability $\mathscr{C}^s$) in cascade with a *stable* (but not necessarily asymptotically stable) subplant with *nonpositive* real eigenvalues. The neighbourhood $\mathscr{N}$ can again be attained from any state $x \in \mathscr{C} = \{x : x^s \in \mathscr{C}^s\}$; however, it is necessary to select (details omitted) an appropriate *non-zero* control in order to attain $\mathscr{N}$ whenever the stable subplant is not asymptotically stable.

and, evaluating (6.93) yields the result

$$\xi(x) = \exp(-2\lambda T_2^s)\beta(x)$$

$$= x_1 - \frac{x_3}{\lambda^2} + \left[x_2 + \frac{x_3}{\lambda}\right]T_2^s - \frac{\text{sgn}(\Xi^s)}{a\lambda^2}$$

$$\times \left[\frac{3}{2\lambda}\exp(-2\lambda T_1^s) - \frac{3}{4\lambda}\exp(-2\lambda T_2^s) - \frac{3}{4\lambda}\right.$$

$$\left. + \tfrac{1}{2}T_2^s - (T_2^s - T_1^s)\exp(-2\lambda T_1^s)\right] \tag{6.106}$$

which, on substituting for T_i^s from (6.6), becomes

$$\xi(x) = x_1 - \frac{x_3}{\lambda^2} - \left(\frac{x_2}{\lambda} + \frac{x_3}{\lambda^2}\right)\ln[1 - \lambda(ax_3\,\text{sgn}(\Xi^s) + 2R(x^s))]$$

$$- \frac{\text{sgn}(\Xi^s)}{a\lambda^3}\left(\tfrac{3}{2}[1 - \lambda(ax_3\,\text{sgn}(\Xi^s) + R(x^s))]^2\right.$$

$$- \tfrac{3}{4}[1 - \lambda(ax_3\,\text{sgn}(\Xi^s) + 2R(x^s)]^2$$

$$- \tfrac{3}{4} - \tfrac{1}{2}\ln[1 - \lambda(ax_3\,\text{sgn}(\Xi^s) + 2R(x^s))]$$

$$+ [1 - \lambda(ax_3\,\text{sgn}(\Xi^s) + R(x^s))]^2$$

$$\left.\times \ln\left[\frac{1 - \lambda(ax_3\,\text{sgn}(\Xi^s) + 2R(x^s))}{1 - \lambda(ax_3\,\text{sgn}(\Xi^s) + R(x^s))}\right]\right) \tag{6.107}$$

Here $R(x^s)$ and Ξ^s $(= \Xi^s(x^s))$ are defined by (6.101b) and (6.101c) and, with ξ: $\mathbb{R}^3 \to \mathbb{R}$ defined by (6.107), the time-optimal switching function is again given by (6.102).

(ii) $\lambda > 0$: If the plant to be controlled is unstable, i.e. $\lambda > 0$, then the time-optimal switching function is again given by (6.102) and (6.107) *but now restricted to the domain of null controllability* $\mathscr{C} \subset \mathbb{R}^3$ which is again given by (6.103) with $\lambda_3 = \lambda$.

6.9.3 Case 3 : repeated eigenvalues : $\lambda_1 = \lambda_3 = \lambda \neq 0; \lambda_2 = 2\lambda$

(i) $\lambda < 0$: Setting

$$P = \begin{bmatrix} 1 & 0 & 1 \\ 0 & 1 & \lambda \\ 0 & -\lambda & 0 \end{bmatrix} \tag{6.108}$$

then $P^{-1}\exp(At)P = \exp(\Lambda t)$ where, in this case,

$$\Lambda = \begin{bmatrix} \lambda & 1 & 0 \\ 0 & \lambda & 0 \\ 0 & 0 & 2\lambda \end{bmatrix} \quad \text{and} \quad \exp(\Lambda t) = \begin{bmatrix} \exp(\lambda t) & t\exp(\lambda t) & 0 \\ 0 & \exp(\lambda t) & 0 \\ 0 & 0 & \exp(2\lambda t) \end{bmatrix} \tag{6.109}$$

Evaluating (6.93) yields the result

$$\xi(x) = \exp(-\lambda T_2^s(x^s))\beta(x)$$
$$= x_1 - \frac{x_2}{\lambda} - \frac{x_3}{\lambda^2} - \frac{x_3}{\lambda} T_2^s + \frac{\operatorname{sgn}(\Xi^s)}{a\lambda^3}$$
$$\times [\lambda T_2^s - 2\lambda(T_2^s - T_1^s)\exp(-\lambda T_1^s)] \tag{6.110}$$

which, on substituting for T_i^s from (6.6), becomes

$$\xi(x) = x_1 - \frac{x_2}{\lambda} - \frac{x_3}{\lambda^2} + \frac{x_3}{\lambda^2}\ln[1 - \lambda(ax_3 \operatorname{sgn}(\Xi^s) + 2R(x^s))]$$
$$- \frac{\operatorname{sgn}(\Xi^s)}{a\lambda^3}\Bigg(\ln[1 - \lambda(ax_3 \operatorname{sgn}(\Xi^s) + 2R(x^s))]$$
$$- 2[1 - \lambda(ax_3 \operatorname{sgn}(\Xi^s) + R(x^s))]$$
$$\times \ln\left[\frac{1 - \lambda(ax_3 \operatorname{sgn}(\Xi^s) + 2R(x^s))}{1 - \lambda(ax_3 \operatorname{sgn}(\Xi^s) + R(x^s))}\right]\Bigg) \tag{6.111}$$

where $R(x^s)$ and $\Xi^s = \Xi^s(x^s)$ are defined as in (6.101). With $\xi: \mathbb{R}^3 \to \mathbb{R}$ defined by (6.111), the time-optimal switching function is given by (6.102).

(ii) $\lambda > 0$: When restricted to $\mathscr{C} \subset \mathbb{R}^3$, (6.102) and (6.111), again define the time-optimal switching function $\Xi: \mathscr{C} \to \mathbb{R}$ where the domain of null controllability is given by (6.103) with $\lambda_3 = \lambda$.

6.10 System (*e*): Subsystem III-plus-lag; $\lambda_1 < 0, \lambda_2 = 0, \lambda_3 < 0$

The system A matrix in this case is

$$A = \begin{bmatrix} \lambda_1 & 1 & 0 \\ 0 & 0 & 1 \\ 0 & 0 & \lambda_3 \end{bmatrix}; \qquad \lambda_1 < 0, \lambda_3 < 0 \tag{6.112}$$

and all states are null controllable, i.e. $\mathscr{C} = \mathbb{R}^3$. Two possibilities of distinct and repeated eigenvalues arise which are separately studied.

6.10.1 Distinct eigenvalues: $\lambda_1 \neq \lambda_3$

Setting

$$P = \begin{bmatrix} 1 & 1 & 1 \\ 0 & -\lambda_1 & -(\lambda_1 - \lambda_3) \\ 0 & 0 & -\lambda_3(\lambda_1 - \lambda_3) \end{bmatrix} \tag{6.113}$$

then, as before,

$$P^{-1}\begin{bmatrix}\beta(x)\\0\\0\end{bmatrix}=\begin{bmatrix}\beta(x)\\0\\0\end{bmatrix}$$

and $P^{-1}\exp(At)P=\exp(\Lambda t)$, where

$$\Lambda=P^{-1}AP=\begin{bmatrix}\lambda_1 & 0 & 0\\0 & 0 & 0\\0 & 0 & \lambda_3\end{bmatrix}\text{ and}$$

$$\exp(\Lambda t)=\begin{bmatrix}\exp(\lambda_1 t) & 0 & 0\\0 & 1 & 0\\0 & 0 & \exp(\lambda_3 t)\end{bmatrix} \tag{6.114}$$

Evaluating (6.93) in this case yields the result

$$\begin{aligned}\xi(x)&=\exp(-\lambda_1 T_2^s)\beta(x)\\&=x_1+\frac{x_2}{\lambda_1}+\frac{x_3}{\lambda_1(\lambda_1-\lambda_3)}-\frac{\operatorname{sgn}(\Xi^s)}{a\lambda_1^2(\lambda_1-\lambda_3)}\\&\quad\times[\![\exp(-\lambda_1 T_2^s)-2\exp(-\lambda_1 T_1^s)+1]\!]\end{aligned} \tag{6.115}$$

where $\Xi^s=\Xi^s(x^s)$ and $T_i^s=T_i^s(x^s)$ are given by (6.8) and (6.9). Substituting for T_i^s in (6.115) gives

$$\begin{aligned}\xi(x)&=x_1+\frac{x_2}{\lambda_1}+\frac{x_3}{\lambda_1(\lambda_1-\lambda_3)}-\frac{\operatorname{sgn}(\Xi^s)}{a\lambda_1^2(\lambda_1-\lambda_3)}\\&\quad\times[\![\exp[a\lambda_1(\lambda_3 x_2-x_3)\operatorname{sgn}(\Xi^s)][H(x^s)^2-2H(x^s)]+1]\!]\end{aligned} \tag{6.116a}$$

where

$$\begin{aligned}&H(x^s)=\\&\left[1+\sqrt{1-[1-a\lambda_3 x_3\operatorname{sgn}(\Xi^s)]\exp[-a\lambda_3(\lambda_3 x_2-x_3)\operatorname{sgn}(\Xi^s)]}\right]^{\lambda_1/\lambda_3}\end{aligned} \tag{6.116b}$$

and

$$\Xi^s=\Xi^s(x^s)=\begin{cases}\xi^s(x^s)=x_2-\dfrac{x_3}{\lambda_3}-\dfrac{\operatorname{sgn}(x_3)}{a\lambda_3^2}\ln(1-a\lambda_3|x_3|); & \xi^s(x^s)\neq 0\\ x_3; \qquad \xi^s(x^s)=0\end{cases} \tag{6.116c}$$

With ξ: $\mathbb{R}^3 \to \mathbb{R}$ and Ξ^s: $\mathbb{R}^2 \to \mathbb{R}$ defined by (6.116), the overall time-optimal switching function Ξ: $\mathbb{R}^3 \to \mathbb{R}$ is given by

$$\Xi(x) = \begin{cases} \xi(x); & \xi(x) \neq 0 \\ \Xi^s(x^s); & \xi(x) = 0 \end{cases} \tag{6.117}$$

6.10.2 Repeated eigenvalues : $\lambda_1 = \lambda_3 = \lambda < 0$

Setting

$$P = \begin{bmatrix} 1 & 0 & 1 \\ 0 & 1 & -\lambda \\ 0 & \lambda & 0 \end{bmatrix} \tag{6.118}$$

then $P^{-1} \exp(At)P = \exp(\Lambda t)$, where

$$\Lambda = P^{-1}AP = \begin{bmatrix} \lambda & 1 & 0 \\ 0 & \lambda & 0 \\ 0 & 0 & 0 \end{bmatrix} \quad \text{and}$$

$$\exp(\Lambda t) = \begin{bmatrix} \exp(\lambda t) & t\exp(\lambda t) & 0 \\ 0 & \exp(\lambda t) & 0 \\ 0 & 0 & 1 \end{bmatrix} \tag{6.119}$$

Evaluating (6.93) in this case yields the result

$$\begin{aligned} \xi(x) &= \exp(-\lambda T_2^s)\beta(x) \\ &= x_1 + \frac{x_2}{\lambda} - \frac{x_3}{\lambda^2} + \frac{x_3}{\lambda}(T_2^s) \\ &\quad - \frac{\operatorname{sgn}(\Xi^s)}{a\lambda}\left[\frac{2}{\lambda^2}[\exp(-\lambda T_2^s) - 1]\right. \\ &\quad - \frac{4}{\lambda^2}[\exp(-\lambda T_2^s) - \exp(-\lambda T_1^s)] \\ &\quad \left.- \frac{2}{\lambda}(T_2^s - T_1^s)\exp(-\lambda T_1^s) + \frac{T_2^s}{\lambda}\right] \end{aligned} \tag{6.120}$$

where Ξ^s and T_i^s are given (6.8) and (6.9) with $\lambda_3 = \lambda$.

Substituting for T_i^s in (6.120)

$$\begin{aligned} \xi(x) &= x_1 + \frac{x_2}{\lambda} - \frac{x_3}{\lambda^2} - \frac{x_3}{\lambda} \\ &\quad \times \left[a(\lambda x_2 - x_3)\operatorname{sgn}(\Xi^s) + \frac{2}{\lambda}\ln(H(x^s))\right] \\ &\quad - \frac{\operatorname{sgn}(\Xi^s)}{a\lambda^3}[2H(x^s)\exp[a\lambda(\lambda x_2 - x_3)\operatorname{sgn}(\Xi^s)] \\ &\quad \times [\ln(H(x^s)) - H(x^s) + 2] \\ &\quad - 2\ln(H(x^s)) - a\lambda(\lambda x_2 - x_3)\operatorname{sgn}(\Xi^s) - 2] \end{aligned} \tag{6.121a}$$

where

$$H(x^s) = 1 + \sqrt{1 - [1 - a\lambda x_3 \operatorname{sgn}(\Xi^s)] \exp[-a\lambda(\lambda x_2 - x_3) \operatorname{sgn}(\Xi^s)]} \tag{6.121b}$$

and

$$\Xi^s = \Xi^s(x^s) = \begin{cases} \xi^s(x^s) = x_2 - \dfrac{x_3}{\lambda} - \dfrac{\operatorname{sgn}(x_3)}{a\lambda^2} \ln(1 - a\lambda |x_3|); & \xi^s(x^s) \neq 0 \\ x_3; & \xi^s(x^s) = 0 \end{cases} \tag{6.121c}$$

6.11 System (*f*): Subsystem IV-plus-lag: $\lambda_1 < 0, \lambda_2 = 0, \lambda_3 > 0$

It may be seen that the domain of null controllability $\mathscr{C} \subset \mathbb{R}^3$ in this case is given by

$$\mathscr{C} = \{x: x^s \in \mathscr{C}^s\} \tag{6.122a}$$

where, from (6.12),

$$\mathscr{C}^s = \left\{x^s: |x_3| < \frac{1}{a\lambda_3}\right\} \tag{6.122b}$$

i.e. the domain of null controllability is the open convex region of the state space $\mathbb{R}^3$ contained between the two parallel planes $P_1 = \{x: x_3 = 1/a\lambda_3\}$ and $P_2 = \{x: x_3 = -1/a\lambda_3\}$. Restricting x to this domain of null controllability and proceeding as in Subsection 6.10.1 yields

$$\xi(x) = x_1 + \frac{x_2}{\lambda_1} + \frac{x_3}{\lambda_1(\lambda_1 - \lambda_3)} - \frac{\operatorname{sgn}(\Xi^s)}{a\lambda_1^2(\lambda_1 - \lambda_3)} \times [\exp(-\lambda_1 T_2^s) - 2\exp(-\lambda_1 T_1^s) + 1] \tag{6.123}$$

where $\Xi^s: \mathscr{C}^s \to \mathbb{R}$ and $T_i^s \to [0, \infty)$ are given by (6.10) and (6.11). Substituting for T_i^s in (6.123) yields $\xi: \mathscr{C} \to \mathbb{R}$ in the explicit form

$$\xi(x) = x_1 + \frac{x_2}{\lambda_1} + \frac{x_3}{\lambda_1(\lambda_1 - \lambda_3)} - \frac{\operatorname{sgn}(\Xi^s)}{a\lambda_1^2(\lambda_1 - \lambda_3)} \times [\exp[a\lambda_1(\lambda_3 x_2 - x_3) \operatorname{sgn}(\Xi^s)][H(x^s)^2 - 2H(x^s) + 1]]$$

where

$$H(x^s) = [1 - \sqrt{1 - [1 - a\lambda_3 x_3 \operatorname{sgn}(\Xi^s)] \exp[-a\lambda_3(\lambda_3 x_2 - x_3) \operatorname{sgn}(\Xi^s)]}]^{\lambda_1/\lambda_3} \qquad x^s \in \mathscr{C}^s \tag{6.124b}$$

and

$$\Xi^s = \Xi^s(x^s) = \begin{cases} \xi^s(x^s) = x_2 - \dfrac{x_3}{\lambda_3} - \dfrac{\text{sgn}\,(x_3)}{a\lambda_3^2} \ln\,(1 - a\lambda_3 |x_3|); \; \xi^s(x^s) \neq 0 \\ x_3; \qquad \xi^s(x^s) = 0 \end{cases}$$

$$x^s \in \mathscr{C}^s \qquad (6.124c)$$

Chapter 7

Fourth-order time-optimal control system synthesis

7.1 Introduction

The investigations of the previous two chapters are extended to linear fourth-order single-input systems with real eigenvalues in the present chapter. Again, explicit algebraic forms for the time-optimal feedback controls (and associated switching hypersurfaces) are derived. The systems to be considered have the general structure of Fig. 4.3, viz.

$$\left.\begin{aligned} &\dot{x}_1 = \lambda_1 x_1 + x_2; \qquad \dot{x}_2 = \lambda_2 x_2 + x_3; \\ &\dot{x}_3 = \lambda_3 x_3 + x_4; \qquad \dot{x}_4 = \lambda_4 x_4 + \frac{u}{a}; \qquad |u| \leq 1; \qquad a > 0 \end{aligned}\right\} \tag{7.1a}$$

or

$$\dot{x} = Ax + bu; \qquad |u| \leq 1 \tag{7.1b}$$

where

$$A = \begin{bmatrix} \lambda_1 & 1 & 0 & 0 \\ 0 & \lambda_2 & 1 & 0 \\ 0 & 0 & \lambda_3 & 1 \\ 0 & 0 & 0 & \lambda_4 \end{bmatrix}; \qquad b = \begin{bmatrix} 0 \\ 0 \\ 0 \\ \frac{1}{a} \end{bmatrix} \tag{7.1c}$$

Again, recalling the synthesis technique based on the optimality of the strategy of Gulko *et al.* as outlined in Section 4.5, the derivation of the time-optimal feedback control presupposes knowledge of the third-order subsystem switching function Ξ^s, and the dependence of the subsystem switching times $t_1 = T_1^s(x^s)$, $t_2 = T_2^s(x^s)$ and final time $t_3 = T_3^s(x^s)$ on the initial state $x^s \in \mathscr{C}^s$ for the third-order subsystem defined, on elimination of the first member of the serially decomposed system equation set (7.1a), as

$$\left.\begin{aligned} \dot{x}_2 = \lambda_2 x_2 + x_3; \qquad \dot{x}_3 = \lambda_3 x_3 + x_4;& \\ \dot{x}_4 = \lambda_4 x_4 + \frac{u}{a}; \qquad |u| \leq 1; \qquad a > 0& \\ x^s = (x_2, x_3, x_4)' \in \mathscr{C}^s \subseteq \mathbb{R}^3& \end{aligned}\right\} \tag{7.2}$$

Third-order systems of the form (7.2) were studied in Chapter 6, and the associated third-order time-optimal switching functions were derived in a number of cases; however, it was possible to express the optimal switching and final times explicitly in terms of the initial state only in two third-order cases, viz. (i) triple integrator system and (ii) system with eigenvalues in simple ratio. Consequently, it proves possible to solve the time-optimal feedback control problem for the fourth-order system (7.1) only for those systems which exhibit one or other of the aforementioned third-order subsystems.

7.2 Third-order subsystems

7.2.1 Subsystem I: triple integrator, $\lambda_2 = \lambda_3 = \lambda_4 = 0$

Interpreting the results of Section 6.4, the optimal subsystem switching function is given by

$$\Xi^s = \Xi^s(x^s) = \begin{cases} \xi^s(x^s) = x_2 + a x_3 x_4 \operatorname{sgn}(\Xi^{ss}) + \dfrac{a^2 x_4^3}{3} + a^{-1} & \\ \quad \times [a x_3 \operatorname{sgn}(\Xi^{ss}) + \frac{1}{2}(a x_4)^2]^{3/2} \operatorname{sgn}(\Xi^{ss}); & \xi^s(x^s) \neq 0 \\ \Xi^{ss}(x_3, x_4); \qquad \xi^s(x^s) = 0 & \end{cases} \tag{7.3a}$$

where $\Xi^{ss}: \mathbb{R}^2 \to \mathbb{R}$ is the optimal switching function for the double integrator subsystem: $\dot{x}_3 = x_4; \dot{x}_4 = u/a$, viz.

$$\Xi^{ss}(x_3, x_4) = \begin{cases} \xi^{ss}(x_3, x_4) = x_3 + \frac{1}{2} a x_4 |x_4|; & \xi^{ss}(x_3, x_4) \neq 0 \\ x_4; & \xi^{ss}(x_3, x_4) = 0 \end{cases} \tag{7.3b}$$

Interpreting the results of Section 6.5, the optimal subsystem switching times T_1^s, T_2^s and final time T_3^s, (T_i^s: $\mathbb{R}^3 \to [0, \infty)$) are given by

$$T_3^s(x^s) = \begin{cases} S(x^s) + a x_4 \operatorname{sgn}(\xi^s(x^s)); & \xi^s(x^s) \neq 0 \\ T_2^{ss}(x_3, x_4); & \xi^s(x^s) = 0 \end{cases} \tag{7.4a}$$

$$T_2^s(x^s) = \begin{cases} \frac{3}{4} S(x^s) + \gamma_2(x^s)(S(x^s))^{-1} + a x_4 \operatorname{sgn}(\xi^s(x^s)); & \xi^s(x^s) \neq 0 \\ T_1^{ss}(x_3, x_4); & \xi^s(x^s) = 0 \end{cases} \tag{7.4b}$$

$$T_1^s(x^s) = \begin{cases} \frac{1}{4} S(x^s) + \gamma_2(x^s)(S(x^s))^{-1} + a x_4 \operatorname{sgn}(\xi^s(x^s)); & \xi^s(x^s) \neq 0 \\ T_1^{ss}(x_3, x_4); & \xi^s(x^s) = 0 \end{cases} \tag{7.4c}$$

where

$$S(x^s) = \sqrt{\tfrac{1}{4}\tilde{y} + 4\gamma_2}\ \mathrm{sgn}\ (\gamma_1) + [4\gamma_2 - \tfrac{1}{4}\tilde{y} + \sqrt{\tfrac{1}{4}\tilde{y}^2 + 16\gamma_2^2}]^{1/2} \tag{7.5a}$$

$$\tilde{y} = \tilde{y}(x^s) = \begin{cases} 8[f_1(x^s) + \sqrt{f_2(x^s)f_3(x^s)}]^{1/3} \\ \quad + 8[f_1(x^s) - \sqrt{f_2(x^s)f_3(x^s)}]^{1/3} - \tfrac{16}{3}\gamma_2(x^s); & \text{if } x^s \notin \Sigma_1 \\ \tfrac{16}{3}\gamma_2(x^s)\left[\cos\left(\dfrac{2\pi + \omega}{3}\right) - 1\right]; \\ \omega = \omega(x^s) = \tan^{-1}\left(\dfrac{\sqrt{|f_2(x^s)f_3(x^s)|}}{f_1(x^s)}\right); & \text{if } x^s \in \Sigma_1 \end{cases} \tag{7.5b}$$

with the range of the function $\tan^{-1}$ taken as $[0, \pi)$ with $\omega = \pi/2$ for $f_1(x^s) = 0$ and

$$f_1(x^s) = (\gamma_1(x^s))^2 - \tfrac{26}{27}(\gamma_2(x^s))^3 \tag{7.5c}$$

$$f_2(x^s) = (\gamma_1(x^s))^2 - (\gamma_2(x^s))^3 \tag{7.5d}$$

$$f_3(x^s) = (\gamma_1(x^s))^2 - \tfrac{25}{27}(\gamma_2(x^s))^3 \tag{7.5e}$$

$$\gamma_1(x^s) = ax_2\ \mathrm{sgn}\ (\xi^s) + a^2x_3x_4 + \tfrac{1}{3}(ax_4)^3\ \mathrm{sgn}\ (\xi^s) \tag{7.5f}$$

$$\gamma_2(x^s) = ax_3\ \mathrm{sgn}\ (\xi^s) + \tfrac{1}{2}(ax_4)^2 \tag{7.5g}$$

$$\Sigma_1 = \{x^s: (\gamma_1(x^s))^2 - (\gamma_2(x^s))^3 \le 0; (\gamma_1(x^s))^2 - \tfrac{25}{27}(\gamma_2(x^s))^3 > 0\} \tag{7.5h}$$

where $\xi^s: \mathbb{R}^3 \to \mathbb{R}$ is defined in (7.3a). The functions $T_i^{ss}: \mathbb{R}^2 \to [0, \infty)$ are the optimal switching and final times for the double integrator subsystem $\dot{x}_3 = x_4$, $\dot{x}_4 = u/a$, which, from (5.16), become

$$T_1^{ss}(x_3, x_4) = ax_4\ \mathrm{sgn}\ (\Xi^{ss}) + \sqrt{ax_3\ \mathrm{sgn}\ (\Xi^{ss}) + \tfrac{1}{2}(ax_4)^2} \tag{7.6a}$$

$$T_2^{ss}(x_3, x_4) = ax_4\ \mathrm{sgn}\ (\Xi^{ss}) + 2\sqrt{ax_3\ \mathrm{sgn}\ (\Xi^{ss}) + \tfrac{1}{2}(ax_4)^2} \tag{7.6b}$$

with $\Xi^{ss} = \Xi^{ss}(x_3, x_4)$ defined by (7.3b).

7.2.2 Subsystem II: Eigenvalues in simple ratio: $\lambda_2 = 3\lambda$; $\lambda_3 = 2\lambda$; $\lambda_1 = \lambda \neq 0$

Interpreting the results of Section 6.7, it may be concluded that the subsystem switching function $\Xi^s: \mathscr{C}^s \subseteq \mathbb{R}^3 \to \mathbb{R}$ in this case is given by (7.3), identical to the triple integrator subsystem (restricted to the subsystem domain of null controllability $\mathscr{C}^s \subset \mathbb{R}^3$ when $\lambda > 0$); while the functions $T_i^s: \mathscr{C}^s \subseteq \mathbb{R}^3 \to [0, \infty)$ are given by

$$T_3^s(x^s) = \begin{cases} -\dfrac{1}{\lambda}\ln[1 - \lambda[S(x^s) + ax_4\ \mathrm{sgn}\ (\xi^s)]]; & \xi^s = \xi^s(x^s) \neq 0 \\ T_2^{ss}(x_3, x_4); & \xi^s = \xi^s(x^s) = 0 \end{cases} \tag{7.7a}$$

$$T_2^s(x^s) = \begin{cases} -\dfrac{1}{\lambda} \ln [\![1 - \lambda[\tfrac{3}{4}S(x^s) + \gamma_2(x^s)(S(x^s))^{-1} + ax_4 \operatorname{sgn}(\xi^s)]]\!]; & \xi^s = \xi^s(x^s) \neq 0 \\ T_1^{ss}(x_3, x_4); & \xi^s = \xi^s(x^s) = 0 \end{cases} \tag{7.7b}$$

$$T_1^s(x^s) = \begin{cases} -\dfrac{1}{\lambda} \ln [\![1 - \lambda[\tfrac{1}{4}S(x^s) + \gamma_2(x^s)(S(x^s))^{-1} + ax_4 \operatorname{sgn}(\xi^s)]]\!]; & \xi^s = \xi^s(x^s) \neq 0 \\ T_1^{ss}(x_3, x_4); & \xi^s = \xi^s(x^s) = 0 \end{cases} \tag{7.7c}$$

with S, γ_2 and ξ^s defined as in (7.3) and (7.5) and where T_i^{ss} are the switching and final times for the second-order subsystem: $\dot{x}_3 = 2\lambda x_3 + x_4$; $\dot{x}_4 = \lambda x_4 + u/a$, i.e.

$$\begin{aligned} T_2^{ss} &= T_2^{ss}(x_3, x_4) \\ &= -\frac{1}{\lambda} \ln [\![1 - \lambda[ax_4 \operatorname{sgn}(\Xi^{ss}) + 2\sqrt{ax_3 \operatorname{sgn}(\Xi^{ss}) + \tfrac{1}{2}(ax_4)^2}]]\!] \end{aligned} \tag{7.8a}$$

$$\begin{aligned} T_1^{ss} &= T_1^{ss}(x_3, x_4) \\ &= -\frac{1}{\lambda} \ln [\![1 - \lambda[ax_4 \operatorname{sgn}(\Xi^{ss}) + \sqrt{ax_3 \operatorname{sgn}(\Xi^{ss}) + \tfrac{1}{2}(ax_4)^2}]]\!] \end{aligned} \tag{7.8b}$$

Finally, from (6.86) and (6.88), the subsystem domain of null controllability may be defined as

$$\mathscr{C}^s = \begin{cases} \mathbb{R}^3; & \lambda < 0 \\ \left\{x^s\colon S(x^s) + ax_4 \operatorname{sgn}(\xi^s) < \dfrac{1}{\lambda};\ \xi^s = \xi^s(x^s) \neq 0\right\} \cup \\ \left\{x^s\colon ax_4 \operatorname{sgn}(\Xi^{ss}) + 2\sqrt{ax_3 \operatorname{sgn}(\Xi^{ss}) + \tfrac{1}{2}(ax_4)^2} < \dfrac{1}{\lambda};\ \xi^s(x^s) = 0\right\}; & \lambda > 0 \end{cases} \tag{7.9}$$

7.3 Fourth-order time-optimal synthesis

Again recalling the general approach of Subsection 4.5.2, the time-optimal feedback control for the fourth-order system (7.1) may be explicitly obtained in those cases where λ_2, λ_3, λ_4 are such that they define one or other of the above third-order subsystems. Specifically, from (4.49), the time-optimal switching function $\Xi : \mathscr{C} \subseteq \mathbb{R}^4 \to \mathbb{R}$ is given by

$$\Xi(x) = \begin{cases} \xi(x); & \xi(x) \neq 0 \\ \Xi^s(x^s); & \xi(x) = 0 \end{cases} \tag{7.10}$$

where, from (4.48), $\xi(x) = \exp(-\lambda_1 T_3^s(x^s))\beta(x)$ and

$$\begin{bmatrix} \beta(x) \\ 0 \\ 0 \\ 0 \end{bmatrix} = \exp(AT_3^s) \left[\!\left[x - \operatorname{sgn}(\Xi^s) \left[\int_0^{T_1^s} \exp(-As)b\, ds - \int_{T_1^s}^{T_2^s} \exp(-As)b\, ds + \int_{T_2^s}^{T_3^s} \exp(-As)b\, ds \right] \right]\!\right] \tag{7.11}$$

with Ξ^s and T_1^s, T_2^s, T_3^s defined as for Subsystem I or Subsystem II of Section 7.2 as appropriate. As in the previous chapter, determination of the function β in (7.11) is sometimes facilitated by the introduction of a nonsingular matrix P such that

$$P^{-1} \begin{bmatrix} \beta(x) \\ 0 \\ 0 \\ 0 \end{bmatrix} = \begin{bmatrix} \beta(x) \\ 0 \\ 0 \\ 0 \end{bmatrix}$$

which transforms (7.11) into an equivalent but more convenient form

$$\begin{bmatrix} \beta(x) \\ 0 \\ 0 \\ 0 \end{bmatrix} = \exp(\Lambda T_3^s) \left[\!\left[P^{-1}x - \operatorname{sgn}(\Xi^s) \times \left[\int_0^{T_1^s} \exp(-\Lambda s)P^{-1}b\, ds - \int_{T_1^s}^{T_2^s} \exp(-\Lambda s)P^{-1}b + \int_{T_2^s}^{T_3^s} \exp(-\Lambda s)P^{-1}b\, ds \right] \right]\!\right] \tag{7.12}$$

where $\Lambda = P^{-1}AP$ is a Jordan matrix.

On the above basis, the time-optimal switching functions may be determined as in Ryan (1977b), for the following systems (where $G(s)$ denotes the transfer function from the Laplace transformed input u to output x_1):

(*a*) Fourth-order integrator: $G(s) = \dfrac{1}{as^4}$

(*b*) Eigenvalues in simple ratio:

$$G(s) = \frac{1}{a(s-4\lambda)(s-3\lambda)(s-2\lambda)(s-\lambda)}; \qquad \lambda \neq 0$$

(*c*) Subsystem I and first-order element in series:

$$G(s) = \frac{1}{as^3(s-\lambda)}; \qquad \lambda \neq 0$$

(*d*) Subsystem II and single integrator in series:

$$G(s) = \frac{1}{as(s-3\lambda)(s-2\lambda)(s-\lambda)}; \qquad \lambda \neq 0$$

(*e*) Subsystem II and first-order element in series:

$$G(s) = \frac{1}{a(s-\lambda_1)(s-3\lambda)(s-2\lambda)(s-\lambda)}; \qquad \lambda_1, \lambda \neq 0$$

7.4 System (*a*): Fourth-order integrator

In this case, the system eigenvalues are zero, $\lambda_i = 0$, $i = 1, 2, 3, 4$. The state equations are

$$\dot{x}_1 = x_2; \qquad \dot{x}_2 = x_3; \qquad \dot{x}_3 = x_4; \qquad \dot{x}_4 = \frac{u}{a} \tag{7.13}$$

Clearly, the third-order subsystem exhibited by system (7.13) corresponds to Subsystem I of Section 7.2. For system (7.13),

$$\exp(At) = \begin{bmatrix} 1 & t & \dfrac{t^2}{2} & \dfrac{t^3}{6} \\ 0 & 1 & t & \dfrac{t^2}{2} \\ 0 & 0 & 1 & t \\ 0 & 0 & 0 & 1 \end{bmatrix}$$

and evaluating (7.11) yields

$$\xi(x) = \beta(x) = x_1 + x_2 T_3^s + \tfrac{1}{2}x_3(T_3^s)^2 + \tfrac{1}{6}x_4(T_3^s)^3 - \frac{\text{sgn}\,(\Xi^s)}{24a}\left[(T_3^s)^4 - 2(T_3^s - T_1^s)^4 + 2(T_3^s - T_2^s)^4\right] \quad (7.14)$$

Substituting for T_1^s, T_2^s, T_3^s (given by (7.4)) and rearranging terms yields

$$\xi(x) = \alpha + \beta S + \tfrac{35}{96}\gamma S^2 + \tfrac{1}{4}\gamma^2\,\text{sgn}\,(\Xi^s) - \frac{\gamma^3}{6S^2} - \frac{S^4}{64}\,\text{sgn}\,(\Xi^s) \quad (7.15a)$$

where

$$\left.\begin{aligned} \alpha &= ax_1 + a^2x_2x_4\,\text{sgn}\,(\Xi^s) + \tfrac{1}{2}a^3x_3x_4^2 + \tfrac{1}{8}(ax_4)^4\,\text{sgn}\,(\Xi^s) \\ \beta &= ax_2 + a^2x_3x_4\,\text{sgn}\,(\Xi^s) + \frac{(ax_4)^3}{3} \\ \gamma &= ax_3 + \tfrac{1}{2}(ax_4)^2\,\text{sgn}\,(\Xi^s) \\ S &= S(x^s) \end{aligned}\right\} \quad (7.15b)$$

with Ξ^s and $S(x^s)$ defined as in (7.3) and (7.5). It remains to interpret the term $\gamma^3/6S^2$ for the special case of $S = 0$. If $S = 0$ it may easily be verified that the state point must lie on the unique p-path or the unique n-path which leads directly to the state origin. Furthermore, under such conditions it may be verified that $\text{sgn}\,(\Xi^s) = \text{sgn}\,(x_4)$ and $ax_3 = -\tfrac{1}{2}(ax_4)^2\,\text{sgn}\,(x_4) \Rightarrow \gamma = 0$. Now γ is of order S^2 and hence $\gamma^3/6S^2 = (\gamma^2/6)(\gamma/S^2) \to 0$ as $\gamma, S \to 0$.

With $\xi\colon \mathbb{R}^4 \to \mathbb{R}$ and $\Xi^s\colon \mathbb{R}^3 \to \mathbb{R}$ defined by (7.15) and (7.3), respectively, the time-optimal switching function is given by (7.10). For a gain parameter value $a = 1$, the time-optimal response, with initial state $x(0) = (1, 0, 0, 0)'$ and obtained by digital simulation of the complete system under the feedback control $u(x) = -\text{sgn}\,(\Xi(x))$, is shown in Fig. 7.1. Such digital simulation of the feedback

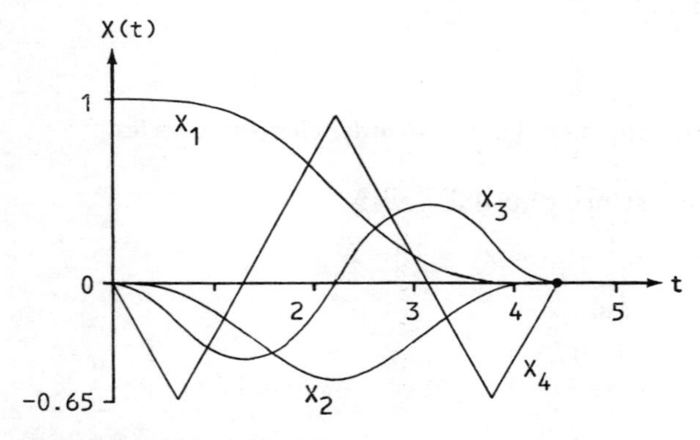

Fig. 7.1 *Time-optimal response of system (a)*
a = 1, *x*(0) = (1, 0, 0, 0)′

system (also used to produce the time-optimal responses of Figs. 7.3 to 7.6) provides a valuable check on the correctness of the associated switching function.

7.5 System (*b*): Eigenvalues in simple ratio

The eigenvalues λ_i in system equations (7.1) are now assumed to be in the simple ratio 4 : 3 : 2 : 1, i.e.

$$\dot{x}_1 = 4\lambda x_1 + x_2; \qquad \dot{x}_2 = 3\lambda x_2 + x_3; \qquad \dot{x}_3 = 2\lambda x_3 + x_4;$$
$$\dot{x}_4 = \lambda x_4 + \frac{u}{a}; \qquad |u| \leq 1; \qquad a > 0 \tag{7.16}$$

with $\lambda =$ constant $\neq 0$.

Assuming initially that $\lambda < 0$, i.e. that the plant to be controlled is stable, it may be concluded from theorem 4.3.1 that all states are null controllable, i.e. $\mathscr{C} = \mathbb{R}^4$. Furthermore, the system equivalence property of Section 3.6 can be invoked to establish that the time-optimal feedback control is independent of the parameter λ and is identical to that of the fourth-order integrator system (system (*a*)) of the previous section. In particular, the time-optimal feedback control is given by

$$u(x) = -\operatorname{sgn}(\Xi(x)); \qquad \Xi(x) = \begin{cases} \xi(x); & \xi(x) \neq 0 \\ \Xi^s(x^s); & \xi(x) = 0 \end{cases} \tag{7.17}$$

with $\xi\colon \mathbb{R}^4 \to \mathbb{R}$ and $\Xi^s\colon \mathbb{R}^3 \to \mathbb{R}$ defined by (7.15) and (7.3).

If the plant to be controlled is unstable, i.e. if $\lambda > 0$, then the time-optimal feedback control is again given by (7.17), but now restricted to the domain of null controllability $\mathscr{C}$. However, in this case it is not clear how to characterise the set $\mathscr{C}$ explicitly.

7.6 System (*c*): Subsystem I and first-order element in series

In this case the system eigenvalues satisfy

$$\lambda_1 = \lambda \neq 0; \qquad \lambda_2 = \lambda_3 = \lambda_4 = 0 \tag{7.18}$$

The state equations are

$$\dot{x}_1 = \lambda x_1 + x_2; \qquad \dot{x}_2 = x_3; \qquad \dot{x}_3 = x_4; \qquad \dot{x}_4 = \frac{u}{a} \tag{7.19}$$

The third-order subsystem exhibited by system (7.19) corresponds to Subsystem I of Subsection 7.2.1, as depicted in Fig. 7.2*a*.

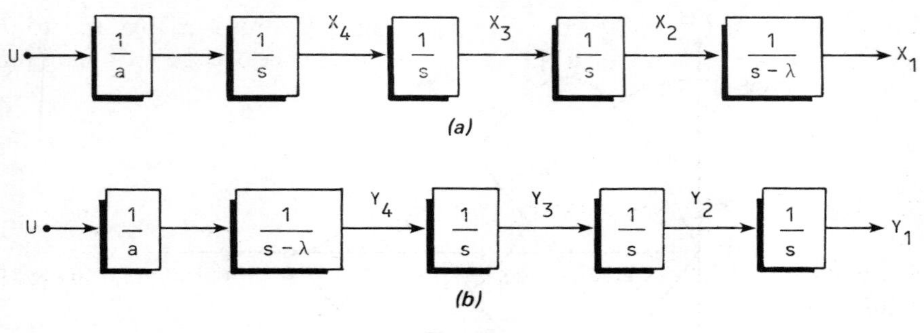

Fig. 7.2

Assuming initially that $\lambda < 0$ (i.e. stable plant), evaluating (7.11) yields

$$\begin{aligned}
\xi(x) &= \exp(-\lambda T_3^s)\beta(x) \\
&= x_1 + \frac{x_2}{\lambda}[1 - \exp(-\lambda T_3^s(x^s))] \\
&\quad + \frac{x_3}{\lambda}\left[\frac{1}{\lambda} - \exp(-\lambda T_3^s(x^s))\left[T_3^s(x^s) + \frac{1}{\lambda}\right]\right] \\
&\quad + \frac{x_4}{\lambda}\left[\frac{1}{\lambda^2} - \exp(-\lambda T_3^s(x^s))\left[\tfrac{1}{2}(T_3^s(x^s))^2 + \frac{1}{\lambda}T_3^s(x^s) + \frac{1}{\lambda^2}\right]\right] \\
&\quad - \frac{\operatorname{sgn}(\Xi^s)}{a\lambda}\left[\frac{1}{\lambda^3}[1 - 2\exp(-\lambda T_1^s(x^s)) + 2\exp(-\lambda T_2^s(x^s))\right. \\
&\quad - \exp(-\lambda T_3^s(x^s))] \\
&\quad - \frac{1}{\lambda^2}\exp(-\lambda T_3^s(x^s))[T_3^s(x^s) - 2T_2^s(x^s) + 2T_1^s(x^s)] \\
&\quad - \frac{1}{2\lambda}\exp(-\lambda T_3^s(x^s)) \\
&\quad \times [\{T_3^s(x^s)\}^2 - 2\{T_3^s(x^s) - T_1^s(x^s)\}^2 + 2\{T_3^s(x^s) - T_2^s(x^s)\}^2] \\
&\quad - \tfrac{1}{6}\exp(-\lambda T_3^s(x^s)) \\
&\quad \left.\times [\{T_3^s(x^s)\}^3 - 2\{T_3^s(x^s) - T_1^s(x^s)\}^3 + 2\{T_3^s(x^s) - T_2^s(x^s)\}^3]\right]
\end{aligned} \tag{7.20}$$

where $\Xi^s: \mathbb{R}^3 \to \mathbb{R}$ and $T_i^s: \mathbb{R}^3 \to [0, \infty)$ are given by (7.3) and (7.4).

With $\xi: \mathbb{R}^4 \to \mathbb{R}$ and $\Xi^s: \mathbb{R}^3 \to \mathbb{R}$ defined by (7.20) and (7.3), respectively, the time-optimal switching function $\Xi: \mathbb{R}^4 \to \mathbb{R}$ is given by (7.10). For system parameter values $a = 2$ and $\lambda = -\frac{1}{2}$, Fig. 7.3 depicts the time-optimal response,

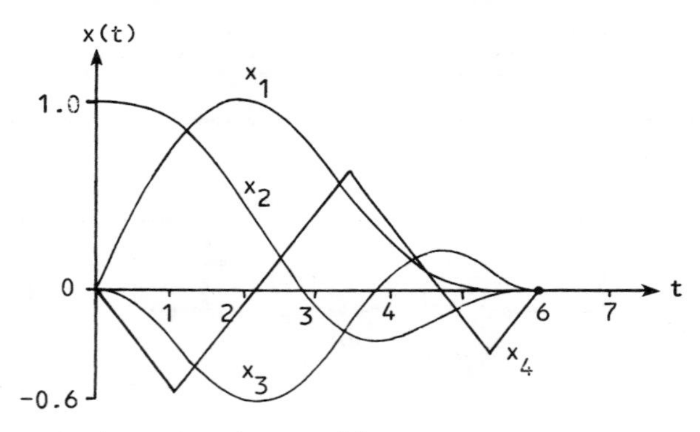

Fig. 7.3 *Time-optimal response of system (c)*
$a = 2, \lambda = -\frac{1}{2}, x(0) = (0, 1, 0, 0)'$

with initial state $x(0) = (0, 1, 0, 0)'$, generated by the feedback control $u(x) = -\operatorname{sgn}(\Xi(x))$.

If the plant to be controlled is *unstable*, i.e. if $\lambda > 0$, then (7.20) is again valid *when restricted to the domain of null controllability* $\mathscr{C}$ which can be characterised as follows.

Under the linear state transformation

$$\begin{bmatrix} y_1 \\ y_2 \\ y_3 \\ y_4 \end{bmatrix} = \begin{bmatrix} 1 & 0 & 0 & 0 \\ \lambda & 1 & 0 & 0 \\ \lambda^2 & \lambda & 1 & 0 \\ \lambda^3 & \lambda^2 & \lambda & 1 \end{bmatrix} \begin{bmatrix} x_1 \\ x_2 \\ x_3 \\ x_4 \end{bmatrix} \tag{7.21}$$

the system representation is of the form depicted in Fig. 7.2*b* which, with $\lambda > 0$, consists of an unstable first-order sybsystem $(\dot{y}_4 = \lambda y_4 + u/a)$ in cascade with a triple integrator subsystem. In terms of the y state vector, the domain of null controllability $\mathscr{C}$ is comprised of the set of vectors y for which the fourth component y_4 lies in the domain of null controllability of the unstable first-order subsystem, viz. $|y_4| < 1/a\lambda$ (see (5.6) and footnote on page 188). Hence, in view of (7.21), the domain of null controllability for $\lambda > 0$ may be expressed, in terms of the original x state vector, as

$$\mathscr{C} = \left\{ x: |\lambda^3 x_1 + \lambda^2 x_2 + \lambda x_3 + x_4| < \frac{1}{a\lambda} \right\}; \qquad \lambda > 0 \tag{7.22}$$

i.e. the region of the x state space lying between the parallel planes $\lambda^3 x_1 + \lambda^2 x_2 + \lambda x_3 + x_4 = \pm 1/a\lambda$.

7.7 System (*d*): Subsystem II and single integrator in series

In this case the system eigenvalues are given by

$$\lambda_1 = 0; \qquad \lambda_2 = 3\lambda; \qquad \lambda_3 = 2\lambda; \qquad \lambda_4 = \lambda \qquad (\lambda = \text{constant} \neq 0) \tag{7.23}$$

It is assumed initially that the plant to be controlled is stable, i.e. $\lambda < 0$, so that all states are null controllable ($\mathscr{C} = \mathbb{R}^4$).

Considerable simplification of the analysis is achieved by the introduction of the diagonalising transformation matrix

$$P = \begin{bmatrix} 1 & 1 & 1 & 1 \\ 0 & 3\lambda & 2\lambda & \lambda \\ 0 & 0 & -2\lambda^2 & -2\lambda^2 \\ 0 & 0 & 0 & 2\lambda^3 \end{bmatrix} \tag{7.24}$$

with the properties

$$P^{-1} \begin{bmatrix} \beta(x) \\ 0 \\ 0 \\ 0 \end{bmatrix} = \begin{bmatrix} \beta(x) \\ 0 \\ 0 \\ 0 \end{bmatrix} \tag{7.25a}$$

and

$$P^{-1} \exp(At)P = \exp(P^{-1}APt) = \exp(\Lambda t) \tag{7.25b}$$

where

$$\Lambda = P^{-1}AP = \begin{bmatrix} 0 & 0 & 0 & 0 \\ 0 & 3\lambda & 0 & 0 \\ 0 & 0 & 2\lambda & 0 \\ 0 & 0 & 0 & \lambda \end{bmatrix} \quad \text{and}$$

$$\exp(\Lambda t) = \begin{bmatrix} 1 & 0 & 0 & 0 \\ 0 & \exp(3\lambda t) & 0 & 0 \\ 0 & 0 & \exp(2\lambda t) & 0 \\ 0 & 0 & 0 & \exp(\lambda t) \end{bmatrix} \tag{7.25c}$$

Combining (7.12), (7.24) and (7.25) yields the result

$$\xi(x) = \beta(x) = x_1 - \frac{x_2}{3\lambda} + \frac{x_3}{6\lambda^2} - \frac{x_4}{6\lambda^3} + \frac{\operatorname{sgn}(\Xi^s)}{6a\lambda^3} \times [T_3^s(x^s) - 2T_2^s(x^s) + 2T_1^s(x^s)] \tag{7.26}$$

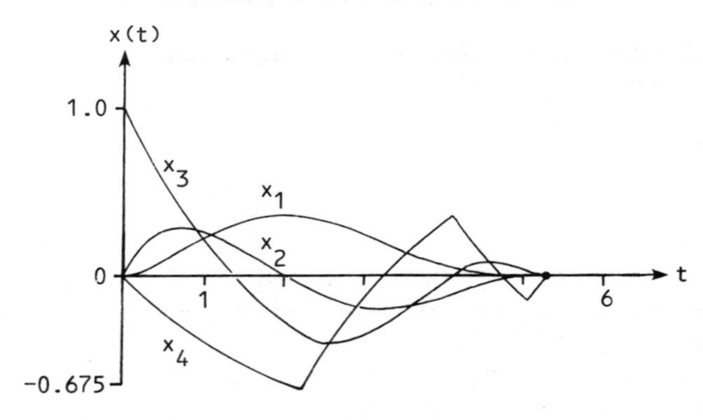

Fig. 7.4 *Time-optimal response of system (d)*
$a = 2, \lambda = -\frac{1}{2}, x(0) = (0, 0, 1, 0)'$

where Ξ^s: $\mathbb{R}^3 \to \mathbb{R}$ and T_i^s: $\mathbb{R}^3 \to [0, \infty)$ are given by (7.3) and (7.7). Again, with ξ: $\mathbb{R}^4 \to \mathbb{R}$ and Ξ^s: $\mathbb{R}^3 \to \mathbb{R}$ defined by (7.26) and (7.3), respectively, the time-optimal switching function Ξ: $\mathbb{R}^4 \to \mathbb{R}$ is given by (7.10). For system parameter values $a = 2$ and $\lambda = -\frac{1}{2}$, Fig. 7.4 depicts the time-optimal response, with initial state $x(0) = (0, 0, 1, 0)'$, generated by the feedback control $u(x) = -\text{sgn}(\Xi(x))$.

Now suppose that the plant to be controlled is unstable, i.e. $\lambda > 0$. In this case, (7.26) is again valid when restricted to the domain of null controllability $\mathscr{C}$ which may be expressed as

$$\mathscr{C} = \{x\colon x^s = (x_2, x_3, x_4)' \in \mathscr{C}^s\} \tag{7.27}$$

where $\mathscr{C}^s$ is the subsystem domain of null controllability defined by (7.9) with $\lambda > 0$.

7.8 System (*e*): Subsystem II and first-order element in series

In this case the system eigenvalues satisfy†

$$\lambda_1 \neq 0; \qquad \lambda_2 = 3\lambda; \qquad \lambda_3 = 2\lambda; \qquad \lambda_4 = \lambda \qquad (\lambda = \text{constant} \neq 0) \tag{7.28}$$

It is initially assumed that the plant to be controlled is stable, i.e. λ_1, $\lambda < 0$ so that all states are null controllable ($\mathscr{C} = \mathbb{R}^4$). The case of distinct eigenvalues and the three possible cases of repeated eigenvalues will be treated separately.

† Note that the specific case of (7.28) for which $\lambda_1 = 4\lambda$ has previously been considered in Section 7.5.

7.8.1 Stable system (i) : distinct eigenvalues $\lambda_1 \neq \lambda_2, \lambda_3, \lambda_4$
Introducing the diagonalising matrix

$$P = \begin{bmatrix} 1 & 1 & 1 & 1 \\ 0 & -(\lambda_1 - 3\lambda) & -(\lambda_1 - 2\lambda) & -(\lambda_1 - \lambda) \\ 0 & 0 & \lambda(\lambda_1 - 2\lambda) & 2\lambda(\lambda_1 - \lambda) \\ 0 & 0 & 0 & -2\lambda^2(\lambda_1 - \lambda) \end{bmatrix} \tag{7.29}$$

with the properties (7.25a) and (7.25b) where, in this case,

$$\Lambda = \begin{bmatrix} \lambda_1 & 0 & 0 & 0 \\ 0 & 3\lambda & 0 & 0 \\ 0 & 0 & 2\lambda & 0 \\ 0 & 0 & 0 & \lambda \end{bmatrix} \quad \text{and}$$

$$\exp(\Lambda t) = \begin{bmatrix} \exp(\lambda_1 t) & 0 & 0 & 0 \\ 0 & \exp(3\lambda t) & 0 & 0 \\ 0 & 0 & \exp(2\lambda t) & 0 \\ 0 & 0 & 0 & \exp(\lambda t) \end{bmatrix} \tag{7.30}$$

then (7.12) yields the result

$$\begin{aligned} \xi(x) &= \exp(\lambda_1 T_3^s(x^s))\beta(x) \\ &= x_1 + \frac{x_2}{(\lambda_1 - 3\lambda)} + \frac{x_3}{(\lambda_1 - 3\lambda)(\lambda_1 - 2\lambda)} \\ &\quad + \frac{x_4}{(\lambda_1 - 3\lambda)(\lambda_1 - 2\lambda)(\lambda_1 - \lambda)} \\ &\quad + \frac{\operatorname{sgn}(\Xi^s)}{a\lambda_1(\lambda_1 - 3\lambda)(\lambda_1 - 2\lambda)(\lambda_1 - \lambda)} \\ &\quad \times [\exp(-\lambda_1 T_3^s(x^s)) - 2\exp(-\lambda_1 T_2^s(x^s)) \\ &\quad + 2\exp(-\lambda_1 T_1^s(x^s)) - 1] \end{aligned} \tag{7.31}$$

where $\Xi^s\colon \mathbb{R}^3 \to \mathbb{R}$ and $T_i^s\colon \mathbb{R}^3 \to [0, \infty)$ are defined by (7.3) and (7.7).
Under the associated time-optimal feedback control

$$u(x) = -\operatorname{sgn}(\Xi(x)); \quad \Xi(x) = \begin{cases} \xi(x); & \xi(x) \neq 0 \\ \Xi^s(x^s); & \xi(x) = 0 \end{cases}$$

Fig. 7.5 depicts the computed time-optimal response from an initial state $x(0) = (0, 0, 0, 1)'$ for system parameter values $a = 2, \lambda_1 = -0{\cdot}5, \lambda = -0{\cdot}1$.

7.8.2 Stable system (ii) : repeated eigenvalues; $\lambda_1 = \lambda_2 = 3\lambda < 0$
Introducing the matrix

$$P = \begin{bmatrix} 1 & 0 & 1 & 1 \\ 0 & 1 & -\lambda & -2\lambda \\ 0 & 0 & \lambda^2 & 4\lambda^2 \\ 0 & 0 & 0 & -4\lambda^3 \end{bmatrix} \qquad (7.32)$$

with properties (7.25a) and (7.25b) where, in this case,

$$\Lambda = \begin{bmatrix} 3\lambda & 1 & 0 & 0 \\ 0 & 3\lambda & 0 & 0 \\ 0 & 0 & 2\lambda & 0 \\ 0 & 0 & 0 & \lambda \end{bmatrix} \quad \text{and}$$

$$\exp(\Lambda t) = \begin{bmatrix} \exp(3\lambda t) & t\exp(3\lambda t) & 0 & 0 \\ 0 & \exp(3\lambda t) & 0 & 0 \\ 0 & 0 & \exp(2\lambda t) & 0 \\ 0 & 0 & 0 & \exp(\lambda t) \end{bmatrix} \qquad (7.33)$$

then (7.12) yields the result

$$\begin{aligned} \xi(x) &= \exp(-3\lambda T_3^s(x^s))\beta(x) \\ &= x_1 - \frac{x_3}{\lambda^2} - \frac{3x_4}{4\lambda^3} + \left[x_2 + \frac{x_3}{\lambda} + \frac{x_4}{2\lambda^2}\right] T_3^s(x^s) \\ &\quad - \frac{\operatorname{sgn}(\Xi^s(x^s))}{6a\lambda^3}\Bigg[\Bigg[T_3^s(x^s) - 2[T_3^s(x^s) - T_1^s(x^s)]\exp(-3\lambda T_1^s(x^s)) \\ &\quad + 2[T_3^s(x^s) - T_2^s(x^s)]\exp(-3\lambda T_2^s(x^s)) \\ &\quad + \frac{11}{6\lambda}[\exp(-3\lambda T_3^s(x^s)) - 2\exp(-3\lambda T_2^s(x^s)) \\ &\quad + 2\exp(-3\lambda T_1^s(x^s)) - 1]\Bigg]\Bigg] \end{aligned} \qquad (7.34)$$

where $\Xi^s \colon \mathbb{R}^3 \to \mathbb{R}$ and $T_i^s \colon \mathbb{R}^3 \to [0, \infty)$ are given by (7.3) and (7.7). Under the associated time-optimal feedback control

$$u(x) = -\operatorname{sgn}(\Xi(x)); \qquad \Xi(x) = \begin{cases} \xi(x); & \xi(x) \neq 0 \\ \Xi^s(x^s); & \xi(x) = 0 \end{cases}$$

Fig. 7.6 depicts the computed time-optimal response from an initial state $x(0) = (1, 0, 0, -1)'$ for system parameter values $a = 2$ and $\lambda = -0{\cdot}1$.

7.8.3 Stable system (iii) : repeated eigenvalues; $\lambda_1 = \lambda_3 = 2\lambda < 0$
Following a similar procedure to that of the previous case, the function ξ:

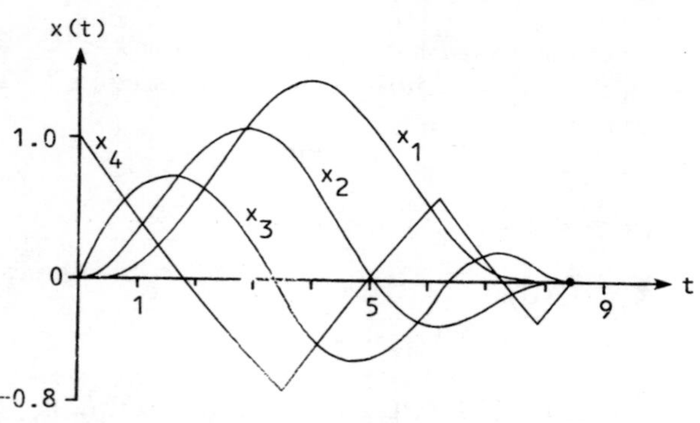

Fig. 7.5 *Time-optimal response of system (e)*
$a = 2, \lambda_1 = -0{\cdot}5, \lambda = -0{\cdot}1, x(0) = (0, 0, 0, 1)'$

$\mathbb{R}^4 \rightarrow \mathbb{R}$ is found to be

$$\xi(x) = x_1 - \frac{x_2}{\lambda} - \frac{x_3}{\lambda^2} - \left[\frac{x_3}{\lambda} + \frac{x_4}{\lambda^2}\right] T_3^s(x^s)$$
$$+ \frac{\text{sgn}\,(\Xi^s(x^s))}{2a\lambda^3}$$
$$\times \Big[\!\Big[T_3^s(x^s) - 2[T_3^s(x^s) - T_1^s(x^s)] \exp\,(-2\lambda T_1^s(x^s))$$
$$+ 2[T_3^s(x^s) - T_2^s(x^s)] \exp\,(-2\lambda T_2^s(x^s))$$
$$+ \frac{1}{2\lambda} [\exp\,(-2\lambda T_3^s(x^s)) - 2 \exp\,(-2\lambda T_2^s(x^s))$$
$$+ 2 \exp\,(-2\lambda T_1^s(x^s)) - 1] \Big]\!\Big] \tag{7.35}$$

where Ξ^s: $\mathbb{R}^3 \rightarrow \mathbb{R}$ and T_i^s: $\mathbb{R}^3 \rightarrow [0, \infty)$ are again given by (7.3) and (7.7).

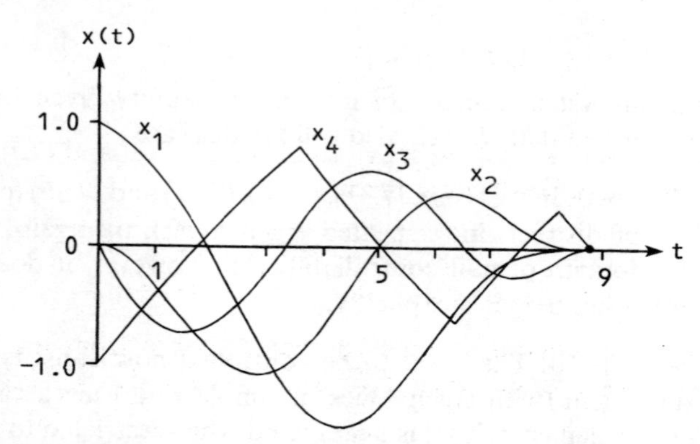

Fig. 7.6 *Time-optimal response of system (e)*
$a = 2, \lambda_1 = -0{\cdot}3, \lambda = -0{\cdot}1, x(0) = (1, 0, 0, -1)'$

7.8.4 Stable system (iv): repeated eigenvalues; $\lambda_1 = \lambda_4 = \lambda < 0$

Again, following a similar procedure to that of Subsection 7.8.2, the function $\xi\colon \mathbb{R}^4 \to \mathbb{R}$ is found to be

$$\xi(x) = x_1 - \frac{x_2}{2\lambda} + \frac{x_3}{2\lambda^2} + \frac{3x_4}{4\lambda^3} + \frac{x_4}{2\lambda^2} T_3^s(x^s)$$
$$- \frac{\operatorname{sgn}(\Xi^s(x^s))}{2a\lambda^3}$$
$$\times \Bigg[T_3^s(x^s) - 2[T_3^s(x^s) - T_1^s(x^s)] \exp(-\lambda T_1^s(x^s))$$
$$+ 2[T_3^s(x^s) - T_2^s(x^s)] \exp(-\lambda T_2^s(x^s))$$
$$- \frac{1}{2\lambda} [\exp(-\lambda T_3^s(x^s)) - 2\exp(-\lambda T_2^s(x^s))$$
$$+ 2\exp(-\lambda T_1^s(x^s)) - 1] \Bigg] \tag{7.36}$$

where $\Xi^s\colon \mathbb{R}^3 \to \mathbb{R}$ and $T_i^s\colon \mathbb{R}^3 \to [0, \infty)$ are again given by (7.3) and (7.7).

7.8.5 Unstable plants

Finally, the cases in which the plant to be controlled is unstable will be briefly discussed.

Case (i): $\lambda_1 < 0$; $\lambda > 0$: In this case, the plant to be controlled consists of an unstable third-order subsystem (with eigenvalues in simple ratio) in cascade with a stable first-order element. When restricted to the domain of null controllability

$$\mathscr{C} = \{x\colon x^s \in \mathscr{C}^s\}$$

where $\mathscr{C}^s$ is the subsystem domain of null controllability given by (7.9) with $\lambda > 0$, it may be verified that (7.31) is also valid in this case.

Case (ii): $\lambda_1 > 0$; $\lambda > 0$: In this case, (7.31), (7.34), (7.35) and (7.36) (corresponding to the cases of distinct and repeated eigenvalues) are again valid when restricted to the domain of null controllability $\mathscr{C}$; however, it does not seem possible to characterise the set $\mathscr{C}$ explicitly.

Case (iii): $\lambda_1 > 0$; $\lambda < 0$: The plant to be controlled now consists of a third-order stable subsystem (with eigenvalues in simple ratio) in cascade with an unstable first-order element; (7.31) is again valid when restricted to the domain of null controllability $\mathscr{C}$ which can be characterised as follows.

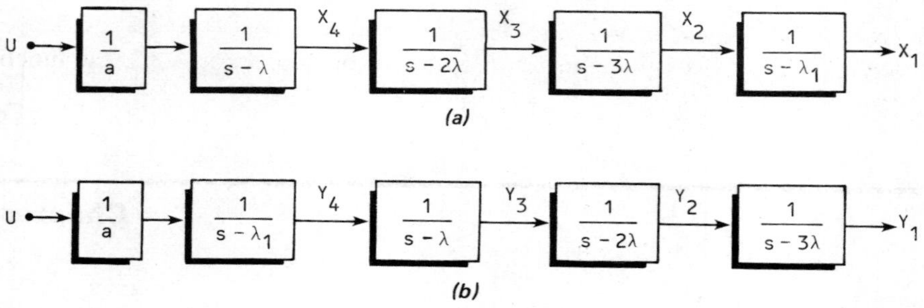

U → $\frac{1}{a}$ → $\frac{1}{s-\lambda}$ → X_4 → $\frac{1}{s-2\lambda}$ → X_3 → $\frac{1}{s-3\lambda}$ → X_2 → $\frac{1}{s-\lambda_1}$ → X_1

(a)

U → $\frac{1}{a}$ → $\frac{1}{s-\lambda_1}$ → Y_4 → $\frac{1}{s-\lambda}$ → Y_3 → $\frac{1}{s-2\lambda}$ → Y_2 → $\frac{1}{s-3\lambda}$ → Y_1

(b)

Fig. 7.7

Under the state transformation

$$\begin{bmatrix} y_1 \\ y_2 \\ y_3 \\ y_4 \end{bmatrix} = \begin{bmatrix} 1 & 0 & 0 & 0 \\ (\lambda_1 - 3\lambda) & 1 & 0 & 0 \\ (\lambda_1 - 3\lambda)(\lambda_1 - 2\lambda) & (\lambda_1 - 2\lambda) & 1 & 0 \\ (\lambda_1 - 3\lambda)(\lambda_1 - 2\lambda)(\lambda_1 - \lambda) & (\lambda_1 - 2\lambda)(\lambda_1 - \lambda) & (\lambda_1 - \lambda) & 1 \end{bmatrix} \begin{bmatrix} x_1 \\ x_2 \\ x_3 \\ x_4 \end{bmatrix} \tag{7.37}$$

the system representation is transformed from that of Fig. 7.7*a* to that of Fig. 7.7*b*. In terms of the y state vector, the domain of null controllability for $\lambda_1 > 0$ may be expressed as (see (5.20))

$$\mathscr{C} = \left\{ y: |y_4| < \frac{1}{a\lambda_1} \right\}$$

which, in terms of the original x state vector, becomes (via (7.37))

$$\mathscr{C} = \left\{ x: |(\lambda_1 - 3\lambda)(\lambda_1 - 2\lambda)(\lambda_1 - \lambda)x_1 + (\lambda_1 - 2\lambda)(\lambda_1 - \lambda)x_2 + (\lambda_1 - \lambda)x_3 + x_4| < \frac{1}{a\lambda_1} \right\} \tag{7.38}$$

i.e. the region of the x-state space lying between the parallel planes

$$(\lambda_1 - 3\lambda)(\lambda_1 - 2\lambda)(\lambda_1 - \lambda)x_1 + (\lambda_1 - 2\lambda)(\lambda_1 - \lambda)x_2 + (\lambda_1 - \lambda)x_3 + x_4 = \pm \frac{1}{a\lambda_1}.$$

Chapter 8

Fuel-optimal control problems

Having devoted the previous three chapters exclusively to problems of time-optimal synthesis, attention is now focused on saturating control systems optimised with respect to other performance criteria. Feedback controls which minimise various cost functionals are derived in this and the following chapter. In particular, three classes of optimal control problems are investigated:

(i) The fuel-optimal control problem and some of its variants are discussed in the present chapter; the optimal synthesis is obtained in a variety of cases. Recall that one such optimal synthesis has been obtained in the lunar soft landing example of Subsection 1.3.1.

(ii) The quadratic-cost control problem for saturating systems (with its characteristic feature of singular subarcs) is studied in Chapter 9.

(iii) Minimisation of nonquadratic cost functionals are also investigated in Chapter 9, viz. integral costs with integrands inhomogeneous in system states, and minmax control systems.

8.1 Fuel-optimal control problems

Linear, single-input systems of the familiar form

$$\dot{x} = Ax + bu, \qquad |u(t)| \leq 1; \qquad x(t) \in \mathbb{R}^n \tag{8.1}$$

only will be considered,† with the usual assumption that (A, b) in a controllable pair, i.e. rank $[b : Ab : \cdots : A^{n-1}b] = n$. It is assumed that the magnitude of the

† For a comprehensive treatment of the fuel-optimal control problem (and its many variants) in a more general setting, the reader is referred to the text by Athans and Falb (1966) which collates many of the earlier results in the area (e.g. Athanassiades, 1963; Flügge-Lotz and Marbach, 1963; Foy, 1963; Athans, 1964; Athans and Canon, 1964; Meditch, 1964a). Related results are also contained in Grimmell (1967) (existence results), Rootenberg (1974) (sensitivity to parameter variation) and Nishimura (1980) (application to VTOL aircraft).

control $|u(t)|$ is proportional to the rate of flow of fuel at time t, so that the functional

$$J(u) = \int_0^{t_f} |u(t)| \, dt \tag{8.2}$$

is a measure of the fuel expended over the control interval $[0, t_f]$. Recalling that free end-time problems (t_f free) only are the concern of this volume, the basic fuel-optimal regulator problem can be stated as follows: determine an admissible control $u^*(t)$, $0 \le t \le t_f$, which generates a trajectory $x(\cdot)$ from the initial state $x(0) = x^0$ to the origin $x(t_f) = 0$ at the free terminal time t_f with minimum fuel expenditure, i.e. such that (8.2) is minimised. It will be shown that the problem as stated above is ill posed in the sense that questions of existence and uniqueness of solutions arise. This is to be expected as the following simple observation shows. Suppose A is an asymptotically stable matrix (all eigenvalues with negative real parts) then, by using the zero control $u \equiv 0$, the initial state can be transferred arbitrarily close to the origin (by the natural system dynamics) with zero fuel expenditure; to actually attain the origin, an appropriate non-zero control must be used. However, the fuel expended can be made arbitrarily small at the expense of elapsed time, by waiting a sufficiently long time for the natural system dynamics to take the state arbitrarily close to the origin before employing some control action. In other words, zero fuel expenditure is a lower bound on the cost which can be approached arbitrarily closely (but never exactly attained). In this way, fuel consumption can in general be reduced at the expense of increased response times which suggests that, while fuel minimisation may be the primary concern, it is frequently necessary to incorporate response time in the problem formulation in order to avoid the impracticalities of large control intervals $[0, t_f]$.

The Hamiltonian for the basic problem is

$$H(x, u, \eta) = \langle \eta, Ax \rangle + \langle \eta, bu \rangle - |u| \tag{8.3}$$

and, in accordance with the maximum principle (theorem 2.4.1), if the control $u^*(\cdot)$ is optimal, generating trajectory $x(\cdot)$ from $x(0) = x^0$ to $x(t_f) = 0$, then there exists a solution $\eta(t) = \exp(-A^T t)\eta^0$, $0 \le t \le t_f$, of the adjoint equation

$$\dot{\eta} = -A^T \eta \tag{8.4}$$

such that

$$H(x(t), u^*(t), \eta(t)) = \max_{|u| \le 1} H(x(t), u, \eta(t)) \qquad \text{a.a.} \qquad t \in [0, t_f] \tag{8.5}$$

and, since t_f is free,

$$H(x(t), u^*(t), \eta(t)) = 0 \qquad \text{a.a.} \qquad t \in [0, t_f] \tag{8.6}$$

Combining (8.3) and (8.5), the H-maximising control satisfies

$$u^*(t) = \begin{cases} \operatorname{sgn} \langle \eta(t), b \rangle; & \text{if } |\langle \eta(t), b \rangle| > 1 \\ 0; & \text{if } |\langle \eta(t), b \rangle| < 1 \end{cases} \tag{8.7a}$$

$$u^*(t) \in \begin{cases} [0, 1]; & \text{if } \langle \eta(t), b \rangle = 1 \\ [-1, 0]; & \text{if } \langle \eta(t), b \rangle = -1 \end{cases} \tag{8.7b}$$

Note that the control $u^*(\cdot)$ is uniquely (almost everywhere) determined by (8.7) provided that $|\langle \eta(t), b \rangle| \neq 1$ almost everywhere, in which case the problem is said to be *normal* or nonsingular. Conversely, if the following condition holds on some finite interval $I \subseteq [0, t_f]$

$$\textit{Singular condition:} \qquad |\langle \eta(t), b \rangle| = 1 \qquad \text{for all } t \in I \tag{8.8}$$

then the control $u^*(\cdot)$ is no longer uniquely (almost everywhere) defined by (8.7a) but instead satisfies the inclusion (8.7b); the problem is then said to be *singular*.

Clearly, if (8.8) is to hold on the finite interval I, then all derivatives of the scalar product $\langle \eta(t), b \rangle = \langle \exp(-A^T t)\eta^0, b \rangle$ must also vanish on the interior of I, whence

$$\frac{d^k}{dt^k} \langle \eta(t), b \rangle = (-1)^k \langle \eta(t), A^k b \rangle = 0, \ \forall\, t \in \operatorname{int}(I);\ k = 0, 1, 2, \ldots$$

and, in particular,

$$\eta(t)'[Ab : A^2 b : \cdots : A^n b] = 0 \ \forall\, t \in \operatorname{int}(I) \tag{8.9}$$

But (8.8) implies that $\eta(t) \neq 0$ and hence, for (8.9) to hold, the $n \times n$ matrix $[Ab : A^2 b : \cdots : A^n b]$ must be singular, i.e.

$$\det[Ab : A^2 b : \cdots : A^n b] = \det[A] \det[b : Ab : \cdots : A^{n-1} b] = 0$$

But, by the initial assumption that the pair (A, b) is controllable, $\det[b : Ab : \cdots : A^{n-1} b] \neq 0$ and so $\det[A] = 0$. These results are summarised in the following lemma.

Lemma 8.1.1: If (A, b) is a controllable pair and if the fuel-optimal problem is singular, then A must be a singular matrix ($\det[A] = 0$); conversely, if (A, b) is a controllable pair, and if A is a nonsingular matrix ($\det[A] \neq 0$) then the fuel-optimal control problem is nonsingular or normal.

8.2 Fuel-optimal control of the double integrator system

To illustrate the questions of existence and uniqueness which can arise in the basic fuel-optimal control problem, the theory is applied to a double integrator system in this section. This example has also been studied by Athans and Falb (1966) whose treatment is closely followed here.

Consider the following fuel-optimal regulator problem:

$$\text{minimise:} \qquad J(u) = \int_0^{t_f} |u(t)| \; dt; \qquad t_f \text{ free}$$

$$\text{subject to:} \quad \begin{cases} \dot{x}_1 = x_2; \quad \dot{x}_2 = \dfrac{u}{a}; \quad |u(t)| \leq 1 \\ x(0) = x^0 = \begin{bmatrix} x_1^0 \\ x_2^0 \end{bmatrix}; \quad x(t_f) = 0 \end{cases} \tag{8.10}$$

In this case,

$$A = \begin{bmatrix} 0 & 1 \\ 0 & 0 \end{bmatrix} \quad \text{and} \quad b = \begin{bmatrix} 0 \\ \dfrac{1}{a} \end{bmatrix}.$$

From (8.7), a fuel-optimal control $u^*(t), 0 \leq t \leq t_f$ (if such a control exists), must satisfy

$$u^*(t) = \begin{cases} \text{sgn } \eta_2(t); & \text{if } |\eta_2(t)| > a \\ 0; & \text{if } |\eta_2(t)| < a \end{cases} \tag{8.11a}$$

$$u^*(t) \in \begin{cases} [0, 1]; & \text{if } \eta_2(t) = a \\ [-1, 0]; & \text{if } \eta_2(t) = -a \end{cases} \tag{8.11b}$$

where

$$\eta_2(t) = \eta_2^0 - \eta_1^0 t, \qquad 0 \leq t \leq t_f. \tag{8.11c}$$

Note that, for the system under consideration, the matrix A is singular so that, in view of lemma 8.1.1, the possibility of singular solutions cannot be discounted.

8.2.1 Candidate singular controls

In particular, if $\eta_1^0 = 0$ and $|\eta_2^0| = a$ then $\eta_1(t) = 0$ and $|\eta_2(t)| = a$ for all $t \in [0, t_f]$. Defining $U^+ \subset U$ as the set of admissible *non-negative* control functions $u(t)$, $0 \leq t \leq t_f$, with values in $[0, +1]$, i.e.

$$U^+ = \{u \in U: 0 \leq u(t) \leq 1\}$$

then, the control

$$u^*(t) = \text{sgn } (\eta_2^0) \, u(t), \qquad 0 \leq t \leq t_f; \qquad u(\cdot) \in U^+; \qquad |\eta_2^0| = a \tag{8.12}$$

is a candidate singular control, generating a state trajectory $x(\cdot)$, satisfying the necessary condition (8.5). Moreover,

$$
\begin{aligned}
H(x(t), u^*(t), \eta(t)) &= \eta_1(t)x_2(t) + \eta_2(t)\frac{u^*(t)}{a} - |u^*(t)| \\
&= |u^*(t)| - |u^*(t)| \\
&= 0 \ \forall \ t \in [0, t_f].
\end{aligned}
$$

so that the candidate control also satisfies condition (8.6).

8.2.2 *Candidate normal controls*

On the other hand, if

$$\eta_1^0 \neq 0 \Rightarrow \eta_1(t) = \eta_1^0 \neq 0, \qquad 0 \leq t \leq t_f$$

then the function

$$\eta_2(t) = \eta_2^0 - \eta_1^0 t$$

varies linearly with t and, in particular, $|\eta_2(t)| = a$ at two isolated instants at most. In other words, if $\eta_1^0 \neq 0$ the fuel-optimal control u^* (if it exists) is uniquely determined (almost everywhere) by the relation

$$u^*(t) = \begin{cases} \operatorname{sgn}(\eta_2(t)); & |\eta_2(t)| > a \\ 0; & |\eta_2(t)| < a \end{cases} \tag{8.13}$$

Since η_2 varies linearly in t, it immediately follows that only the control sequences $\{\pm 1, 0, \mp 1\}$, and the subsequences $\{\pm 1\}$, $\{0\}$, $\{0, \mp 1\}$, $\{\pm 1, 0\}$ thereof, are candidates for normal fuel-optimal controls satisfying (8.5), where the notation $\{\pm 1, 0, \mp 1\}$ denotes a constant control value ± 1 on a finite interval followed by a finite interval of zero-valued control, followed, in turn, by a finite interval of ∓ 1 valued control.

8.2.3 *Infimum of the value function*

As before, the value function $V(x)$ (defined on the region of the state space for which an optimal solution exists) is the minimum cost (fuel expended) incurred on an optimal trajectory from $x(0) = x$ to $x(t_f) = 0$. It will now be shown that, for the problem at hand,

$$V(x) = V(x_1, x_2) \geq a|x_2| \tag{8.14}$$

Specifically, given a starting point $x(0) = x = (x_1, x_2)'$ and any admissible control $u(t)$, $0 \leq t \leq t_f$, which generates a trajectory $x(\cdot)$ to the origin $x(t_f) = 0$, then, as $\dot{x}_2 = u/a$,

$$x_2(t_f) = 0 = x_2 + \frac{1}{a}\int_0^{t_f} u(t)\, dt$$

and so

$$a\,|x_2| = \left|\int_0^{t_f} u(t)\,dt\right|$$

$$\leq \int_0^{t_f} |u(t)|\;dt = J(u)$$

i.e. the cost of *every* trajectory to the origin is bounded from below by the non-negative quantity $a\,|x_2|$, where x_2 is the second component of the initial state vector. Hence the cost $V(x_1, x_2)$ of an optimal trajectory from $x = (x_1, x_2)$ must also be bounded below by $a\,|x_2|$. To see that $a\,|x_2|$ is actually the *greatest* lower bound for $V(x_1, x_2)$ it suffices to show that, for at least one initial state $x \neq 0$, the minimum cost $a\,|x_2|$ is exactly achieved. Consider, for example, a starting point x on the curve $\Gamma = \{x\colon x_1 + \tfrac{1}{2}ax_2|x_2| = 0\}$, which corresponds to the time-optimal switching curve for the double integrator, so that, by the results of Subsection 5.2.1(i), the time-optimal trajectory to the origin is generated by the constant control $u^* \equiv -\text{sgn}\,(x_2)$ of minimum duration $t_f = a\,|x_2|$. The fuel consumed on this (minimum-time) path to the origin is given by

$$J(u) = \int_0^{t_f} |u(t)|\;dt$$

$$= \int_0^{a|x_2|} dt$$

$$= a\,|x_2|$$

i.e. the minimum possible value of cost is incurred and hence the quantity $a\,|x_2|$ is indeed the infimum of the value function $V(x_1, x_2)$. Hence, any admissible control $u^*(t)$, $0 \leq t \leq t_f$ generating a trajectory $x(\cdot)$ to the origin which incurs a cost $a\,|x_2(0)|$ must be optimal.

8.2.4 $x \in \Gamma$ *: Optimal* $\{\pm 1\}$ *control sequence trajectories*

In the previous section it was shown that all initial states $x \in \Gamma$ on the time-optimal switching curve

$$\Gamma = \{x\colon \xi(x) = x_1 + \tfrac{1}{2}ax_2|x_2| = 0\} \tag{8.15}$$

depicted in Fig. 8.1, can be transferred to the origin via the constant control

$$u^* \equiv -\text{sgn}\,(x_2); \qquad x \in \Gamma \tag{8.16}$$

and, since each of these time-optimal trajectories incurs the minimum possible value of cost, viz. $a|x_2|$, each must also be a fuel-optimal trajectory. Uniqueness of the fuei-optimal trajectory from $x \in \Gamma$ follows immediately from the uniqueness of the time-optimal control, i.e. for initial states $x \in \Gamma$ the fuel-optimal and time-optimal solutions coincide.

Referring to Fig. 8.1, the following partition of the state plane $\mathbb{R}^2 = \Gamma \cup \Sigma^- \cup \Sigma^0_- \cup \Sigma^+ \cup \Sigma^0_+$ is introduced where

$$\begin{aligned}
\Sigma^- &= \{x\colon \xi(x) > 0;\ x_2 \geq 0\} \\
\Sigma^0_- &= \{x\colon \xi(x) > 0;\ x_2 < 0\} \\
\Sigma^+ &= \{x\colon \xi(x) < 0;\ x_2 \leq 0\} \\
\Sigma^0_+ &= \{x\colon \xi(x) < 0;\ x_2 > 0\}
\end{aligned} \tag{8.17}$$

and initial states in each of these regions are considered. It will be shown that for $x \in \Sigma^0_- \cup \Sigma^0_+$ infinitely many fuel-optimal controls exist (nonuniqueness); while for $x \in \Sigma^+ \cup \Sigma^-$ a fuel-optimal control does *not* exist.

8.2.5 $x \in \Sigma^0_- \cup \Sigma^0_+$: *nonuniqueness of fuel-optimal controls*

It will be shown that, from an initial state $x \in \Sigma^0_- \cup \Sigma^0_+$, infinitely many (singular) optimal paths to the origin exist, each of which incurs the same minimum value of cost $a|x_2|$. Recall that, if a singular or normal control $u^*(t), 0 \leq t \leq t_f$, satisfying (8.12) or (8.13), can be found which generates a trajectory $x(\cdot)$ from $x(0) = (x_1, x_2)' \in \Sigma^0_- \cup \Sigma^0_+$ to the origin $x(t_f) = 0$ such that the value of cost is given by $a|x_2|$, then $u^*(\cdot)$ is a fuel-optimal control. The claim that there are many such controls is verified by noting that infinitely many admissible non-negative controls $u^+ \in U^+$ exist with $u^*(t) = \text{sgn}\,(x_2)u^+(t), 0 \leq t \leq t_f$, for which

$$\begin{aligned}
x_1(t_f) &= 0 = x_1 + x_2 t_f - \frac{\text{sgn}\,(x_2)}{a}\int_0^{t_f}\int_0^t u^+(s)\,ds\,dt \\
x_2(t_f) &= 0 = x_2 - \frac{\text{sgn}\,(x_2)}{a}\int_0^{t_f} u^+(t)\,dt
\end{aligned} \tag{8.18}$$

with

$$\begin{aligned}
J(u) &= \int_0^{t_f} |u^*(t)|\,dt = \int_0^{t_f} |u^+(t)|\,dt \\
&= a|x_2|
\end{aligned} \tag{8.19}$$

For example, the piecewise constant control

$$u^*(t) = -\text{sgn}\,(x_2)u^+(t) \tag{8.20a}$$

where

$$u^+(t) = \begin{cases} 0 < K < 1; & 0 \leq t < t_s \\ \quad 1; & t_s \leq t \leq t_f \end{cases} \tag{8.20b}$$

of total duration

$$t_f = \frac{a|x_2|}{K} - \sqrt{(1-K)\left[\left(\frac{a|x_2|}{K}\right)^2 - \frac{2a|x_1|}{K}\right]} \tag{8.20c}$$

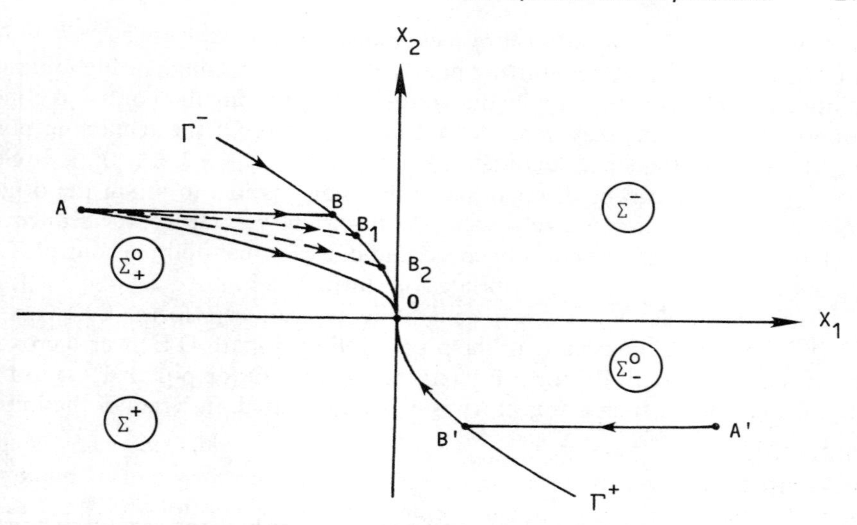

Fig. 8.1 *Regions Σ^+, Σ^-, Σ^0_+, Σ^0_- and nonunique optimal trajectories*

and with the single control function discontinuity occurring at

$$t_s = [t_f - a|x_2|](1 - K)^{-1} \tag{8.20d}$$

generates a trajectory to the origin which incurs the minimum cost $a|x_2|$ for all values of

$$0 < K \leq \frac{(a|x_2|)^2}{2a|x_1|},$$

i.e.

$$\text{if} \quad x \in \Sigma^0_- \cup \Sigma^0_+, \quad \text{then} \quad J(u^*) = V(x_1, x_2) = a|x_2| \ \forall \ K \in \left(0, \frac{(a|x_2|)^2}{2a|x_1|}\right] \tag{8.21}$$

Each of these piecewise constant controls are optimal singular fuel-optimal controls satisfying (8.12). Fig. 8.1 depicts several optimal paths, AB_1O, AB_2O, AO, generated by such piecewise constant singular controls each of which incurs the minimum cost $a|x_2|$. The path ABO or (A′B′O) of Fig. 8.1 is generated by a fuel-optimal nonsingular control function consisting of a zero control interval from A to B (or A′ to B′) followed by a -1 control interval from B to O (or a $+1$ interval from B′ to O); such controls correspond to the $\{0, \pm 1\}$ sequence control candidates satisfying (8.13). Again, it is easily verified that such $\{0, \pm 1\}$ control sequence trajectories exist from all points $x \in \Sigma^0_+ \cup \Sigma^0_-$, each of which incurs the minimum cost value $a|x_2|$. This example clearly illustrates that fuel-optimal controls are, in general, nonunique. Questions of existence will now be studied.

8.2.6 $x \in \Sigma^+ \cup \Sigma^-$: nonexistence of fuel-optimal control

It is easily verified that, for starting points $x \in \Sigma^+ \cup \Sigma^-$, singular fuel-optimal solutions do not exist (essentially due to the fact that a singular control is either nonpositive or nonnegative and cannot change sign). Of the remaining non-singular control sequence candidates, viz. $\{\pm 1, 0, \mp 1\}$, $\{\pm 1, 0\}$, $\{0\}$, it is clear that $\{\pm 1, 0\}$, $\{0\}$ can be disregarded as it is not possible to attain the origin from a non-zero state on a zero valued control interval. Hence, the sequences $\{\pm 1, 0, \mp 1\}$, only, can be considered as candidates for fuel-optimal control. The sequence $\{+1, 0, -1\}$ is appropriate for starting points $x \in \Sigma^+$ and $\{-1, 0, +1\}$ is appropriate for $x \in \Sigma^-$, generating typical trajectories DEFO and D′E′F′O of Fig. 8.2 consisting of the p-path DE (or n-path D′E′) which crosses the x_1 axis, the o-path EF (or E′F′) and the n-path FO (or p-path F′O) to the origin. The cost of such a trajectory is easily calculated, in terms of the initial state $x = (x_1, x_2) \in \Sigma^+ \cup \Sigma^-$, as

$$J(u) = a[|x_2| + 2\epsilon] \tag{8.22}$$

where $\epsilon > 0$ is the modulus of the x_2 component at the first and second switches, i.e. the x_2 component of the switch points E and F (or E′ and F′). Note that

$$\lim_{\epsilon \to 0} J(u) = a|x_2|$$

so that, by switching sufficiently close to the x_1-axis, the minimum possible value of cost can be approached arbitrarily closely; on the other hand, note that the duration of the *o*-path EF or E′F′ is given, in terms of the initial state $x \in \Sigma^+ \cup \Sigma^-$, as

$$\tau = \frac{1}{\epsilon}[x_1 \operatorname{sgn}(x_2) + \tfrac{1}{2}ax_2^2] - a\epsilon \tag{8.23}$$

so that $\tau \to \infty$ as $\epsilon \to 0$. Hence, the minimum possible value of cost can be approached arbitrarily closely at the expense of an unbounded increase in time to the origin. Consequently, for each $x \in \Sigma^+ \cup \Sigma^-$ an admissible control satisfying (8.13) can be constructed which generates a trajectory to the origin incurring a cost arbitrarily close to the minimum possible value $a|x_2|$; however, this minimum value can never be attained exactly. Referring to Fig. 8.2, in the limit $\epsilon \to 0$ the point E (or E′) will lie on the x_1-state axis and, on reaching this point, the state will remain stationary and the origin never attained; if the control switch is delayed by an arbitrarily small amount to allow the state to penetrate to the other side of the x_1-axis, then the origin is subsequently attained and the overall cost is arbitrarily close to the minimum possible value. Hence, for starting points $x \in \Sigma^+ \cup \Sigma^-$, fuel-optimal controls do not exist.

8.2.7 Discussion

The above example serves to illustrate the questions of existence and uniqueness which can arise in problems of fuel-optimal control (see Kishi, 1963; Athans, 1966, for related results). In summary, it has been shown that

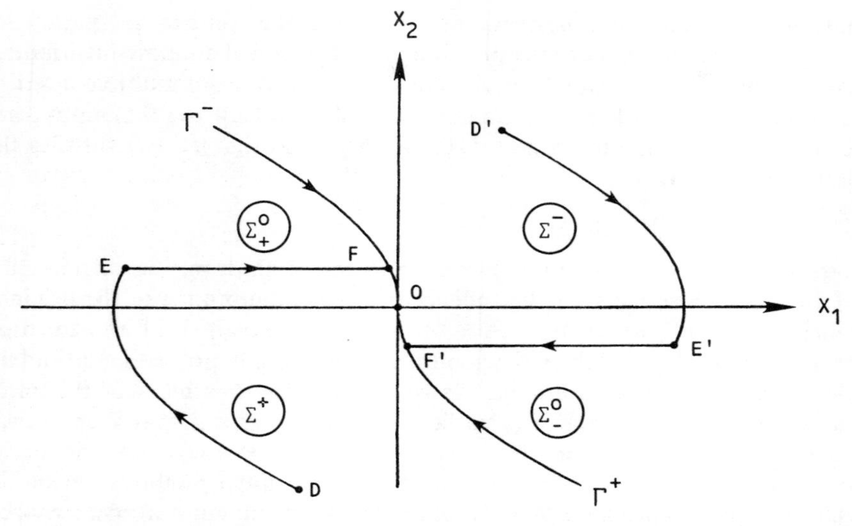

Fig. 8.2 $\{\mp 1, 0, \pm 1\}$ *control sequence trajectories*

(*a*) for an initial state $x \in \Sigma^+ \cup \Sigma^-$, a fuel-optimal trajectory to the origin does not exist, although the infimum of the cost can be approached arbitrarily closely at the expense of times to the origin which tend to infinity;
(*b*) for an initial state $x \in \Sigma_+^0 \cup \Sigma_-^0$ infinitely many finite-time fuel-optimal paths to the origin exist, each of which incurs the same minimum possible value of cost;
(*c*) only for initial states $x \in \Gamma$ on the time-optimal switching curve Γ do *unique* fuel-optimal paths to the origin exist, in which case the fuel-optimal and time-optimal solutions are identical.

The nonuniqueness of fuel-optimal controls is not a serious practical drawback, in fact, it might be regarded as an asset which permits a choice in the realisation of the optimal control. On the other hand, the nonexistence results pose serious practical difficulties; even though the minimum possible cost can be approached arbitrarily closely, the approximately optimal control is of doubtful use in view of the large times necessary to reach the origin. Hence, from a practical view-point, time to the origin should be incorporated in a modified problem formulation. For example, in Sarma and Prasad (1972) (see also Sarma *et al.*, 1969) the theory of differential games is exploited to solve a bicriterion optimal control problem with time and fuel minimisation as twin objectives. Many other modifications of the basic fuel-optimal problem have been proposed in the literature (see e.g. Flügge-Lotz and Marbach, 1963; Athans, 1964; Athans and Falb, 1966; Mih Yin and Grimmell, 1968; Zach, 1972). One modification is to fix the end time t_f, i.e. allowing the control to act over a *prescribed* interval $[0, t_f]$. However, in this case the optimal feedback solution (when it exists) exhibits the unattractive feature of time-variation, involving the 'time-to-go' parameter $\tau = t_f - t$, even though both the plant and cost functional are time-invariant. A second approach is to bound the time to the origin by some multiple $\beta \geq 1$ of the minimum time in the sense that an optimal trajectory $x(t)$, $0 \leq t \leq t_f$, has the property that, at each time $t \in [0, t_f]$ the time-to-go $(t_f - t)$ satisfies the relation

$$t_f - t \leq \beta T^*(x(t)); \qquad \beta \geq 1$$

where $T^*(x(t))$ denotes the minimum time to the origin from the current state $x(t)$. In cases where the state dependence of the minimum time to the origin is explicitly known (i.e. when a closed-form characterisation of the function $T^*: \mathscr{C} \to [0, \infty)$ is available) the modified fuel-optimal problem may admit time-invariant feedback solutions. However, from the treatment of the linear time-optimal control problem of earlier chapters, it is clear that such closed-form expressions for $T^*(\cdot)$ are few in number; this places a severe restriction on the applicability of the second (bounded response time) modification of the basic fuel-optimal problem. A third approach is to minimise a linear combination of elapsed time and consumed fuel (time-fuel-optimal control); this approach circumvents many of the aforementioned difficulties and is exclusively adopted in the sequel.

8.3 Time-fuel-optimal control

Consider again the controllable, autonomous, linear, single-input system

$$\dot{x} = Ax + bu; \qquad |u(t)| \leq 1; \qquad x(t) \in \mathbb{R}^n \tag{8.24}$$

A control $u(t)$, $0 \leq t \leq t_f$, is sought which transfers (8.24) from $x(0) = x^0$ to the state origin $x(t_f) = 0$ such that the cost functional

$$J(u) = \int_0^{t_f} [\mu + (1 - \mu)|u(t)|]\, dt; \qquad \mu \in (0, 1]; \qquad t_f \text{ free} \tag{8.25}$$

is minimised. The parameter μ determines the relative weighting of response time and consumed fuel, with $\mu = 1$ corresponding to the time-optimal control problem.

The Hamiltonian for this time-fuel-optimal control problem is

$$H(x, u, \eta) = \langle \eta, Ax \rangle + \langle \eta, b \rangle u - [\mu + (1 - \mu)|u|] \tag{8.26}$$

and, in accordance with the maximum principle (theorem 2.4.1), if the control $u^*(\cdot)$ is cost-minimising, generating an optimal trajectory $x(\cdot)$ from $x(0) = x^0$ to $x(t_f) = 0$, then there exists a solution $\eta(t) = \exp(-A^T t)\eta^0$, $0 \leq t \leq t_f$, of the adjoint equation

$$\dot{\eta} = -A^T \eta \tag{8.27}$$

such that

$$H(x(t), u^*(t), \eta(t)) = \max_{|u| \leq 1} H(x(t), u, \eta(t)) \qquad \text{a.a.} \quad t \in [0, t_f] \tag{8.28}$$

and, since t_f is free,

$$H(x(t), u^*(t), \eta(t)) = 0 \qquad \text{a.a.} \quad t \in [0, t_f] \tag{8.29}$$

The H-maximising control is uniquely (almost everywhere) determined by the relation

$$u^*(t) = \begin{cases} \operatorname{sgn} \langle \eta(t), b \rangle; & |\langle \eta(t), b \rangle| > 1 - \mu \\ 0; & |\langle \eta(t), b \rangle| < 1 - \mu \end{cases} \tag{8.30}$$

provided that the problem is nonsingular, i.e. provided that $|\langle \eta(t), b \rangle| \neq 1 - \mu$ almost everywhere. That the time-fuel-optimal control problem under consideration is indeed nonsingular is easily established.

Lemma 8.3.1 : The time-fuel-optimal regulator problem (8.24)–(8.25) is nonsingular and the time-fuel-optimal control (if it exists) is unique.

Proof
Suppose the problem is singular, then

$$|\langle \eta(t), b \rangle| = 1 - \mu \qquad \text{on some finite interval } I \subseteq [0, t_f] \tag{8.31}$$

and all derivatives of $\langle \eta(t), b \rangle$ must vanish on the interior of I, viz.

$$\langle \eta(t), A^k b \rangle = 0 \qquad k = 1, 2, \ldots \tag{8.32}$$

Combining (8.31) and (8.32),

$$\eta(t)'[b : Ab : \cdots : A^{n-1}b] = [\pm(1 - \mu), 0, \ldots, 0] \tag{8.33}$$

Now, by the controllability assumption on the pair (A, b), the matrix $[b : \cdots : A^{n-1}b]$ is invertible so that (8.33) has unique constant solution

$$\begin{aligned} \eta(t)' &= [\pm(1 - \mu), 0, \ldots, 0][b : Ab : \cdots : A^{n-1}b]^{-1} \\ &= \text{constant} \qquad \forall\, t \in \operatorname{int}(I) \end{aligned}$$

which, in turn, implies that

$$\dot{\eta}(t) = -A^T \eta(t) = 0 \qquad \forall\, t \in \operatorname{int}(I) \tag{8.34}$$

Now, from (8.26) and (8.31),

$$\begin{aligned} H(x(t), u^*(t), \eta(t)) &= \langle \eta(t), Ax(t) \rangle - \mu \\ &= \langle A^T \eta(t), x(t) \rangle - \mu \\ &= -\mu \qquad \forall\, t \in \operatorname{int}(I) \end{aligned}$$

But $\mu \neq 0$ which contradicts (8.29). Hence supposition (8.31) is false, the problem is strictly nonsingular and the optimal control, if it exists, is uniquely determined (almost everywhere) by (8.30), taking the values ± 1, 0.

8.4 Time-fuel-optimal control of the double integrator system

The system equations are

$$\dot{x}_1 = x_2; \qquad \dot{x}_2 = \frac{u}{a}; \qquad |u| \le 1; \qquad a > 0 \tag{8.35a}$$

or

$$\dot{x} = Ax + bu; \qquad |u| \le 1; \qquad A = \begin{bmatrix} 0 & 1 \\ 0 & 0 \end{bmatrix}; \qquad b = \begin{bmatrix} 0 \\ \frac{1}{a} \end{bmatrix} \tag{8.35b}$$

and, from the theory of Section 8.3, if $u^*(t)$, $0 \le t \le t_f$, generates a trajectory $x(\cdot)$ from $x(0) = x^0$ to the origin $x(t_f) = 0$ while minimising the cost functional

$$J(u) = \int_0^{t_f} [\mu + (1 - \mu)|u(t)|]\, dt; \qquad t_f \text{ free} \tag{8.36}$$

then u^* is uniquely determined (almost everywhere) as

$$u^*(t) = \begin{cases} \text{sgn } \langle \eta(t), b \rangle = \text{sgn } (\eta_2(t)); & |\eta_2(t)| > a(1 - \mu) \\ 0; & |\eta_2(t)| < a(1 - \mu) \end{cases} \tag{8.37}$$

where

$$\begin{bmatrix} \eta_1(t) \\ \eta_2(t) \end{bmatrix} = \eta(t) = \exp(-A^T t)\eta^0 = \begin{bmatrix} \eta_1^0 \\ \eta_2^0 - \eta_1^0 t \end{bmatrix} \tag{8.38}$$

for some initial adjoint vector $\eta^0 \ne 0$.

By lemma 8.3.1, $|\eta_2(t)| \ne a(1 - \mu)$ almost everywhere so that the function $|\eta_2(t)| = |\eta_2^0 - \eta_1^0 t|$ can take the value $a(1 - \mu)$ at two isolated instants at most and the control sequences $\{\mp 1, 0, \pm 1\}$, $\{\pm 1\}$, $\{0, \pm 1\}$ only can be optimal, each of which will be treated separately.

8.4.1 $\{\pm 1\}$-control sequences

Clearly, these must correspond to the symmetric pair of switchless trajectories Γ^+ and Γ^- which go directly to the origin under the constant controls $u \equiv +1$ and $u \equiv -1$, respectively. These trajectories constitute the time-optimal switching curve

$$\Gamma = \Gamma^+ \cup \Gamma^- = \{x: \xi(x) = x_1 + \tfrac{1}{2} a x_2 |x_2| = 0\} \tag{8.39}$$

as shown in Fig. 8.3.

8.4.2 $\{0, \pm 1\}$-control sequences

These generate state trajectories which go to Γ under a zero-control input (o-path) followed by a $\{\pm 1\}$-control generated path (p- or n-path) in Γ to the

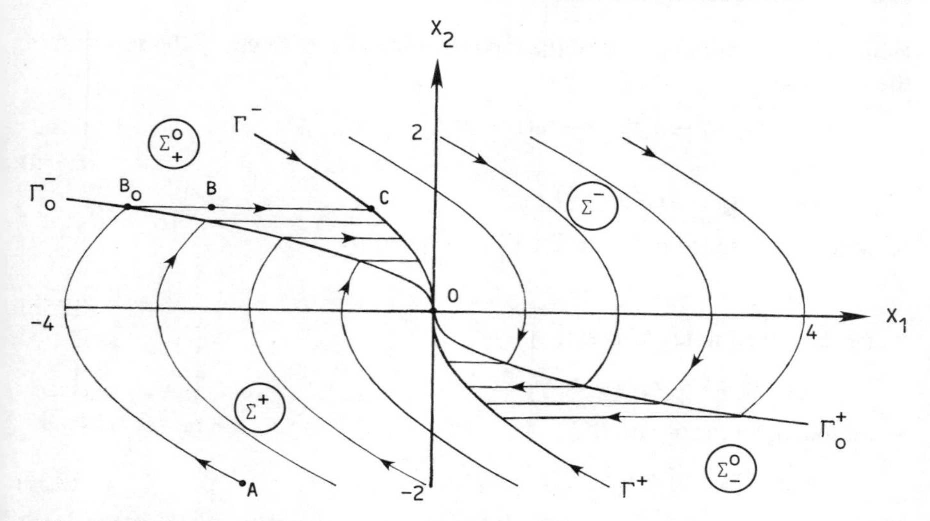

Fig. 8.3 *Time-fuel-optimal synthesis for double integrator system with* $a = 1$ *and* $\mu = \frac{1}{2}$

origin. Consider one such path BCO of Fig. 8.3 generated by a $\{0, -1\}$ control sequence. In the case of the fuel-optimal control problem of Section 8.2, the duration t_0 of the o-path BC (generated by a zero control $u \equiv 0$) was unbounded, i.e. B could be translated an arbitrary distance into the left half-plane. However, for the time-fuel-optimal control problem under consideration, the duration t_{BC} of the o-path is bounded as will now be shown. From (8.37), for a $\{0, -1\}$ control sequence, the function $\eta_2(t) = \eta_2^0 - \eta_1^0 t$, $0 \leq t \leq t_f$ must have the general form depicted in Fig. 8.4 with

$$\left.\begin{aligned} &-a(1-\mu) < \eta_2^0 \leq a(1-\mu) \\ \text{and} \quad & \\ &\eta_1^0 > 0 \end{aligned}\right\} \tag{8.40}$$

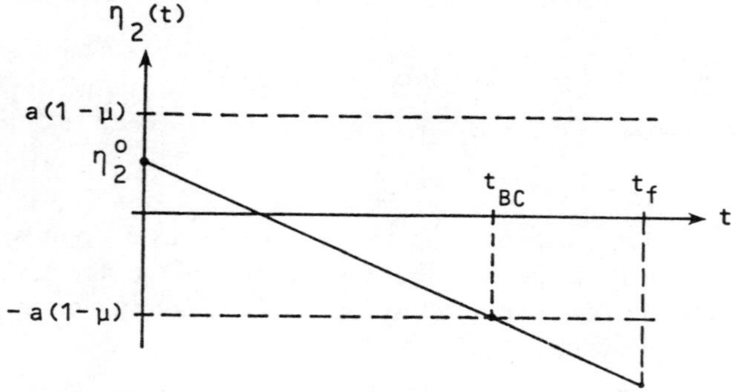

Fig. 8.4 *Typical* η_2 *function which generates a* $\{0, -1\}$ *control sequence*

Hence, the duration $t_{BC} > 0$ of the 0-control interval is given by the solution of the equation

$$\eta_2(t_{BC}) = \eta_2^0 - \eta_1^0 t_{BC} = -a(1-\mu)$$

i.e.

$$t_{BC} = [a(1-\mu) + \eta_2^0]/\eta_1^0$$

Now the state coordinates at $C \in \Gamma^-$ are given by

$$C: (-\tfrac{1}{2}ax_2^2, x_2); \qquad x_2 > 0$$

while the coordinates at B satisfy

$$B: (-(\tfrac{1}{2}ax_2^2 + x_2 t_{BC}), x_2)$$

From (8.29), on the o-path BC

$$H = \eta_1^0 x_2 - \mu = 0$$

so that

$$\eta_1^0 = x_2^{-1}\mu$$

and

$$t_{BC} = \mu^{-1}[a(1-\mu) + \eta_2^0]x_2$$

which, in view of (8.40), takes its maximum value $t_{BC} = t_{B_0C}$ for $\eta_2^0 = a(1-\mu)$, i.e.

$$t_{B_0C} = 2a\mu^{-1}(1-\mu)x_2 \tag{8.41}$$

Hence, from C the o-path can be extended backwards only as far as the point B_0 with coordinates

$$(-\tfrac{1}{2}ax_2^2[1 + 4\mu^{-1}(1-\mu)], x_2)$$

Thus, to every endpoint $C \in \Gamma^-$ of a o-path there (uniquely) corresponds a point B_0 such that the duration of the o-path B_0C is the maximum possible value t_{B_0C}; such a o-path will be referred to as *maximal*. The set of all maximal o-path starting points B_0 constitutes the set Γ_0^- of Fig. 8.3 with the characterisation

$$\Gamma_0^- = \{x: x_1 + \tfrac{1}{2}a\mu^{-1}(4-3\mu)x_2^2 = 0; x_2 > 0\} \tag{8.42}$$

By symmetry, the set

$$\Gamma_0^+ = \{x: x_1 - \tfrac{1}{2}a\mu^{-1}(4-3\mu)x_2^2 = 0; x_2 < 0\} \tag{8.43}$$

may be constructed, corresponding to the set of starting points of $\{0, +1\}$ control sequence trajectories such that the duration of the zero control interval takes its maximum value. The composition of Γ_0^+, Γ_0^- and the origin yields what is in effect a second switch curve Γ_0, viz.

$$\Gamma_0 = \Gamma_0^+ \cup \Gamma_0^- \cup \{0\} = \{x: \xi_0(x) = x_1 + \tfrac{1}{2}a\mu^{-1}(4-3\mu)x_2|x_2| = 0\} \tag{8.44}$$

Defining the region

$$\Sigma^0 = \Sigma^0_+ \cup \Sigma^0_- = \{x\colon \xi(x) < 0;\ \xi_0(x) \geq 0\}$$
$$\cup \{x\colon \xi(x) > 0;\ \xi_0(x) \leq 0\} \tag{8.45a}$$

where

$$\xi(x) = x_1 + \tfrac{1}{2}ax_2|x_2|; \qquad \xi_0(x) = x_1 + \tfrac{1}{2}a\mu^{-1}(4 - 3\mu)x_2|x_2| \tag{8.45b}$$

then a non-zero state $x \in \Sigma^0_+$ may be steered to the origin using a control sequence $\{0, +1\}$ while the sequence $\{0, -1\}$ is appropriate for initial states $x \in \Sigma^0_-$.

8.4.3 $\{\mp 1, 0, \pm 1\}$*-control sequences*

On a full control sequence $\{\mp 1, 0, \pm 1\}$ trajectory the duration of the o-path must be maximal (see e.g. Fig. 8.3), so that the $\mp 1 \rightarrow 0$ switch must occur on Γ_0 and the $0 \rightarrow \pm 1$ switch must occur on Γ. Defining the regions

$$\Sigma^+ = \{x\colon \xi(x) < 0;\ \xi_0(x) < 0\} \tag{8.46a}$$

and

$$\Sigma^- = \{x\colon \xi(x) > 0;\ \xi_0(x) > 0\} \tag{8.46b}$$

then, for an initial state $x \in \Sigma^+$, the origin is attained using a $\{+1, 0, -1\}$ control sequence as on trajectory AB_0CO of Fig. 8.3; while for $x \in \Sigma^-$ the sequence $\{-1, 0, +1\}$ is appropriate.

8.4.4 Optimal control synthesis

The overall optimal state portrait† is depicted in Fig. 8.3. The optimal feedback control may be explicitly expressed as

$$u = u(x) = \begin{cases} +1; & x \in \Sigma^+ \cup \Gamma^+ \\ 0; & x \in \Sigma^0 = \Sigma^0_+ \cup \Sigma^0_- \\ -1; & x \in \Sigma^- \cup \Gamma^- \end{cases} \tag{8.47}$$

where $\Gamma^+, \Gamma^-, \Sigma^0, \Sigma^+, \Sigma^-$ are given by (8.39), (8.45) and (8.46).

8.4.5 Value function

Finally, the minimum cost of a trajectory from an arbitrary starting point x will be calculated.

Suppose initially that the starting point x lies in the region Σ^+ (or Σ^-) so that the full control sequence $\{+1, 0, -1\}$ (or $\{-1, 0, +1\}$) is optimal. Denoting the

† Strictly speaking, optimality of the trajectories cannot be immediately concluded insofar as unique trajectories satisfying the necessary conditions of the maximum principle only have been constructed; unlike the linear time-optimal problem of earlier chapters, the maximum principle may not be sufficient for optimality for the problem at hand. However, the synthesis of Fig. 8.3 constitutes a regular synthesis in the sense of Boltyanskii (see Section 2.7) and hence is optimal.

durations of the +1, 0, −1 (or −1, 0, +1) control intervals by $\tau_1 = \tau_1(x)$, $\tau_2 = \tau_2(x)$, $\tau_3 = \tau_3(x)$ respectively, it follows from (8.36) that the minimum cost of the trajectory from x is given by

$$V(x) = \mu[\tau_1(x) + \tau_2(x) + \tau_3(x)] + (1 - \mu)[\tau_1(x) + \tau_3(x)]$$

i.e.

$$V(x) = \tau_1(x) + \mu\tau_2(x) + \tau_3(x); \qquad x \in \Sigma^+ \cup \Sigma^-$$

It remains to determine the functions $\tau_i: \mathbb{R}^2 \to [0, \infty)$. The time τ_1 from the starting point to the first switch point on Γ_0 is easily calculated by integrating the state equations from $x(0) = x$ to $x(\tau_1) = x^1 = (x_1^1, x_2^1)' \in \Gamma_0$ under the constant control $u = -\operatorname{sgn}(\xi(x))$ and solving the resulting equations for τ_1. This procedure yields the result

$$\tau_1 = \tau_1(x) = ax_2 \operatorname{sgn}(\xi) + \sqrt{\left(\frac{\mu}{2-\mu}\right)(ax_1 \operatorname{sgn}(\xi) + \tfrac{1}{2}(ax_2)^2)} \tag{8.48a}$$

where

$$\xi = \xi(x) = x_1 + \tfrac{1}{2}ax_2|x_2| \tag{8.48b}$$

Now, similar to (8.41), the duration of the subsequent o-path (in Σ^0) from $x^1 = (x_1^1, x_2^1)' \in \Gamma_0$ to $x^2 = (x_1^2, x_2^2)' \in \Gamma$ is given by

$$\tau_2 = 2a\mu^{-1}(1 - \mu)|x_2^1| = 2a\mu^{-1}(1 - \mu)\left[x_2 - \operatorname{sgn}(\xi(x))\frac{\tau_1}{a}\right]$$

i.e.

$$\tau_2 = \tau_2(x) = 2\mu^{-1}(1 - \mu)\sqrt{\left(\frac{\mu}{2-\mu}\right)(ax_1 \operatorname{sgn}(\xi) + \tfrac{1}{2}(ax_2)^2)} \tag{8.49}$$

Clearly, the duration τ_3 of the subsequent path (in Γ) from $x^2 = (x_1^2, x_2^2)' \in \Gamma$ to the origin is given by

$$\tau_3 = a|x_2^2|$$
$$= a|x_2^1| \quad \text{(since the } x_2 \text{ state component is unchanged on a o-path)}$$

i.e.

$$\tau_3 = \tau_3(x) = \frac{\tau_2(x)}{2\mu^{-1}(1 - \mu)} \tag{8.50}$$

Combining the above results yields the cost of a trajectory from $x \in \Sigma^+ \cup \Sigma^-$ as

$$V(x) = ax_2 \operatorname{sgn}(\xi) + 2(2 - \mu)\sqrt{\left(\frac{\mu}{2-\mu}\right)(ax_1 \operatorname{sgn}(\xi) + \tfrac{1}{2}(ax_2)^2)}; \qquad x \in \Sigma^+ \cup \Sigma^-$$

Suppose now that the starting point $x \neq 0$ lies in the region Σ^0 so that the control sequence $\{0, -1\}$ or $\{0, +1\}$ is optimal. A straightforward calculation gives the minimum cost of the trajectory in this case as

$$V(x) = \left(\frac{2-\mu}{2}\right) a|x_2| - \mu x_1 x_2^{-1}; \qquad x \in \Sigma^0$$

Lastly, if the initial state x lies on the curve Γ, then the cost of the trajectory is given by $V(x) = a|x_2|$. Summarising, the minimum cost of a trajectory from a non-zero initial state $x \neq 0$ is given by

$$V(x) = \begin{cases} ax_2 \operatorname{sgn}(\xi) + 2(2-\mu)\sqrt{\left(\frac{\mu}{2-\mu}\right)(ax_1 \operatorname{sgn}(\xi) + \frac{1}{2}(ax_2)^2)}; \\ \qquad\qquad\qquad\qquad x \in \Sigma^+ \cup \Sigma^- \\ \left(\frac{2-\mu}{2}\right) a|x_2| - \mu x_1 x_2^{-1}; \qquad x \in \Sigma^0 \\ a|x_2|; \qquad x \in \Gamma \end{cases} \tag{8.51}$$

with $V(x) = 0$ at $x = 0$.

8.5 Time-fuel-optimal control of system with eigenvalues in simple ratio

Suppose now that the system equations are

$$\frac{dx_1}{dt} = 2\lambda x_1 + x_2; \qquad \frac{dx_2}{dt} = \lambda x_2 + \frac{u}{a}; \qquad |u| \leq 1;$$
$$a > 0; \qquad \lambda = \text{constant} \neq 0 \tag{8.52}$$

i.e. the plant eigenvalues are in the simple ratio 2 : 1. The adjoint equations are

$$\frac{d\eta_1}{dt} = -2\lambda\eta_1; \qquad \frac{d\eta_2}{dt} = -\lambda\eta_2 - \eta_1 \tag{8.53}$$

The time-optimal control problem for this plant was studied in Subsections 5.2.1(v) and 5.2.1(vii); the system equivalence property of Section 3.6 ensures that the time-optimal feedback control is identical to the time-optimal feedback control for the double integrator plant. In the case of time-fuel-minimisation, viz.

$$J(u) = \int_0^{t_f} [\mu + (1-\mu)|u(t)|]\, dt; \qquad \mu \in (0, 1]; \qquad t_f \text{ free}$$

the analysis is again simplified by the adoption of the transformation of variables approach of Section 3.6, whereby the plant may be reduced to an equivalent double integrator plant. However, in contrast to the previous results on time-optimal control, the time-fuel-optimal control for one plant is not mapped by the transformation into the time-fuel-optimal for the equivalent plant. The optimal control (if it exists) is uniquely (almost everywhere) determined by

$$u(t) = \begin{cases} \operatorname{sgn}(\eta_2(t)); & |\eta_2(t)| > a(1-\mu) \\ 0; & |\eta_2(t)| < a(1-\mu) \end{cases} \tag{8.54}$$

and since λ is a real-valued parameter, it follows from (8.53) and (8.54) that the control sequences $\{\pm 1\}$, $\{0, \pm 1\}$, $\{\mp 1, 0, \pm 1\}$ only can be optimal. As in the previous example, the time-fuel-optimal synthesis will be obtained via an explicit partitioning of the state space into regions $\Sigma^+, \Sigma^0, \Sigma^-$ wherein the control takes the values $+1, 0, -1$, respectively. The boundaries of these regions form the optimal switch curves Γ and Γ_0 where Γ denotes the set of $0 \to \pm 1$ control switches and Γ_0 denotes the set of $\mp 1 \to 0$ control switches. The ensuing analysis is essentially that of Ryan (1980a).

8.5.1 Transformation of variables

The following transformations are introduced

$$w_1(t) = \exp(-2\lambda t)x_1(t); \qquad w_2(t) = \exp(-\lambda t)x_2(t) \tag{8.55a}$$

$$\gamma_1(t) = \exp(2\lambda t)\eta_1(t); \qquad \gamma_2(t) = \exp(\lambda t)\eta_2(t) \tag{8.55b}$$

then, from (8.52) and (8.53),

$$\frac{dw_1}{dt} = \exp(-\lambda t)w_2; \qquad \frac{dw_2}{dt} = \exp(-\lambda t)u \tag{8.56a}$$

$$\frac{d\gamma_1}{dt} = 0; \qquad \frac{d\gamma_2}{dt} = -\exp(-\lambda t)\gamma_1 \tag{8.56b}$$

Now introduce the further transformation

$$s = \mathscr{T}(t) = -\frac{1}{\lambda}[\exp(-\lambda t) - 1];$$

$$\text{with} \qquad t = \mathscr{T}^{-1}(s) = \mathscr{S}(s) = -\frac{1}{\lambda}\ln(1-\lambda s) \tag{8.57}$$

and define the new variables

$$\begin{aligned} y_i(s) &= w_i(\mathscr{S}(s)); \qquad i = 1, 2 \\ v(s) &= u(\mathscr{S}(s)) \\ \psi_i(s) &= \gamma_i(\mathscr{S}(s)); \qquad i = 1, 2 \end{aligned} \tag{8.58}$$

then (8.52) may be written as

$$\frac{dy_1}{ds} = y_2; \qquad \frac{dy_2}{ds} = \frac{v}{a}; \qquad |v| \leq 1 \tag{8.59}$$

with the same initial condition $y(0) = x(0) = x^0$ and the adjoint system becomes

$$\frac{d\psi_1}{ds} = 0; \qquad \frac{d\psi_2}{ds} = -\psi_1 \tag{8.60}$$

i.e. the original system has been transformed into the double integrator system (8.59) with associated adjoint system (8.60).

8.5.2 Case (i): stable plant $\lambda < 0$

The optimal $\{\pm 1\}$ control trajectories clearly must correspond to the unique p-path ($u = +1$) and unique n-path ($u = -1$) which lead to the state origin. For the transformed system (8.59) these paths constitute the set

$$\Gamma = \Gamma^+ \cup \Gamma^- = \{y: \xi(y) = y_1 + \tfrac{1}{2}ay_2|y_2| = 0\} \tag{8.61a}$$

and consequently it may be concluded, from (8.55) and (8.58), that the corresponding set for the original system (8.52) is

$$\Gamma = \Gamma^+ \cup \Gamma^- = \{x: \xi(x) = x_1 + \tfrac{1}{2}ax_2|x_2| = 0\} \tag{8.61b}$$

Γ is depicted in Fig. 8.5 and is comprised of the origin together with the segments Γ^- (n-path) and Γ^+ (p-path) lying in the second and fourth quadrants, respectively. For $x \in \Gamma$ the optimal control is given by

$$u^* = u(x) = -\operatorname{sgn}(x_2); \qquad x \in \Gamma \tag{8.62}$$

The optimal $\{0, \pm 1\}$ control sequence trajectories consist of a o-path ($u = 0$) leading to Γ followed by an optimal $\{\pm 1\}$ control path in Γ leading to the origin. It is easily verified that all o-paths leading to Γ^- (in forwards time) lie in the second quadrant ($x_1 < 0$; $x_2 > 0$) and to the left of Γ^-, while all o-paths leading to Γ^+ (in forwards time) lie in the fourth quadrant ($x_1 > 0$; $x_2 < 0$) and to the right of Γ^+. Hence the set $\Gamma_0 = \Gamma_0^+ \cup \Gamma_0^-$ of optimal $\mp 1 \to 0$ control switches must also lie entirely in the second (i.e. segment Γ_0^-) and fourth (i.e. segment Γ_0^+) quadrants as depicted in Fig. 8.5. Consider now a trajectory starting at point A in Γ_0, i.e. $x(0) = x^A \in \Gamma_0$. Assume that the other switch curve Γ is intersected at B via the o-path AB of duration $t_1 = T_1(x^A)$, i.e. $x(t_1) = x^B \in \Gamma$, then, from (8.57), the corresponding path AB for the transformed system with the same initial condition $y(0) = x^A$ has duration

$$s_1 = S_1(x^A) = \mathcal{T}(T_1(x^A)) = -\frac{1}{\lambda}[\exp(-\lambda T_1(x^A)) - 1] \tag{8.63}$$

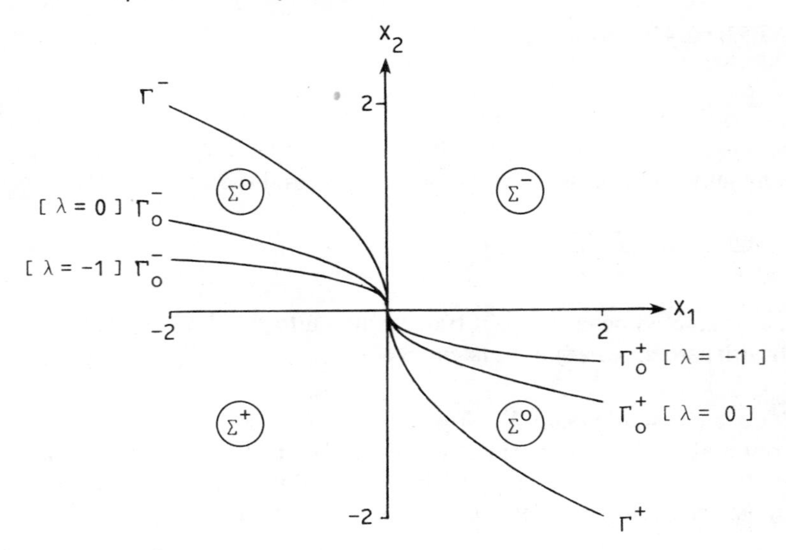

Fig. 8.5 $a = 1$; $\mu = \frac{1}{2}$
Time-fuel-optimal synthesis for stable plant with eigenvalues in simple ratio

Integrating the transformed equations (8.59) along AB (with $v \equiv 0$) yields

$$\begin{aligned} y_1^B &= y_1(s_1) = y_1(0) + y_2(0)s_1 = x_1^A + x_2^A S_1(x^A) \\ y_2^B &= y_2(s_1) = y_2(0) = x_2^A \end{aligned} \tag{8.64a}$$

Hence, from (8.55) and (8.58), under the inverse transformation, the state co-ordinates at B for the original system are

$$\begin{aligned} x_1^B &= \frac{y_1^B}{(1-\lambda s_1)^2} = (x_1^A + x_2^A S_1(x^A))(1 - \lambda S_1(x^A))^{-2} \\ x_2^B &= \frac{y_2^B}{(1-\lambda s_1)} = x_2^A(1 - \lambda S_1(x^A))^{-1} \end{aligned} \tag{8.64b}$$

Now, since $x^B \in \Gamma$,

$$\xi(x^B) = x_1^B + \tfrac{1}{2}a x_2^B |x_2^B| = 0$$

which, in view of (8.64b), implies that

$$x_1^A + x_2^A S_1(x^A) + \tfrac{1}{2}a x_2^A |x_2^A| = 0$$

Writing

$$\xi_0(x) = x_1 + x_2[S_1(x) + \tfrac{1}{2}a|x_2|] \tag{8.65a}$$

and since $A \in \Gamma_0$ is arbitrary, it may be concluded that the $\mp 1 \to 0$ switch curve has the characterisation

$$\Gamma_0 = \{x: \xi_0(x) = 0\} \tag{8.65b}$$

It now remains to define the function $S_1(x)$, which determines the duration s_1 of the maximal zero-control path from $x \in \Gamma_0$ to the switch curve Γ.

Maximum duration S_1 of o-control interval

Consider now the adjoint solution $\eta_2(t)$, $0 \le t \le t_1$, which generates the extremal control over path AB. From (8.54) and (8.62)

$$\eta_2(t_1) = \eta_2^B = -a(1-\mu)\ \mathrm{sgn}\ (x_2^B) = -a(1-\mu)\ \mathrm{sgn}\ (x_2^A)$$

and since $x^A \in \Gamma_0$, at the starting point A

$$\eta_2(0) = \eta_2^A = -\eta_2^B = a(1-\mu)\ \mathrm{sgn}\ (x^A)$$

Also

$$H = \eta_1(2\lambda x_1 + x_2) + \eta_2\left(\lambda x_2 + \frac{u}{a}\right) - \mu - (1-\mu)|u| \equiv 0$$

on an optimal path, so that at the starting point A:

$$\eta_1(0) = \eta_1^A = (\mu - a(1-\mu)\lambda|x_2^A|)(2\lambda x_1^A + x_2^A)^{-1} \triangleq \chi(x^A)$$

Noting that the transformed adjoint system has the same initial condition, i.e. $\psi(0) = \eta^A$, integrating (8.60) over AB yields

$$\begin{aligned}\psi_2(s_1) = \psi_2^B &= \psi_2(0) - \psi_1(0)s_1 \\ &= \eta_2^A - \eta_1^A s_1 \\ &= a(1-\mu)\ \mathrm{sgn}\ (x_2^A) - \chi(x^A)S_1(x^A)\end{aligned} \tag{8.66}$$

Now, from (8.55) and (8.58)

$$\begin{aligned}\psi_2^B &= \eta_2^B(1-\lambda s_1)^{-1} \\ &= -a(1-\mu)\ \mathrm{sgn}\ (x_2^A)(1-\lambda S_1(x^A))^{-1}\end{aligned}$$

which, on substituting in (8.66) and solving the resulting quadratic, yields the single admissible (i.e. real and non-negative) root (omitting the superscript 'A' as the starting point is arbitrary)

$$S_1(x) = \frac{1}{2}\left[\frac{1}{\lambda} + \alpha(x)\right] + \frac{1}{2}\left[\left[\frac{1}{\lambda} + \alpha(x)\right]^2 - \frac{8}{\lambda}\alpha(x)\right]^{1/2} \tag{8.67a}$$

where

$$\begin{aligned}\alpha(x) &= \frac{a(1-\mu)\ \mathrm{sgn}\ (x_2)}{\chi(x)} \\ &= a(1-\mu)\left[\frac{-2\lambda|x_1| + |x_2|}{\mu - a(1-\mu)\lambda|x_2|}\right]\end{aligned} \tag{8.67b}$$

Note that for $\lambda < 0$ and $\mu \in (0, 1]$, α is real-valued and non-negative for all $x \in \mathbb{R}^2$, as is the square root in (8.67a).

With the functions $S_1: \mathbb{R}^2 \to [0, \infty)$ and $\xi_0: \mathbb{R}^2 \to \mathbb{R}$ defined as above, the curve $\Gamma_0 = \{x: \xi_0(x) = 0\}$ partitions the state plane into the two regions $\{x: \xi_0(x) > 0\}$ and $\{x: \xi_0(x) < 0\}$ lying to the right and left of Γ_0, respectively. The regions Σ^+, Σ^0 and Σ^- of positive, zero and negative control may be explicitly characterised as follows:

$$\Sigma^+ = \{x: \xi(x) < 0; \quad \xi_0(x) < 0\} \cup \Gamma^+ \tag{8.68a}$$

$$\Sigma^0 = \{x: \xi(x) < 0; \quad \xi_0(x) \geq 0\} \cup \{x: \xi(x) > 0; \quad \xi_0(x) \leq 0\} \tag{8.68b}$$

$$\Sigma^- = \{x: \xi(x) > 0; \quad \xi(x) > 0\} \cup \Gamma^- \tag{8.68c}$$

where

$$\xi(x) = x_1 + \tfrac{1}{2}ax_2|x_2| \tag{8.68d}$$

$$\xi_0(x) = x_1 + \tfrac{1}{2}x_2[2S_1(x) + a|x_2|] \tag{8.68e}$$

$$2S_1(x) = [\lambda^{-1} + \alpha(x)] + \sqrt{[\lambda^{-1} + \alpha(x)]^2 - 8\lambda^{-1}\alpha(x)} \tag{8.68f}$$

$$\alpha(x) = a(1-\mu)\left[\frac{-2\lambda|x_1| + |x_2|}{\mu - a(1-\mu)\lambda|x_2|}\right] \tag{8.68g}$$

The optimal feedback control is given by

$$u(x) = \begin{cases} +1; & x \in \Sigma^+ \\ 0; & x \in \Sigma^0 \\ -1; & x \in \Sigma^- \end{cases} \tag{8.69}$$

For system parameter values $\lambda = -1$, $a = 1$ and $\mu = \frac{1}{2}$ (i.e. equal weighting of time and fuel) the optimal switch curves Γ_0, Γ and control regions Σ^+, Σ^0, Σ^- are depicted in Fig. 8.5, which also includes, for comparison, the case $\lambda = 0$ (double integrator). If $\mu = 1$, then $S_1(x) = 0$ and the curve Γ_0 coincides with Γ to give the familiar time-optimal synthesis. As $\mu \to 0$ the curve Γ_0 tends to the x_1 state axis which coincides with an eigenvector of the uncontrolled (free) system; consequently, in the limit, for $x(0) \in \Gamma_0$ the origin is approached *exponentially* along this (zero-cost) eigenvector so that (as previously discussed) strictly speaking, a fuel-optimal control does not exist in this case (which of course was excluded at the outset on setting $0 < \mu \leq 1$).

8.5.3 Case (ii) : unstable plant $\lambda > 0$

In the treatment of the time-optimal problem for this unstable plant in Subsection 5.2.1(vii), it was shown that the origin can be obtained only from starting points in the domain of null controllability $\mathscr{C}$, where

$$\mathscr{C} = \left\{x: ax_2 \operatorname{sgn}(\Xi) + 2\sqrt{ax_1 \operatorname{sgn}(\Xi) + \tfrac{1}{2}(ax_2)^2} < \frac{1}{\lambda}\right\} \tag{8.70a}$$

with

$$\Xi = \Xi(x) = \begin{cases} \xi(x) = x_1 + \frac{1}{2}ax_2|x_2|; & \xi(x) \neq 0 \\ x_2; & \xi(x) = 0 \end{cases} \tag{8.70b}$$

It immediately follows that time-fuel-optimal paths to the origin can exist only within this domain of null controllability. With this restriction (i.e. $x \in \mathscr{C}$) the analysis of Subsection 8.5.2 is again valid with the following modifications:

(*a*) with $\lambda > 0$, the time interval $0 \leq t < \infty$ is mapped by (8.57) to the transformed interval $0 \leq s < 1/\lambda$. Hence, it may be verified that the unique admissible root $S_1 \in [0, 1/\lambda)$ of the quadratic equation for the maximum duration of the 0-control interval is given by the difference of the terms of (8.67a), i.e.

$$S_1(x) = \tfrac{1}{2}[\lambda^{-1} + \alpha(x)] - \tfrac{1}{2}\sqrt{[\lambda^{-1} + \alpha(x)]^2 - 8\lambda^{-1}\alpha(x)}; \qquad x \in \mathscr{C} \tag{8.71}$$

with α defined as before but now restricted to $\mathscr{C}$.

(*b*) with $\lambda > 0$, the function S_1 is not real-valued for all $x \in \mathscr{C}$; this necessitates straightforward but lengthy alterations (omitted here) to (8.68) in order to obtain a well defined characterisation of the control regions Σ^+, Σ^0, Σ^- and associated feedback laws.

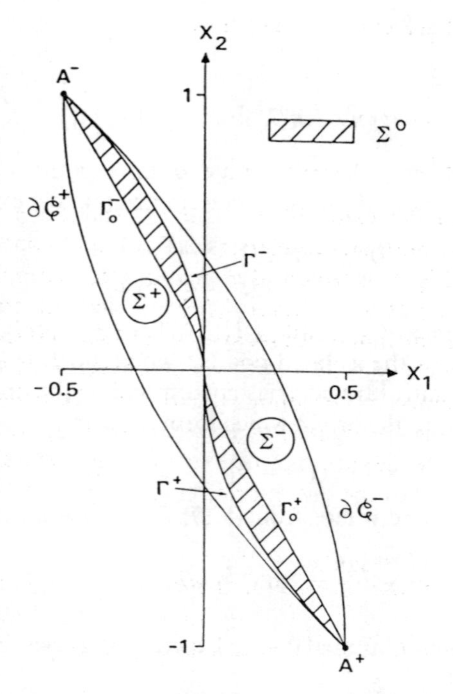

Fig. 8.6 $a = 1$; $\mu = \frac{1}{2}$: domain of null controllability and time-fuel-optimal synthesis for unstable plant with eigenvalues in simple ratio

For system parameter values $\lambda = 1$, $a = 1$ and $\mu = \frac{1}{2}$ the region of controllability $\mathscr{C}$ is depicted in Fig. 8.6, together with the optimal switch curves Γ, Γ_0 and control regions $\Sigma^+, \Sigma^0, \Sigma^-$. If $\mu = 1$, then the boundaries Γ and Γ_0 coincide to give the time-optimal synthesis of Subsection 5.2.1(vii).

8.6 Time-fuel-optimal control of integrator-plus-lag system

Suppose now that the system equations are

$$\dot{x}_1 = x_2; \qquad \dot{x}_2 = \lambda x_2 + \frac{u}{a}; \qquad \lambda < 0; \qquad a > 0; \qquad |u| \leq 1 \tag{8.72a}$$

or

$$\dot{x} = Ax + bu; \qquad A = \begin{bmatrix} 0 & 1 \\ 0 & \lambda \end{bmatrix}; \qquad b = \begin{bmatrix} 0 \\ \frac{1}{a} \end{bmatrix}; \qquad |u| \leq 1 \tag{8.72b}$$

so that the adjoint equations become

$$\dot{\eta}_1 = 0; \qquad \dot{\eta}_2 = -\lambda\eta_2 - \eta_1 \tag{8.73}$$

with solution

$$\left.\begin{aligned} \eta_1(t) &= \eta_1^0 \\ \eta_2(t) &= \exp(-\lambda t)[\eta_2^0 - \eta_1^0\lambda^{-1}(\exp(\lambda t) - 1)] \end{aligned}\right\} \qquad 0 \leq t \leq t_f \tag{8.74}$$

As before, for some $\eta^0 \neq 0$, the time-fuel optimal control steering system (8.72) for $x(0) = x^0$ to $x(t_f) = 0$ is uniquely determined (almost everywhere) by

$$u^*(t) = \begin{cases} \operatorname{sgn}(\eta_2(t)); & |\eta_2(t)| > a(1-\mu) \\ 0; & |\eta_2(t)| < a(1-\mu) \end{cases} \tag{8.75}$$

The control sequences $\{\pm 1\}$, $\{0, \pm 1\}$, $\{\mp 1, 0, \pm 1\}$ only can be optimal with switches occurring at the isolated points (two at most) where $|\eta_2(t)| = a(1-\mu)$. Again the $\{\pm 1\}$ control trajectories correspond to the unique p-path (Γ^+) and unique n-path (Γ^-) to the origin which form the set

$$\Gamma = \Gamma^+ \cup \Gamma^- = \{x\colon \xi(x) = 0\} \tag{8.75a}$$

depicted in Fig. 8.7, and where, from (5.17), $\xi\colon \mathbb{R}^2 \to \mathbb{R}$ is defined as

$$\xi(x) = x_1 - \frac{x_2}{\lambda} - \frac{\operatorname{sgn} x_2}{a\lambda^2} \ln(1 - a\lambda|x_2|) \tag{8.75b}$$

Γ constitutes the set of optimal $0 \to \pm 1$ control switches. For $x \in \Gamma$, the optimal control is given by

$$u(x) = -\operatorname{sgn}(x_2); \qquad x \in \Gamma \tag{8.76}$$

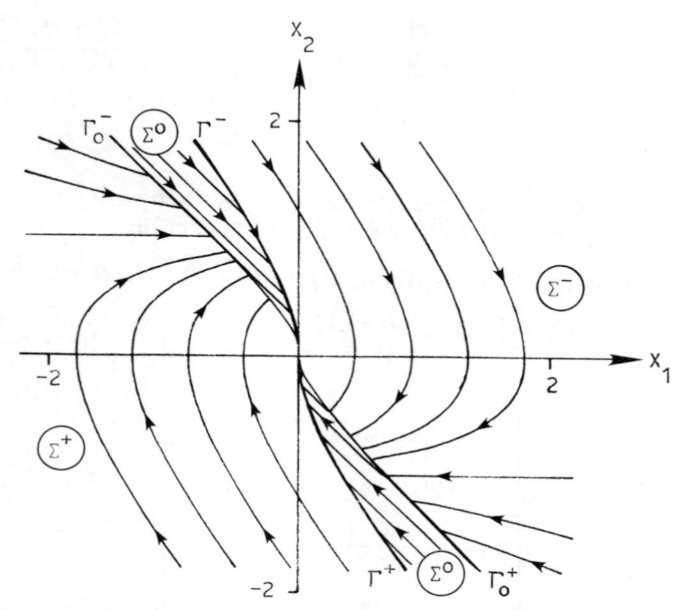

Fig. 8.7 *Time-fuel-optimal synthesis for integrator-plus-lag system with* $a = 1$, $\lambda = -1$, $\mu = \frac{1}{2}$

The set $\Gamma_0 = \Gamma_0^+ \cup \Gamma_0^-$ of optimal $\mp 1 \to 0$ control switches may be calculated, in a similar manner to that of the previous examples, by considering the maximal o-paths AB connecting $x^A \in \Gamma_0$ to $x^B \in \Gamma$. The overall portrait of o-paths (i.e. solutions of (8.72) with $u \equiv 0$) constitutes a family of straight lines (of slope $1/\lambda$) which go to the x_1-axis in forwards time from the upper and lower half-planes. Hence, the o-path AB is a straight line segment, such that the state coordinates have the same polarity at points A and B. The duration t_1 of the maximal o-path AB is easily calculated by considering the associated adjoint solution $\eta_2(t), 0 \leq t \leq t_1$. In particular

$$\eta_2(t_1) = -a(1-\mu)\,\mathrm{sgn}\,(x_2);$$

$$\eta_2^0 = \eta_2(0) = -\eta_2(t_1) = a(1-\mu)\,\mathrm{sgn}\,(x_2) \qquad (8.77)$$

As the Hamiltonian H vanishes identically along an optimal trajectory, evaluating H at $t = 0$ gives

$$H(x^A, u(0), \eta(0)) = \eta_1^0 x_2^A + \eta_2^0 \lambda x_2^A - \mu = 0 \qquad (8.78a)$$

so that

$$\eta_1^0 = \frac{\mu}{x_2^A} - \lambda \eta_2^0 \qquad (8.78b)$$

Combining (8.74), (8.77), (8.78) and omitting the arbitrary superscript A, gives the duration of the maximal o-path AB as a function of the starting point $x \in \Gamma_0$ as

$$t_1 = T_1(x) = -\frac{1}{\lambda} \ln [1 - 2a\lambda\mu^{-1}(1-\mu)|x_2|]; \qquad x \in \Gamma_0 \tag{8.79}$$

It is now straightforward to characterise Γ_0 explicitly by integrating the state equations along the o-path ($u \equiv 0$) AB, yielding the relation

$$x \in \Gamma_0 \Leftrightarrow \exp(AT_1(x))x \in \Gamma \Leftrightarrow \xi(\exp(AT_1(x))x) \triangleq \xi_0(x) = 0 \tag{8.80}$$

i.e. $x \in \Gamma_0$ if and only if $\xi_0(x) = \xi(\exp(AT_1(x))x) = 0$ where $\xi\colon \mathbb{R}^2 \to \mathbb{R}$ is given by (8.75b).

Now

$$\exp(AT_1(x))x = \begin{bmatrix} x_1 + \dfrac{x_2}{\lambda}[\exp(\lambda T_1(x)) - 1] \\ \exp(\lambda T_1(x))x_2 \end{bmatrix} = \begin{bmatrix} x_1 + \dfrac{2a\mu^{-1}(1-\mu)x_2|x_2|}{1 - 2a\lambda\mu^{-1}(1-\mu)|x_2|} \\ \dfrac{x_2}{1 - 2a\lambda\mu^{-1}(1-\mu)|x_2|} \end{bmatrix} \tag{8.81}$$

Hence the function $\xi_0\colon \mathbb{R}^2 \to \mathbb{R}$ is given by

$$\begin{aligned}\xi_0(x) &= \xi(\exp(AT_1(x))x) \\ &= x_1 + \frac{2a\mu^{-1}(1-\mu)x_2|x_2|}{1 - 2a\lambda\mu^{-1}(1-\mu)|x_2|} - \frac{1}{\lambda}\left[\frac{x_2}{1 - 2a\lambda\mu^{-1}(1-\mu)|x_2|}\right] \\ &\quad - \frac{\operatorname{sgn} x_2}{a\lambda^2} \ln [1 - a\lambda|x_2|(1 - 2a\lambda\mu^{-1}(1-\mu)|x_2|)^{-1}]\end{aligned}$$

which simplifies to

$$\begin{aligned}\xi_0(x) = x_1 - \frac{x_2}{\lambda} - \frac{\operatorname{sgn} x_2}{a\lambda^2} \\ \times \ln [1 - a\lambda|x_2|(1 - 2a\lambda\mu^{-1}(1-\mu)|x_2|)^{-1}]\end{aligned} \tag{8.82a}$$

so that

$$\Gamma_0 = \Gamma_0^+ \cup \Gamma_0^- = \{x\colon \xi_0(x) = 0\} \tag{8.82b}$$

Finally, the regions $\Sigma^+, \Sigma^0, \Sigma^-$ may be explicitly characterised as

$$\begin{aligned}\Sigma^+ &= \{x\colon \xi(x) < 0;\quad \xi_0(x) < 0\} \cup \Gamma^+ \\ \Sigma^0 &= \{x\colon \xi(x) < 0;\quad \xi_0(x) \geq 0\} \cup \{x\colon \xi(x) > 0;\quad \xi_0(x) \leq 0\} \\ \Sigma^- &= \{x\colon \xi(x) > 0;\quad \xi_0(x) > 0\} \cup \Gamma^-\end{aligned} \tag{8.83a}$$

where

$$\xi(x) = x_1 - \frac{x_2}{\lambda} - \frac{\operatorname{sgn} x_2}{a\lambda^2} \ln(1 - a\lambda|x_2|)$$

$$\xi_0(x) = x_1 - \frac{x_2}{\lambda} - \frac{\operatorname{sgn} x_2}{a\lambda^2} \times \ln[1 - a\lambda|x_2|(1 - 2a\lambda\mu^{-1}(1-\mu)|x_2|)^{-1}] \tag{8.83b}$$

The optimal feedback control is given by

$$u(x) = \begin{cases} +1; & x \in \Sigma^+ \\ 0; & x \in \Sigma^0 \\ -1; & x \in \Sigma^- \end{cases} \tag{8.84}$$

For parameter values $\lambda = -1$, $a = 1$ and $\mu = \frac{1}{2}$ (i.e. equal weighting of time and fuel), the time-fuel-optimal synthesis is depicted in Fig. 8.7. Again, for $\mu = 1$ the curves Γ and Γ_0 coincide to give the time-optimal synthesis of Subsection 5.2.1(ii).

8.7 Time-fuel-optimal control of integrator plus unstable first-order element

Suppose the system equations are as in Section 8.6 but with the distinction that $\lambda > 0$. It is easy to verify that, in this case, the results of the previous case are again valid when restricted to the domain of null controllability (see Subsection 5.2.1(iii))

$$\mathscr{C} = \left\{x : |x_2| < \frac{1}{a\lambda}\right\} \tag{8.85}$$

For $\lambda = 1 = a$ and $\mu = \frac{1}{2}$, the resulting time-fuel-optimal synthesis is shown in Fig. 8.8.

8.8 Time-fuel-optimal control of the triple integrator

The time-fuel-optimal regulator synthesis will now be derived (see also Ryan 1978b) for the triple integrator system

$$\dot{x}_1 = x_2; \qquad \dot{x}_2 = x_3; \qquad \dot{x}_3 = \frac{u}{a}; \qquad |u| \le 1; \qquad a > 0 \tag{8.86}$$

with the associated cost functional

$$J(u) = \int_0^{t_f} [\mu + (1-\mu)|u(t)|]\, dt; \qquad \mu \in (0, 1]; \qquad t_f \text{ free} \tag{8.87}$$

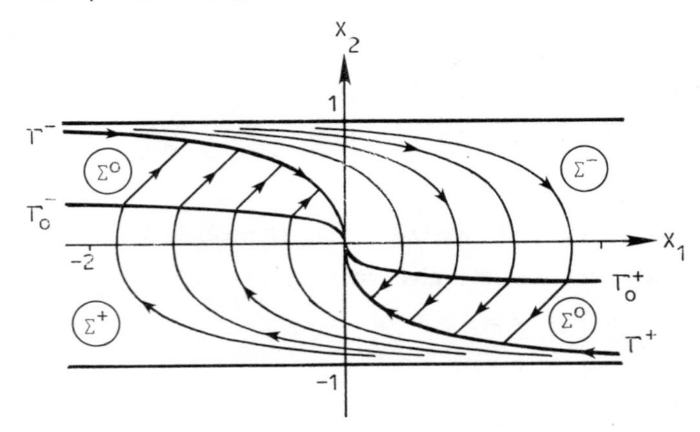

Fig. 8.8 *Domain of null controllability and time-fuel-optimal synthesis for unstable system with $a = 1$, $\lambda = 1$, $\mu = \frac{1}{2}$*

From the theory of Section 8.3, the optimal control u^* is uniquely determined (almost everywhere) as

$$u^*(t) = \begin{cases} \text{sgn}\,(\eta_3(t)); & |\eta_3(t)| > a(1-\mu) \\ 0; & |\eta_3(t)| < a(1-\mu) \end{cases} \tag{8.88}$$

where

$$\eta_3(t) = \eta_3^0 - \eta_2^0 t + \tfrac{1}{2}\eta_1^0 t^2, \qquad 0 \le t \le t_f \tag{8.89a}$$

is a nontrivial solution of the adjoint equation

$$\left.\begin{aligned} \dot{\eta} &= -A^T\eta \\ \eta(0) &= \eta^0 \end{aligned}\right\} \Leftrightarrow \begin{cases} \dot{\eta}_1 = 0; \quad \dot{\eta}_2 = -\eta_1; \quad \dot{\eta}_3 = -\eta_2 \\ \eta_1(0) = \eta_1^0; \quad \eta_2(0) = \eta_2^0; \quad \eta_3(0) = \eta_3^0 \end{cases} \tag{8.89b}$$

By lemma 8.3.1, $|\eta_3(t)| \neq a(1-\mu)$ almost everywhere so that, in view of (8.88) and the *quadratic* function (8.89a), it may be concluded that:

(P1) the control sequences $\{\pm 1\}$, $\{0, \pm 1\}$, $\{\mp 1, 0, \pm 1\}$, $\{\pm 1, 0, \pm 1\}$, $\{0, \mp 1, 0, \pm 1\}$, $\{\pm 1, 0, \mp 1, 0, \pm 1\}$ only can be optimal;

(P2) on an optimal solution with full control sequence $\{\pm 1, 0, \mp 1, 0, \pm 1\}$ the intervals of zero control are of equal duration.

The following results for the double integrator subsystem will be required later.

8.8.1 Double integrator subsystem

From the results of Section 8.4, the subsequences $\{\pm 1\}$, $\{0, \pm 1\}$, $\{\mp 1, 0, \pm 1\}$ only can be optimal for the double integrator subsystem

$$\dot{x}_2 = x_3; \qquad \dot{x}_3 = \frac{u}{a}; \qquad |u| \le 1; \qquad a > 0 \tag{8.90}$$

Moreover, from (8.48), (8.49), (8.50), on an optimal $\{\mp 1, 0, \pm 1\}$ control generated trajectory with initial subsystem state $x^s = (x_2, x_3)'$, the durations τ_i of the successive intervals of constant control are given by

$$\left.\begin{aligned} \tau_1 &= \tau_1(x^s) = ax_3 \operatorname{sgn}(\xi^s) + \tau_3(x^s) \\ &\qquad (\mp 1 \text{ control interval}) \\ \tau_2 &= \tau_2(x^s) = 2\mu^{-1}(1-\mu)\tau_3(x^s) \\ &\qquad (0 \text{ control interval}) \\ \tau_3 &= \tau_3(x^s) = \sqrt{\left(\frac{\mu}{2-\mu}\right)(ax_2 \operatorname{sgn}(\xi^s) + \tfrac{1}{2}(ax_3)^2)} \\ &\qquad (\pm 1 \text{ control interval}) \end{aligned}\right\} \tag{8.91a}$$

where $x^s = (x_2, x_3)'$ is the initial subsystem state and

$$\xi^s = \xi^s(x^s) = x_2 + \tfrac{1}{2}ax_3|x_3| \tag{8.91b}$$

8.8.2 Switching surfaces in (Θ, Φ)*-space*

Recalling the multiple integrator system invariance properties of Section 3.4, it is straightforward to show that the time-fuel-optimal control problem under consideration exhibits property 3.4.3 of special invariance. As a consequence of this property the dynamic behaviour can be analysed in the two-dimensional (Θ, Φ)-space of Section 3.5. Since the control takes the values $+1$, 0, -1 almost everywhere, the optimal vector field (state portrait) is comprised of p-paths in a region Σ^+, o-paths in a region Σ^0, and n-paths in a region Σ^-. The individual portraits of p-paths and n-paths in (Θ, Φ)-space have previously been depicted in Figs. 3.9 and 3.10; the individual portrait of o-paths in (Θ, Φ)-space is shown in Fig. 8.9, where the singular points E^+ and E^- correspond to the positive and negative x_1 state semi-axes, respectively. The separating boundaries Γ_i of the regions Σ^+, Σ^0, Σ^- constitute the optimal switching surfaces in x-state space which map to optimal switching curves in (Θ, Φ)-space. Properties P1 and P2 above, together with the subsystem results (8.91), enable the boundaries Γ_i and the optimal state portrait to be constructed as follows.

Adopting an illustrative cost parameter value $\mu = \frac{1}{2}$ (i.e. equal weighting of time and fuel) and referring to Fig. 8.10, the unique p- and n-paths which lead to the state origin and which correspond to optimal $\{\pm 1\}$ control sequence trajectories map to the points C^+ and C^-. Since optimal $\{\mp 1, 0, \pm 1\}$ trajectories, when projected into the (x_2, x_3) subspace, must coincide with optimal $\{\mp 1, 0, \pm 1\}$ subsystem trajectories, it follows that the set Γ_1 of initial states for which $\{\mp 1, 0, \pm 1\}$ controls are optimal maps to Γ_1 in (Θ, Φ)-space, comprised of the p-path (Γ_1^+) connecting C^+ to A^- and the n-path (Γ_1^-) connecting C^- to A^+ where A^+, A^- are such that the time of transition τ_2 along the o-paths Γ_3^+, Γ_3^- (from A^+ to C^+ and A^- to C^-) satisfies (8.91) for all $x \in \Gamma_1$. By property P2, on

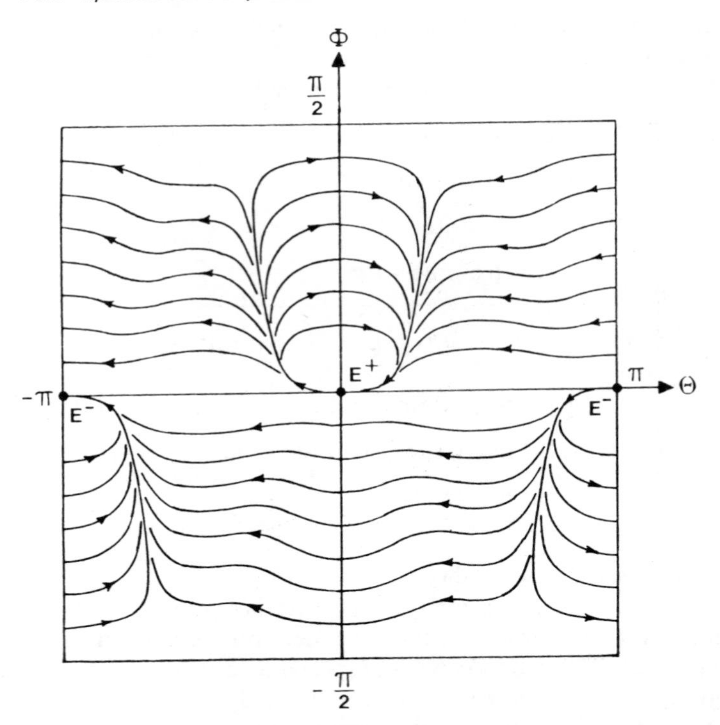

Fig. 8.9 o-*paths*

optimal $\{0, \mp 1, 0, \pm 1\}$ trajectories the duration of the first 0-control interval is at most equal to the duration of the second. Denoting by Γ_2 the set of initial states for which the two 0-control intervals are of equal duration, then for each point $p \in \Gamma_1$ a corresponding point $q \in \Gamma_2$ is generated by tracing, in reverse time from p, a o-path pq of duration τ_2 (satisfying (8.91) at p). This construction yields Γ_2^+ connecting C^+ to B^- and Γ_2^- connecting C^- to B^+. Finally, denoting by $\Gamma_4 = \Gamma_4^+ \cup \Gamma_4^-$ the o-paths connecting B^+ to A^+ and B^- to A^-, it may be concluded that for initial states in the region

(i) Σ_-^0 bounded by $\Gamma_1^+ \cup \Gamma_2^+ \cup \Gamma_4^-$ the sequence $\{0, +1, 0, -1\}$ is optimal;
(ii) Σ_+^0 bounded by $\Gamma_1^- \cup \Gamma_2^- \cup \Gamma_4^+$ the sequence $\{0, -1, 0, +1\}$ is optimal;
(iii) Σ^- lying above the curve $\Gamma_1^- \cup C^- \cup \Gamma_3^- \cup \Gamma_4^- \cup \Gamma_2^+ \cup C^+ \cup \Gamma_3^+$ the sequence $\{-1, 0, +1, 0, -1\}$ is optimal;
(iv) Σ^+ lying below the curve $\Gamma_2^- \cup C^- \cup \Gamma_3^- \cup \Gamma_1^+ \cup C^+ \cup \Gamma_3^+ \cup \Gamma_4^+$ the sequence $\{+1, 0, -1, 0, +1\}$ is optimal.

Writing $\Sigma^0 = \Sigma_+^0 \cup \Sigma_-^0$ and combining the above results, the optimal control is given by

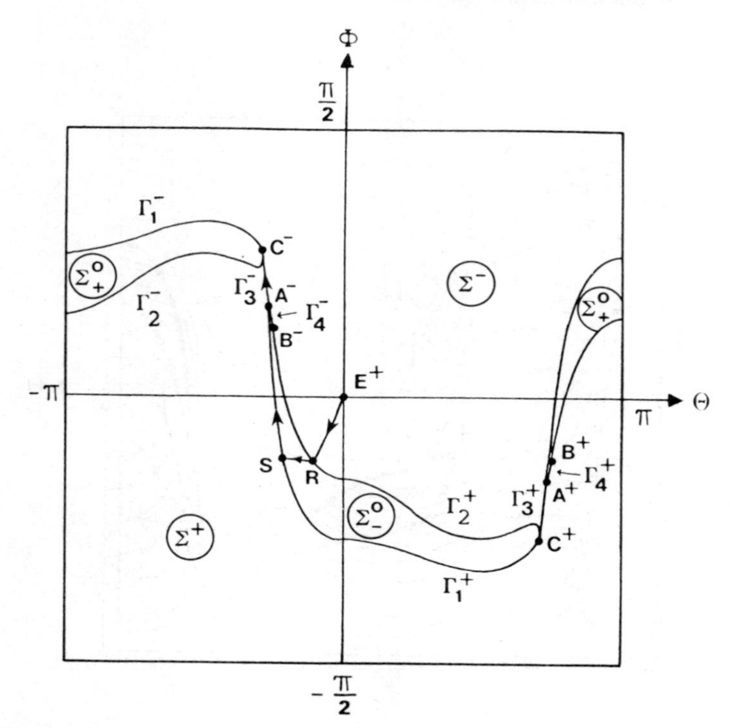

Fig. 8.10 $\mu = \frac{1}{2}$
Time-fuel-optimal switching surfaces in (Θ, Φ)-space

$$u(x) = \begin{cases} +1; & x \in \Sigma^+ \cup \Gamma_1^+ \cup C^+ \\ 0; & x \in \Sigma^0 \cup \Gamma_2 \cup \Gamma_3 \cup \Gamma_4 \\ -1; & x \in \Sigma^- \cup \Gamma_1^- \cup C^- \end{cases} \tag{8.92}$$

Moreover, a straightforward but lengthy algebraic analysis enables the switching boundaries Γ_i and regions $\Sigma^+, \Sigma^0, \Sigma^-$ to be expressed as explicit functions of the state x. These expressions and the resulting explicit feedback characterisation of the control (8.92) are, however, of a high level of complexity and can be found in (Ryan 1977c).

The optimal state portrait is shown in Fig. 8.11. Consider, for example, initial states $x = (x_1, 0, 0)'$ on the x_1 state axis; if $x_1 > 0$, then sequence $\{-1, 0, +1, 0, -1\}$ is optimal and the corresponding state trajectories map to trajectory $E^+RSA^-C^-$ of Fig. 8.10.

A representative value $\mu = \frac{1}{2}$ was adopted in Figs. 8.10 and 8.11. Clearly, similar results are obtained for arbitrary $\mu \in (0, 1]$. Two cases are of particular interest, viz.

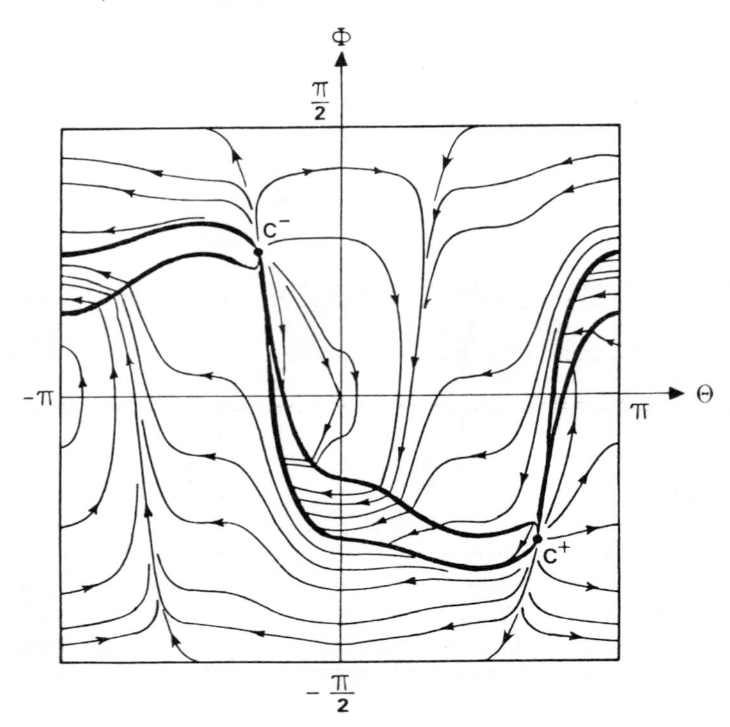

Fig. 8.11 $\mu = \frac{1}{2}$
Time-fuel-optimal state portrait in (Θ, Φ)-space

(*a*) $\mu = 1$ (time-optimal control) for which A^+, B^+ (A^-, B^-) are confluent at C^+ (C^-), i.e. Σ^0 is empty; Γ_1^+, Γ_2^+ coincide with the p-path from C^+ to C^- (and Γ_1^-, Γ_2^- coincide with the n-path from C^- to C^+) giving the time-optimal switching curve and state portrait of Fig. 3.11.
(*b*) $\mu \to 0$, in which case A^+, $B^+ \to E^-$ (A^-, $B^- \to E^+$) and $\Gamma_1^+ \to$ p-path from C^+ to E^+ ($\Gamma_1^- \to$ n-path from C^- to E^-) implying that, in the limit, the x_1 state axis belongs to the o-control region Σ^0. Consequently, on trajectories which start at, or arrive at, a point $x = (x_1, 0, 0)'$, $x_1 \neq 0$, the state will subsequently remain stationary at x and the origin cannot be attained in finite time; this again illustrates that solutions to the fuel-optimal control problem with free end-time (i.e. $\mu = 0$) do not exist in general, as discussed previously in Subsection 8.2.7.

Chapter 9

Minimisation of quadratic and nonquadratic cost functionals

9.1 Introduction

The optimal quadratic cost regulator problem for linear systems with scalar saturating control was considered initially by Letov (1960a, b; 1961) and Fuller (1960a, b, c) and subsequently studied by many authors including Chang (1961), Fuller (1963a; 1964a, b), Wonham (1963), Wonham and Johnson (1964), Grensted and Fuller (1965), Johnson and Wonham (1965), Zachary *et al.* (1965), Zachary (1966a, b), Rohrer and Sobral (1966), Tchamran (1966), Sirisena (1968), Roberts (1973), Bershchanskii (1976), Buelens, Van Rompay and Hellinckx (1978). A characteristic feature of quadratic-cost saturating control problems is the existence of singular solutions† in certain subspaces of the state space $\mathbb{R}^n$, i.e. optimal trajectories can exhibit segments or subarcs along which the Hamiltonian does not depend explicitly on the control u. These singular subarcs may be synthesised by linear feedback control acting over an infinite time horizon while the nonsingular solutions are generated by bang-bang controls. Hence, the optimal control usually has a dual mode character: linear feedback in a singular set and bang-bang elsewhere. However, in special cases (Fuller 1960c, Grensted and Fuller 1965) where one component, only, of the state vector appears in the cost functional, the control problem turns out to be strictly nonsingular; optimal controls are bang-bang and the desired state (origin) is attained in finite time. These nongeneric cases do not exhibit the usual 'dual-mode' features associated with quadratic-cost saturating control problems and have a structure which is more amenable to analysis. Moreover, these special cases provide the key to the nonquadratic-cost regulator problem (Ryan 1979, Ryan and Dorling 1981),

† See, for example, Johnson and Gibson (1963), Johnson (1965), Bell (1975), Bell and Jacobson (1975), Gabasov and Kirillova (1978) and Haas (1978) for general treatments of singular control problems. For computational approaches to the quadratic-cost problem for saturating systems (including fixed end-time problems) see, e.g. Brennan and Roberts (1962), Newmann and Zachary (1965), Edgar and Lapidus (1972), Roberts (1973), Sirisena (1974), Flaherty and O'Malley (1977) (and bibliography therein) and Aly (1978).

studied in Section 9.3, for which the structural properties of the special quadratic-cost problems are preserved. The quadratic-cost problem is considered initially.

9.2 Quadratic-cost regulator problem

The regulator problem to be considered is the following: minimise the quadratic cost functional†

$$J_M(u) = \int_0^{t_f} \langle x(t), Mx(t) \rangle \, dt; \qquad t_f \text{ free} \tag{9.1}$$

subject to

$$\dot{x} = Ax + bu; \qquad x(t) \in \mathbb{R}^n; \qquad |u(t)| \leq 1 \tag{9.2a}$$

$$x(0) = x^0 \tag{9.2b}$$

$$x(t_f) = 0 \tag{9.2c}$$

It is assumed that (A, b) is a controllable pair and of the canonical form

$$A = \begin{bmatrix} 0 & 1 & 0 & \cdots & 0 \\ 0 & 0 & 1 & & 0 \\ \vdots & & & \ddots & \vdots \\ 0 & 0 & 0 & \cdots & 1 \\ a_1 & a_2 & a_3 & \cdots & a_n \end{bmatrix}; \qquad b = \begin{bmatrix} 0 \\ 0 \\ \vdots \\ 0 \\ \frac{1}{a} \end{bmatrix} \qquad (a > 0) \tag{9.2d}$$

i.e. the state (phase) vector x is comprised of the scalar output x_1 and its first $(n-1)$ derivatives of a single-input, single-output system with transfer function

$$G(s) = \frac{1}{a(s^n - a_n s^{n-1} - a_{n-1} s^{n-2} - \cdots - a_2 s - a_1)}$$

Wonham and Johnson (1964) have shown that there is no loss in generality in assuming the matrix M to be diagonal,‡ i.p.

$$M = \text{diag}\,(\mu_1, \mu_2, \ldots, \mu_n)$$

† For the more general formulation in which a quadratic control penalty term is also included in the cost integrand (i.e. $\langle x, Mx \rangle + u^2$) see Letov (1960a, b; 1961), Johnson and Wonham (1965), Tchamran (1966).

‡ If M is not diagonal then the product terms $x_i x_j (i \neq j)$ in the cost integral, associated with the non-zero off-diagonal elements of M, can be removed by integration by parts. In view of the boundary conditions (9.2b, c), this procedure yields the required diagonalised cost integral plus a quadratic term $\langle x^0, Nx^0 \rangle$ which depends only on the initial state x^0 and hence plays no role in determining the optimal control u^*.

In summary, the quadratic-cost regulator problem is to determine an admissible control function $u^*(t)$, $0 \leq t \leq t_f$ (if one exists), which generates a trajectory $x(\cdot)$ from $x(0) = x^0$ to the origin $x(t_f) = 0$ along which the cost (9.1) is minimised. In previous chapters, the 'free' terminal time was assumed finite, i.e. $0 \leq t_f < \infty$. In this section, t_f is taken as belonging to the extended non-negative reals, i.e.

$$0 \leq t_f \leq \infty \tag{9.3}$$

so that if the origin is attained 'at infinity', i.e. $x(\infty) = 0$, then boundary condition (9.2c) is deemed to be satisfied. Alternatively (9.2c) could be rewritten in the form

$$x(t_f) \to 0 \qquad \text{as} \qquad t_f \to \infty$$

Thus exponentially stable state trajectories ($\|x(t)\| \leq Ke^{\alpha t}$; $K > 0$, $\alpha < 0$) may be considered as candidate solutions. In fact, the existence of such trajectories which approach the origin exponentially is a characteristic feature of the quadratic-cost problem as will be seen.

9.2.1 Equivalent quadratic cost functional

Sirisena (1968) (see also Bershchanskii 1976) has established the following result.

Lemma 9.2.1: If the elements μ_i of the matrix $M = \text{diag}\,(\mu_1, \mu_2, \ldots, \mu_n)$ satisfy the relation

$$P(\omega) = \sum_{i=1}^{n} \mu_i \omega^{2(i-1)} = \mu_1 + \mu_2 \omega^2 + \cdots + \mu_n \omega^{2(n-1)} > 0 \;\forall\; \omega \in \mathbb{R} \tag{9.4}$$

then, corresponding to the quadratic cost functional (9.1) there exists an equivalent† quadratic cost functional

$$J(u) = \int_0^{t_f} \langle c, x(t) \rangle^2 \, dt; \qquad t_f \text{ free}; \qquad c \in \mathbb{R}^n \tag{9.5}$$

where the diagonal elements μ_i of M and the components c_i of the real vector $c \in \mathbb{R}^n$ are related via the equations

$$\begin{aligned} \mu_1 &= c_1^2 \\ \mu_i &= c_i^2 + 2\sum_{j=1}^{i-1} (-1)^{i+j} c_j c_{2i-j}; \qquad i = 2, 3, \ldots, n-1 \\ \mu_n &= c_n^2 \end{aligned} \tag{9.6}$$

and where c_{2i-j} is taken as zero whenever $2i - j > n$.

† Equivalent in the sense that if a control u^* minimises one cost functional then the same control u^* minimises the other.

Proof
Integration by parts, together with the boundary conditions (9.2b, c), gives

$$\int_0^{t_f} \langle c, x(t)\rangle^2 \, dt = \int_0^{t_f} \langle x(t), Mx(t)\rangle \, dt + \langle x^0, Nx^0\rangle$$

where the elements of $c = (c_1, c_2, \ldots, c_n)'$ and $M = \text{diag}(\mu_1, \mu_2, \ldots, \mu_n)$ satisfy (9.6) and where N is a matrix which depends only on the vector c. Hence, the quadratic form $\langle x^0, Nx^0\rangle$ depends only on the initial state x^0 and cost vector c and therefore plays no role in determining the optimal control u. Consequently, given a cost functional of form (9.5) with vector c, an equivalent cost functional of form (9.1), which admits the same minimising control, is uniquely defined by (9.6). On the other hand, given an *arbitrary* diagonal matrix M, a *real* vector c satisfying equations (9.6) may *not* exist; it will now be shown that restriction (9.4) on M is sufficient to ensure the existence of at least one real vector $c \in \mathbb{R}^n$ satisfying (9.6). Note initially that (9.4) implies that

$$\mu_1 > 0 \tag{9.7}$$

Suppose that $1 \leq p \leq n$ is the index of the highest non-zero diagonal element of M, i.e.

$$\mu_p \neq 0; \qquad \mu_{p+1} = \mu_{p+2} = \cdots = \mu_n = 0 \tag{9.8}$$

then (9.4) also implies that

$$\mu_p > 0 \tag{9.9}$$

Consider now the polynomial

$$\begin{aligned} \tilde{P}(\beta) = P(i\beta) &= \sum_{j=1}^{p} \mu_j (i\beta)^{2(j-1)} \\ &= \sum_{j=1}^{p} (-1)^{j-1} \mu_j (\beta)^{2(j-1)} \\ &= \mu_1 - \mu_2 \beta^2 + \cdots + (-1)^{p-1} \mu_p \beta^{2(p-1)} \end{aligned} \tag{9.10}$$

which may be factored into the product of two polynomials, viz.

$$\tilde{P}(\beta) = P_1(\beta) P_1(-\beta) \tag{9.11}$$

where

$$P_1(\beta) = \sum_{j=1}^{p} c_j \beta^{j-1} = c_1 + c_2 \beta + \cdots + c_p \beta^{p-1}$$

and where the coefficients c_i (not necessarily real) satisfy (9.6). Let β_j, $j = 1, 2, \ldots, p-1$, denote the roots of $P_1(\beta)$, then the $2(p-1)$ roots of $\tilde{P}(\beta) = P_1(\beta)P_1(-\beta)$ are $\pm\beta_j$, $j = 1, 2, \ldots, p-1$, and grouping the factors $(\beta + \beta_j)(\beta - \beta_j)$, gives

$$\tilde{P}(\beta) = (-1)^{p-1} c_p^2 \prod_{j=1}^{p-1} (\beta^2 - \beta_j^2) \tag{9.12}$$

This in turn gives the representation of the polynomial $P(\omega)$ as

$$P(\omega) = \tilde{P}(i\omega) = c_p^2 \prod_{j=1}^{p-1} (\omega^2 + \beta_j^2) \tag{9.13}$$

Now, (9.13) together with condition (9.4), viz.

$$P(\omega) > 0 \ \forall \text{ real } \omega$$

implies that the β_js cannot be purely imaginary, i.e.

$$\text{Re}(\beta_j) \neq 0 \qquad j = 1, 2, \ldots, p-1 \tag{9.14}$$

In particular, since to every root β_j of $\tilde{P}(\beta) = 0$ there corresponds a root $-\beta_j$, there is no loss in generality in assuming

$$\text{Re}(\beta_j) < 0; \qquad j = 1, 2, \ldots, p-1. \tag{9.15}$$

Now suppose that at least one of the coefficients c_i in (9.11) is *complex*. This supposition, in turn, implies the existence of at least one *complex* root β_k of $P_1(\beta)$ such that its conjugate $\bar{\beta}_k$ is *not* a root, i.e. $P_1(\beta_k) = 0$ and $P_1(\bar{\beta}_k) \neq 0$. Noting that the complex roots of the polynomial $\tilde{P}(\beta) = P_1(\beta)P_1(-\beta)$ (with *real* coefficients μ_i) must occur in conjugate pairs, it immediately follows that, under the above supposition, the complex root β_k must be purely imaginary, i.e. $\text{Re}(\beta_k) = 0$, which clearly contradicts (9.15). Hence, under condition (9.4), all coefficients c_i are *real*.

Finally, without loss of generality, c_1 and c_p may be defined (see (9.6)) to be the *positive* square roots of $\mu_1 > 0$ and $\mu_p > 0$, respectively, which, when combined with (9.15), implies that

$$c_i > 0 \qquad i = 1, 2, \ldots, p \qquad \text{with} \qquad c_{p+1} = c_{p+2} = \cdots = c_n = 0 \tag{9.16}$$

Hence, given a diagonal matrix M with elements μ_i satisfying (9.4), a real vector $(c_1, c_2, \ldots, c_n)' = c \in \mathbb{R}^n$ satisfying (9.6) can be found so that the cost functionals (9.1) and (9.5) are equivalent. This establishes the lemma.

Condition (9.4) will be assumed in the sequel and the equivalent problem formulation based on cost functional (9.5) will be adopted. Note that (9.4) is satisfied by all positive semidefinite matrices M with non-zero leading diagonal element ($\mu_1 > 0$); moreover, (9.4) may also hold for some $\mu_i < 0$, i.e. for some diagonal matrices M which fail to be positive semidefinite. Hence (9.4) represents a relaxation of the usual condition of positive semidefiniteness of the weighting matrix M.

Example 9.2.1 : If

$$M = \begin{bmatrix} \mu_1 & 0 \\ 0 & \mu_2 \end{bmatrix},$$

then, from (9.4) and (9.6), $\mu_1 > 0$ and $\mu_2 \geq 0$ with

$$c = (c_1, c_2)' = (\sqrt{\mu_1}, \sqrt{\mu_2})' \tag{9.17}$$

and the equivalent cost functional (9.5) becomes

$$J(u) = \int_0^{t_f} [\sqrt{\mu_1}\, x_1(t) + \sqrt{\mu_2}\, x_2(t)]^2 \, dt; \qquad t_f \text{ free.}$$

If

$$M = \begin{bmatrix} \mu_1 & 0 & 0 \\ 0 & \mu_2 & 0 \\ 0 & 0 & \mu_3 \end{bmatrix}$$

then, from (9.4)

$$\mu_1 > 0; \qquad \mu_3 \geq 0; \qquad \mu_2 \geq -2\sqrt{\mu_1 \mu_3}$$

and, from (9.6)

$$c = \begin{bmatrix} c_1 \\ c_2 \\ c_3 \end{bmatrix} = \begin{bmatrix} \sqrt{\mu_1} \\ \sqrt{\mu_2 + 2\sqrt{\mu_1 \mu_3}} \\ \sqrt{\mu_3} \end{bmatrix} \tag{9.18}$$

e.g. if $\mu_1 = \mu_3 = 1$ and $\mu_2 = -1$ (i.e. M not positive semidefinite) then the cost functionals

$$J_M(u) = \int_0^{t_f} [x_1(t)^2 - x_2(t)^2 + x_3(t)^2] \, dt$$

and

$$J(u) = \int_0^{t_f} [x_1(t) + x_2(t) + x_3(t)]^2 \, dt$$

are equivalent.

9.2.2 Ideal model and optimal zero-cost trajectories

Suppose the initial state x^0 satisfies

$$\langle c, x^0 \rangle = 0 \tag{9.19}$$

and that an admissible control $u^*(t)$, $0 \leq t \leq t_f$, can be found such that the corresponding trajectory $x(\cdot)$ satisfies

$$\langle c, x(t) \rangle = 0 \qquad 0 \leq t \leq t_f \tag{9.20}$$

then, from (9.5), this trajectory incurs zero cost, $J(u^*) = 0$. Furthermore, differentiating (9.20) $n - p$ times yields the following set of equations which must also be satisfied on $(0, t_f)$

$$\frac{d}{dt}\langle c, x\rangle = c_1x_2 + c_2x_3 + \cdots + c_px_{p+1} = \langle c, Ax\rangle = 0$$

$$\frac{d^2}{dt^2}\langle c, x\rangle = c_1x_3 + c_2x_4 + \cdots + c_px_{p+2} = \langle c, A^2x\rangle = 0 \tag{9.21}$$

$$\vdots$$

$$\frac{d^{n-p}}{dt^{n-p}}\langle c, x\rangle = c_1x_{n-p+1} + c_2x_{n-p+2} + \cdots + c_px_n = \langle c, A^{n-p}x\rangle = 0$$

Thus, combining (9.20) and (9.21), it may be concluded that all zero-cost trajectories must lie in the $(p-1)$-dimensional linear subspace Π corresponding to the intersection of $(n-p+1)$ hyperplanes, viz.

$$\Pi = \{x\colon \langle c, A^ix\rangle = 0; i = 0, 1, 2, \ldots, n-p\} \tag{9.22}$$

Taking the $(n-p+1)$th-derivative of (9.20) yields the relation

$$\begin{aligned}\frac{d^{n-p+1}}{dt^{n-p+1}}\langle c, x\rangle &= \frac{d}{dt}\langle c, A^{n-p}x\rangle\\ &= \langle c, A^{n-p+1}x\rangle + \langle c, A^{n-p}b\rangle u\\ &= 0\end{aligned}$$

which must also hold on a zero-cost trajectory. Hence for zero-cost motion in Π the control must satisfy

$$u = -\langle c, A^{n-p}b\rangle^{-1}\langle c, A^{n-p+1}x\rangle \tag{9.23}$$

and the state equation becomes

$$\begin{aligned}\dot{x} &= Ax - b\langle c, A^{n-p}b\rangle^{-1}\langle c, A^{n-p+1}x\rangle\\ &= [I - \langle c, A^{n-p}b\rangle^{-1}bc'A^{n-p}]Ax\\ &= \tilde{A}x\end{aligned} \tag{9.24}$$

governing the motion in the 'zero-cost' subspace Π. Now, in Π

$$\begin{aligned}\langle c, x\rangle &= c_1x_1 + c_2x_2 + \cdots + c_px_p\\ &= c_p\frac{d^{p-1}}{dt^{p-1}}x_1 + c_{p-1}\frac{d^{p-2}}{dt^{p-2}}x_1\\ &\quad + \cdots + c_2\frac{dx_1}{dt} + c_1x_1\\ &= 0\end{aligned} \tag{9.25}$$

so that (9.25) may be regarded as an idealised 'zero-cost' model whose dynamic behavior the actual system attempts to match by appropriate choice of control input; this match is exact in the zero-cost subspace Π.

Now, as regulation, i.e.

$$x(t) \to 0 \qquad \text{as} \qquad t \to \infty$$

is the primary control objective, the ideal model (9.25) associated with the cost functional should be asymptotically stable, i.e. the roots β_i of the 'ideal model' characteristic equation

$$c_1 + c_2\beta + \cdots + c_p\beta^{p-1} = 0$$

should have negative real parts which is in agreement with (9.15). Note also that this interpretation of the cost functional defining an asymptotically stable ideal model provides additional justification for condition (9.4) which, in view of (9.14), ensures that the ideal model does not have undamped oscillatory modes.

In summary, it has been established that the original cost functional (9.1) (where, without loss of generality, M is now assumed to be diagonal) may be related, via (9.6), to an equivalent cost functional of the form (9.5) provided that the diagonal elements of the original cost weighting matrix M satisfy condition (9.4). The equivalent cost functional (9.5) admits an interpretation in terms of an asymptotically stable model of ideal (zero-cost) motion. If, by appropriate choice of control, the actual system trajectory can match the model exactly, then zero cost is incurred; moreover, as the model is asymptotically stable, the zero-cost system trajectory is optimal. It has been shown that if such optimal 'model-matching' controls exist then the associated zero-cost optimal trajectories lie in the $(p-1)$-dimensional linear subspace

$$\Pi = \{x: \langle c, A^i x\rangle = 0;\ i = 0, 1, 2, \ldots, n-p\} \tag{9.26}$$

Now, a zero-cost trajectory in Π is generated by a control

$$\begin{aligned} u(t) &= -\langle c, A^{n-p}b\rangle^{-1}\langle c, A^{n-p+1}x(t)\rangle \\ &= -\langle c, A^{n-p}b\rangle^{-1}\langle c, A^{n-p+1}\exp(\tilde{A}t)x^0\rangle \end{aligned} \tag{9.27a}$$

Imposing the saturation constraint $|u(t)| \leq 1$ it may be concluded that the set $\Gamma^s \subseteq \Pi$ defined as

$$\Gamma^s = \{x \in \Pi: |\langle c, A^{n-p}b\rangle^{-1}\langle c, A^{n-p+1}\exp(\tilde{A}t)x\rangle| \leq 1, \forall\, t \geq 0\} \tag{9.27b}$$

constitutes the set of starting points from which optimal zero-cost trajectories to the origin exist. That Γ^s is convex is easily verified as follows: let x^A, $x^B \in \Gamma^s$,

then

$$
\begin{aligned}
|\langle c, A^{n-p}b\rangle^{-1}\langle c, A^{n-p+1} &\exp(\tilde{A}t)[\alpha x^A + (1-\alpha)x^B]\rangle| \\
&\le \alpha |\langle c, A^{n-p}b\rangle^{-1}\langle c, A^{n-p+1}\exp(\tilde{A}t)x^A\rangle| \\
&\quad + (1-\alpha)|\langle c, A^{n-p}b\rangle^{-1}\langle c, A^{n-p+1}\exp(\tilde{A}t)x^B\rangle| \\
&\le \alpha + (1-\alpha) = 1
\end{aligned}
$$

so that $\alpha x^A + (1-\alpha)x^B \in \Gamma^s, \forall\, \alpha \in [0, 1]$. Hence Γ^s is a convex subset of Π and clearly lies between the parallel hyperplanes

$$\langle c, A^{n-p+1}x\rangle = \pm\langle c, A^{n-p}b\rangle \tag{9.28}$$

It will now be shown that the optimal zero-cost trajectories in Γ^s correspond to singular solutions of the Pontryagin equations.

9.2.3 *Application of the maximum principle*

The Hamiltonian for the (equivalent) quadratic cost problem is

$$H(x, u, \eta) = \langle \eta, Ax\rangle + \langle \eta, bu\rangle - \langle c, x\rangle^2 \tag{9.29}$$

and, in accordance with the maximum principle (theorem 2.4.1), if $u^*(t)$, $0 \le t \le t_f$, is an optimal control then there exists a solution $\eta(t)$, $0 \le t \le t_f$ of the adjoint equation

$$\dot{\eta} = -\frac{\partial H}{\partial x} = -A^T\eta + 2\langle c, x\rangle c \tag{9.30}$$

such that

$$H(x(t), u^*(t), \eta(t)) = \max_{|u|\le 1} H(x(t), u, \eta(t)) \qquad \text{a.a.} \quad 0 \le t \le t_f \tag{9.31}$$

It follows that $u^*(t)$ is defined by the (familiar) relation

$$
\begin{aligned}
u^*(t) &= \operatorname{sgn}\langle b, \eta(t)\rangle \\
&= \operatorname{sgn}(\eta_n(t)) \qquad \text{if } \eta_n(t) \ne 0
\end{aligned}
\tag{9.32}
$$

However, if $\eta_n(t)$ vanishes on an interval I of positive measure then the Hamiltonian is independent of u and u^* is no longer defined by (9.32). In this case the control is said to be *singular* and the corresponding trajectory $x(t)$, $t \in I$, is termed a *singular subarc*. Finally, as H is time-invariant and the terminal time t_f is free, by corollary 2.4.1, the Hamiltonian takes the value zero a.e. along an optimal (singular or nonsingular) trajectory, i.e.

$$H(x(t), u^*(t), \eta(t)) = 0, \qquad \text{a.a. } 0 \le t \le t_f \tag{9.33}$$

9.2.4 Singular subarcs

From (9.2a) and (9.30), the adjoint equations are

$$\begin{aligned}\dot{\eta}_1 &= -a_1\eta_n + 2\langle c, x\rangle c_1 \\ \dot{\eta}_i &= -\eta_{i-1} - a_i\eta_n + 2\langle c, x\rangle c_i; \qquad i = 2, 3, \ldots, p \\ \dot{\eta}_j &= -\eta_{j-1} - a_j\eta_n; \qquad j = p+1, p+2, \ldots, n\end{aligned} \tag{9.34}$$

where, as before, p is the index of the highest non-zero entry in the vector c, i.e.

$$c = (c_1, c_2, \ldots, c_p, 0, \ldots, 0)'; \qquad c_p \neq 0$$

On a singular subarc, $\eta_n(t) = 0 \;\forall\, t \in I$, so that all derivatives of η_n vanish in the singular interval I. Thus, from (9.34),

$$\begin{aligned}\eta_j(t) &= 0; \qquad j = p, p+1, \ldots, n \quad \forall\, t \in \operatorname{int} I \\ \eta_{p-1}(t) &= 2\langle c, x(t)\rangle c_p\end{aligned} \tag{9.35}$$

and hence, the first $p - 1$ components of the adjoint solution must satisfy the equations

$$\begin{aligned}\dot{\eta}_1 &= c_1c_p^{-1}\eta_{p-1} \\ \dot{\eta}_i &= -\eta_{i-1} + c_ic_p^{-1}\eta_{p-1}, \qquad i = 2, 3, \ldots, p-1\end{aligned} \tag{9.36a}$$

or equivalently

$$\begin{bmatrix}\dot{\eta}_1 \\ \dot{\eta}_2 \\ \dot{\eta}_3 \\ \vdots \\ \dot{\eta}_{p-1}\end{bmatrix} = \begin{bmatrix}0 & 0 & \cdots & 0 & c_1c_p^{-1} \\ -1 & 0 & \cdots & 0 & c_2c_p^{-1} \\ 0 & -1 & \cdots & 0 & c_3c_p^{-1} \\ \vdots & \vdots & \ddots & \vdots & \vdots \\ 0 & 0 & \cdots & -1 & c_{p-1}c_p^{-1}\end{bmatrix}\begin{bmatrix}\eta_1 \\ \eta_2 \\ \eta_3 \\ \vdots \\ \eta_{p-1}\end{bmatrix} \tag{9.36b}$$

which, on solving for η_{p-1}, yields

$$\eta_{p-1}(t) = \sum_{i=1}^{k}\sum_{j=1}^{m_i}\alpha_{ij}t^{j-1}\exp(\gamma_i t); \qquad \alpha_{ij} = \text{constant} \tag{9.37}$$

where γ_i ($i = 1, 2, \ldots, k$), with multiplicity m_i (i.e. $p = \sum_{i=1}^{k} m_i$), are the distinct roots of the characteristic equation of system (9.36), viz.

$$\sum_{i=1}^{p} c_i(-\gamma)^{i-1} = c_1 - c_2\gamma + c_3\gamma^2 - \cdots + (-1)^{p-1}c_p\gamma^{p-1} = 0 \tag{9.38}$$

Now, in view of (9.15), the roots β_i of the equation $\sum_{i=1}^{p} c_i\beta^{i-1} = 0$ have negative real parts and hence the roots γ_i of (9.38) have positive real parts. Combining (9.35) and (9.37) gives

$$2c_p\langle c, x(t)\rangle = \sum_{i=1}^{k}\sum_{j=1}^{m_i}\alpha_{ij}t^{j-1}\exp(\gamma_i t), \qquad \operatorname{Re}(\gamma_i) > 0 \tag{9.39}$$

on a singular subarc. Since Re $(\gamma_i) > 0$, if $\alpha_{ij} \neq 0$ then the singular subarc contains modes which *diverge* from the origin. Intuitively, it is not plausible that such a divergent singular subarc could form part of an optimal trajectory to the origin and consequently *convergent* subarcs only will be investigated. This latter restriction carries the implication that

$$\alpha_{ij} = 0 \qquad \text{for all } i, j$$

which, in turn, yields the result

$$\langle c, x(t) \rangle = 0 \qquad \text{on a convergent singular subarc} \tag{9.40}$$

i.e. all *convergent* singular subarcs lie in the hyperplane $\langle c, x \rangle = 0$ and clearly correspond to the optimal zero-cost trajectories analysed in Subsection 9.2.2. The convex set Γ^s of such optimal, zero-cost, convergent, singular subarcs is given by (9.27) and will henceforth be referred to as the (optimal) *singular set*.

9.2.5 *Nonsingular arcs*

Outside the singular set Γ^s, candidate trajectories are nonsingular and are generated by bang-bang controls of the form (9.32). If the singular set has dimension $(n-1)$ (i.e. if $\mu_n \neq 0 \Leftrightarrow c_n \neq 0$), then a candidate trajectory consists of a nonsingular arc emanating from $x^0 \notin \Gamma^s$ which hits the singular set Γ^s at some point x^s (which may be the origin $x^s = 0$) and subsequently follows a zero-cost singular subarc in Γ^s exponentially to the origin. The field of such nonsingular arcs can be generated by integrating the state and adjoint equations, *in reverse time* $\tau = -t$, from the singular set Γ^s under control (9.32) and with the initial (reverse-time) conditions

$$\left.\begin{aligned} x(0) &\in \Gamma^s \\ \eta(0) &= 0 \\ u^*(0) &= \pm 1 \end{aligned}\right\} \tag{9.41}$$

On the other hand, if the dimension of the singular set is less than $(n-1)$ then it may not be possible to generate the nonsingular arcs by the above 'back-tracking' method; this is due to the fact that, for $p < n$, nonsingular arcs can exist which exhibit an infinite number of control switches in any finite neighbourhood of Γ^s. In effect, nonsingular arcs which impinge directly on the singular set Γ^s (as depicted schematically in Fig. 9.1*a* for an $(n-1)$-dimensional set Γ^s) may not exist when Γ^s has dimension less than $n-1$; instead, the behaviour of Fig. 9.1*b* can occur, in which case the (countable) infinity of switches on the nonsingular arcs resembles a chattering motion about the set Γ^s. The analysis of such motion (see e.g. Marchal 1973) and the junction conditions under which nonsingular and singular arcs can be pieced together (see e.g. McDanell and Powers, 1971; Bershchanskii, 1976, 1979; Bell, 1978; Bell and Boissard, 1979; Lewis, 1980) still present many unresolved difficulties which are beyond the scope of the present chapter. Instead, the characteristic features of the quadratic-cost regulator problem will be illustrated by a number of examples.

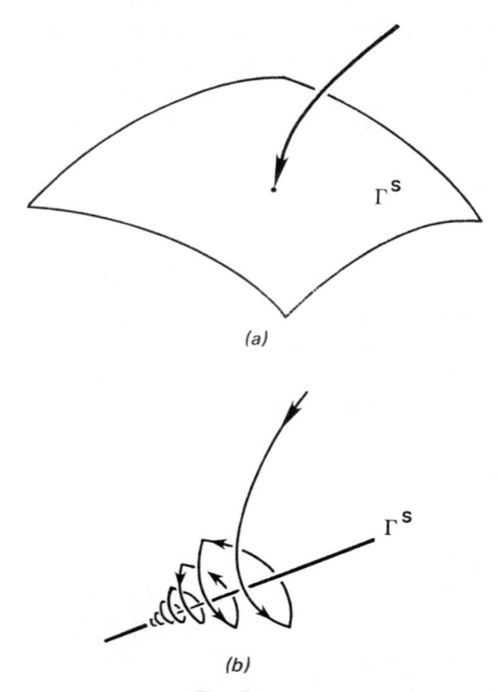

Fig. 9.1

9.2.6 Example 1: Singular set Γ^s coincident with linear subspace Π
Consider the second-order system

$$\dot{x} = Ax + bu; \qquad |u(t)| \leq 1; \qquad x(t) \in \mathbb{R}^2 \tag{9.42a}$$

where

$$A = \begin{bmatrix} 0 & 1 \\ -1 & -2 \end{bmatrix} \quad \text{and} \quad b = \begin{bmatrix} 0 \\ 1 \end{bmatrix} \tag{9.42b}$$

and a quadratic cost functional with $M = I$

$$J_M(u) = \int_0^{t_f} [x_1(t)^2 + x_2(t)^2]\, dt; \qquad t_f \text{ free} \tag{9.43}$$

By lemma 9.2.1, the problem of determining a regulating control which minimises (9.43) is equivalent to minimising (see (9.17)) the cost functional

$$J(u) = \int_0^{t_f} \langle c, x(t) \rangle^2\, dt$$

$$= \int_0^{t_f} [x_1(t) + x_2(t)]^2\, dt \tag{9.44}$$

To see the equivalence of the above minimisation problems, note that (9.44) may be written in the form

$$J(u) = \int_0^{t_f} [x_1(t)^2 + x_2(t)^2]\, dt + 2 \int_0^{t_f} x_1(t)x_2(t)\, dt$$

and integrating the second term (noting that $x_2 = \dot{x}_1$ and $x(0) = x^0$, $x(t_f) = 0$) yields the relation

$$J(u) = J_M(u) - (x_1^0)^2 \tag{9.45}$$

so that a control which minimises J also minimises J_M.

For the problem under consideration, $p = 2 = n$ and, from (9.22), the linear subspace Π is given by

$$\Pi = \{x: \langle c, x \rangle = (x_1 + x_2) = 0\} \tag{9.46}$$

It follows from (9.27a) that the motion in Π is generated by the control

$$\begin{aligned} u(t) &= -\langle c, Ax(t) \rangle; \qquad x(t) \in \Pi \\ &= x_1(t) + x_2(t) \\ &= 0 \quad (\text{since } x_1(t) + x_2(t) = 0 \text{ in } \Pi) \end{aligned}$$

i.e. the optimal singular zero-cost trajectories are generated by *zero* controls. This result is not surprising as the subspace spanned by the eigenvector $v = (1, -1)'$, associated with the eigenvalue $\lambda = -1$ (multiplicity 2) of the system matrix A, coincides with the subspace Π. Hence, for $x^0 \in \Pi$, the origin is approached exponentially along this eigenvector of the free ($u \equiv 0$) system: $\dot{x} = Ax$. Clearly, the condition

$$|u| = |\langle c, b \rangle^{-1} \langle c, Ax \rangle| \leq 1$$

is satisfied for all $x \in \Pi$ and hence the singular set Γ^s coincides with the entire subspace Π, i.e.

$$\Gamma^s = \Pi \tag{9.47}$$

as shown in Fig. 9.2. For $x \in \Gamma^s$ (= Π), the control

$$u = u(x) = 0; \qquad x \in \Gamma^s \tag{9.48}$$

is optimal and the associated state trajectory to the origin incurs zero cost $J(u) = 0$ or, in terms of the original problem formulation, the cost is given by $J_M(u) = (x_1^0)^2$.

To generate the nonsingular (or normal) arcs, the adjoint equations

$$\begin{aligned} \frac{d\eta_1}{dt} &= \eta_2 + 2(x_1 + x_2) \\ \frac{d\eta_2}{dt} &= -\eta_1 + 2\eta_2 + 2(x_1 + x_2) \end{aligned} \tag{9.49}$$

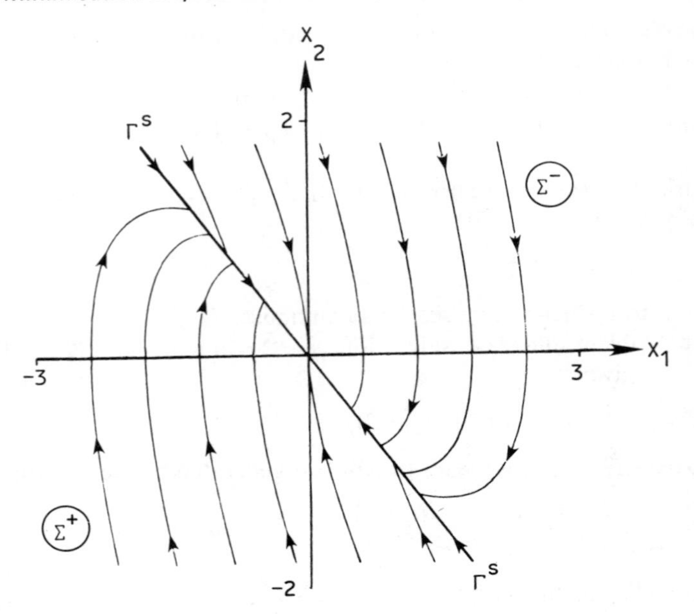

Fig. 9.2 *Example 1*
Optimal synthesis

are integrated in reverse time $\tau = -t$, under the control $u(\tau) = \text{sgn}\,(\eta_2(\tau))$, $\tau > 0$, from Γ^s with initial (reverse-time) conditions

$$\left.\begin{aligned} &x(0) = x^0 \in \Gamma^s \Leftrightarrow x(0) = (x_1^0, \ -x_1^0)' \\ &\eta(0) = 0 \\ &u(0) = u^0 = \pm 1 \end{aligned}\right\} \tag{9.50}$$

It is easily verified that, for $u^0 = +1$, the reverse-time trajectory enters the region $\Sigma^+ = \{x\colon x_1 + x_2 < 0\}$ lying below Π and subsequently remains there; $\eta_2(t)$ increases monotonically from its initial value $\eta_2(0) = 0$ so that no subsequent zeros, and hence no control switches, can occur, i.e. *switchless* nonsingular arcs (generated by *constant* nonsingular controls $u^* \equiv +1$) fill the entire region Σ^+. Similarly, for $u^0 = -1$, a field of switchless nonsingular arcs (generated by constant nonsingular controls $u^* \equiv -1$) can be constructed which fill the region $\Sigma^- = \{x\colon x_1 + x_2 > 0\}$ lying above Π. Hence, the optimal feedback control is given by

$$u = u(x) = \begin{cases} -\text{sgn}\,(x_1 + x_2); & (x_1 + x_2) \neq 0 \\ 0; & (x_1 + x_2) = 0 \end{cases} \tag{9.51}$$

In this case the singular set Γ^s plays the role of a switching line which partitions the state space into the mutually exclusive regions Σ^+ and Σ^- of optimal nonsingular control +1 and −1, respectively. Note that with the sole exception of the pair of switchless nonsingular trajectories which lead directly to the origin, all trajectories contain a singular zero-cost subarc in Π leading to the origin.

9.2.7 Example 2 : Double integrator system

Suppose now that the system matrices are

$$A = \begin{bmatrix} 0 & 1 \\ 0 & 0 \end{bmatrix}; \qquad b = \begin{bmatrix} 0 \\ 1 \end{bmatrix} \tag{9.52}$$

and the cost functional is

$$J_M(u) = \int_0^{t_f} [x_1(t)^2 + \mu x_2(t)^2]\, dt; \qquad t_f \text{ free}; \qquad \mu \geq 0 \tag{9.53}$$

which, by lemma 9.2.1, is equivalent to

$$J(u) = \int_0^{t_f} [x_1(t) + \sqrt{\mu}\, x_2(t)]^2\, dt \tag{9.54}$$

In this case, from (9.26) and (9.27), the singular set Γ^s is a bounded subset of the linear subspace

$$\Pi = \{x\colon x_1 + \sqrt{\mu}\, x_2 = 0\} \tag{9.55a}$$

Specifically,

$$\Gamma^s = \{x\colon x_1 + \sqrt{\mu}\, x_2 = 0; |x_2| \leq \sqrt{\mu}\} \subset \Pi \tag{9.55b}$$

and, for $x \in \Gamma^s$, the optimal control is given by

$$\begin{aligned} u &= -\langle c, b\rangle^{-1}\langle c, Ax\rangle \\ &= -\frac{x_2}{\sqrt{\mu}}; \qquad x \in \Gamma^s \end{aligned} \tag{9.56}$$

As in the previous example, with $\mu \neq 0$ (the special case of $\mu = 0$ will be treated separately in lemma 9.2.2), the normal arcs can be constructed by integrating the state and adjoint equations in reverse time $\tau = -t$, under the control $u(\tau) = \operatorname{sgn} \eta_2(\tau)$ $(\tau > 0)$, from Γ^s with initial (reverse-time) conditions

$$\begin{aligned} &x(0) = x^0 \in \Gamma^s \Leftrightarrow x^0 = (-\sqrt{\mu}\, x_2^0, x_2^0)', \ |x_2^0| \leq \sqrt{\mu} \\ &\eta(0) = 0 \\ &u(0) = u^0 = \pm 1 \end{aligned} \tag{9.57}$$

However, in contrast to the previous example, the solution $\eta_2(\tau)$, $\tau > 0$ exhibits an infinity of isolated zeros, each corresponding to a control switch. The state coordinates at the first of these switches are easily calculated as follows. Suppose the first switch occurs at (reverse) time $\tau = \tau_1$, so that $u(\tau) = u^0$ $(= \pm 1)$

$0 \le \tau \le \tau_1$. Integrating the state equations (in reverse time) under this control from Γ^s, i.e. from $x^0 = (-\sqrt{\mu}\, x_2^0,\ x_2^0)'$ yields

$$\left.\begin{aligned} x_1(\tau) &= -\sqrt{\mu}\, x_2^0 - x_2^0 \tau + \tfrac{1}{2}u^0\tau^2 \\ x_2(\tau) &= x_2^0 - u^0\tau \end{aligned}\right\} \quad 0 \le \tau \le \tau_1 \tag{9.58}$$

Now, the (reverse-time) adjoint equations are

$$\frac{d\eta_1}{d\tau} = -2\langle c, x\rangle c_1 = -2[x_1 + \sqrt{\mu}\, x_2]$$

$$\frac{d\eta_2}{d\tau} = \eta_1 - 2\langle c, x\rangle c_2 = \eta_1 - 2\sqrt{\mu}\, [x_1 + \sqrt{\mu}\, x_2]$$

with solution, from (9.57) and (9.58),

$$\begin{aligned} \eta_1(\tau) &= -2\int_0^\tau [x_1(\sigma) + \sqrt{\mu}\, x_2(\sigma)]\, d\sigma \\ &= 2\int_0^\tau [(x_2^0 + \sqrt{\mu}\, u^0)\sigma - \tfrac{1}{2}u^0\sigma^2]\, d\sigma \\ &= (x_2^0 + \sqrt{\mu}\, u^0)\tau^2 - \tfrac{1}{3}u^0\tau^3; \qquad 0 \le \tau \le \tau_1 \end{aligned} \tag{9.59a}$$

$$\begin{aligned} \eta_2(\tau) &= \int_0^\tau \eta_1(\sigma)\, d\sigma + \sqrt{\mu}\, \eta_1(\tau) \\ &= (\sqrt{\mu}\, x_2^0 + \mu u^0)\tau^2 + \tfrac{1}{3}x_2^0\tau^3 - \tfrac{1}{12}u^0\tau^4; \qquad 0 \le \tau \le \tau_1 \end{aligned} \tag{9.59b}$$

But, at the first switch point (occurring at $\tau = \tau_1 > 0$) $\eta_2(\tau_1) = 0$ and hence, from (9.59b),

$$(\sqrt{\mu}\, x_2^0 u^0 + \mu) + \tfrac{1}{3}(x_2^0 u^0)\tau_1 - \tfrac{1}{12}\tau_1^2 = 0$$

which admits the unique positive real root

$$\tau_1 = 2[\![(x_2^0 u^0) + \sqrt{(x_2^0 u^0)^2 + 3[\sqrt{\mu}(x_2^0 u^0) + \mu]}]\!] \tag{9.60}$$

Hence, from (9.58), the state $(x_1^1,\ x_2^1)' = x^1 = x(\tau_1)$ at the first reverse-time switch is given, in terms of the starting point $x^0 \in \Gamma^s$ and initial value of control u^0, as

$$\begin{aligned} x_1^1 = x_1(\tau_1) &= u^0[\![6\mu + 5\sqrt{\mu}\,(x_2^0 u^0) + 2(x_2^0)^2 \\ &\qquad + 2(x_2^0 u^0)\sqrt{(x_2^0)^2 + 3[\sqrt{\mu}\,(x_2^0 u^0) + \mu]}]\!] \\ x_2^1 = x_2(\tau_1) &= -u^0[\![(x_2^0 u^0) + 2\sqrt{(x_2^0)^2 + 3[\sqrt{\mu}\,(x_2^0 u^0) + \mu]}]\!] \end{aligned} \tag{9.61}$$

As x_2^0 ranges over the admissible interval $-\sqrt{\mu} \le x_2^0 \le \sqrt{\mu}$ and (*a*) if $u^0 = +1$, then the point x^1 traces a curve Γ_1^+ which abuts one end of the singular set Γ^s at the point with coordinates $(\mu,\ -\sqrt{\mu})$; while (*b*) if $u^0 = -1$, then x^1 traces a curve Γ_1^-, which abuts the other end $(-\mu,\ \sqrt{\mu})$ of the singular set Γ^s. The

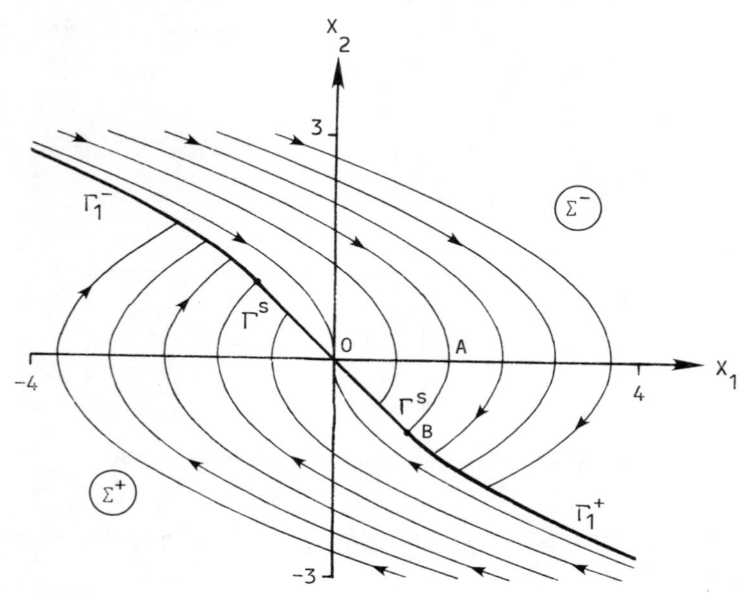

Fig. 9.3 *Example 2*
Optimal synthesis for $\mu = 1$ (singular case)

composite curve

$$\Gamma_1 = \Gamma_1^+ \cup \Gamma^s \cup \Gamma_1^-$$

with end points $\pm\{\mu(13 + 2\sqrt{7}), -\sqrt{\mu}(1 + 2\sqrt{7})\}$ constitutes a bounded segment of the overall switching curve Γ. To construct the complete curve Γ the above analysis must be extended to the second, third, etc. (reverse-time) switch points.

For a weighting parameter $\mu = 1$, the singular set Γ^s, the sets Γ_1^+ and Γ_1^-, and the optimal state vector field are depicted in Fig. 9.3. With the exception of the symmetric pair of nonsingular arcs which terminate in the p- or n-path leading directly to the origin, all state trajectories hit the set Γ^s at some point $x \neq 0$ from where the origin is approached exponentially. For example, consider the trajectory ABO of Fig. 9.3. From the initial point A with coordinates $(\frac{3}{2}, 0)$, the nonsingular n-path AB of finite duration is initially followed until the end-point $(1, -1)$ of Γ^s is reached, from where the exponentially stable singular subarc BO is followed to the origin. The optimal control which generates the trajectory ABO is shown in Fig. 9.4 which clearly illustrates the dual mode character of the optimal quadratic-cost control. Finally, note that for $\mu = 0 \Leftrightarrow p = 1$, the singular set Γ^s collapses to the state origin, the problem becomes nonsingular and the optimal control is bang-bang and satisfies (9.32) almost everywhere. The solution in this case was initially derived by Fuller (1960c) (see also Fuller,

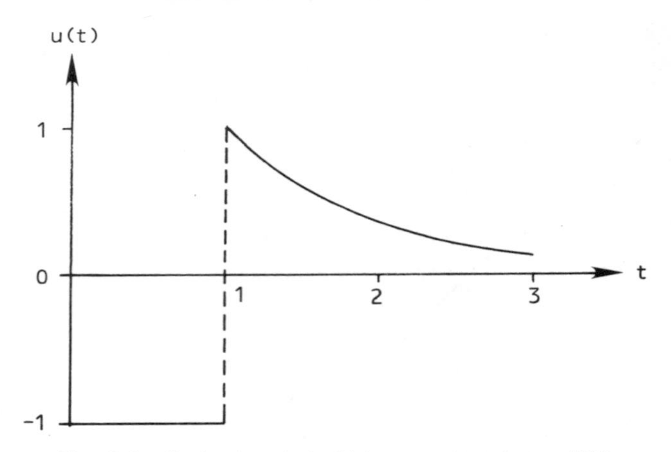

Fig. 9.4 *Optimal control which generates trajectory* ABO

1963a; 1964a, b; 1966; Wonham, 1963) who established the following property of the optimal system.

Lemma 9.2.2: If $\mu = 0$, every optimal trajectory is a *constant-ratio trajectory* generated by an optimal nonsingular control which contains a countable infinity of switches such that, if (x_1, x_2) are the state coordinates at one such switch, then $(-\rho^2 x_1, -\rho x_2)$ are the state coordinates at the next switch where

$$\rho = \tfrac{1}{4}[3 + \sqrt{33} - \sqrt{26 + 6\sqrt{33}}] \tag{9.62}$$

Moreover, while the number of control switches is infinite, the origin is always attained in finite time $t_f < \infty$.

Proof

For detailed proof of the lemma see Fuller (1960c; 1963a; 1964a, b; 1966) and Wonham (1963). Here a partial verification of the lemma only is given; specifically, it will be shown that a solution $\eta(t)$, $0 \le t \le t_f$ of the adjoint equation exists such that the extremal control $u^*(t) = \operatorname{sgn}(\eta_2(t))$ does indeed generate a finite-time constant-ratio trajectory to the origin with constant-ratio parameter (9.62). Recall that constant-ratio trajectories have already been discussed in relation to sensitivity studies of time-optimal systems in Subsections 5.2.3 and 6.6.2. For $\mu = 0$, the state and adjoint equations become

$$\begin{aligned} \dot{x}_1 &= x_2; \qquad \dot{x}_2 = \operatorname{sgn}(\eta_2) \qquad (\eta_2(t) \neq 0 \text{ almost everywhere}) \\ \dot{\eta}_1 &= 2x_1; \qquad \dot{\eta}_2 = -\eta_1 \end{aligned} \tag{9.63}$$

Let (x_1, x_2) and $(\eta_1, 0)$ be the state and adjoint coordinates at an arbitrary switch point of the conjectured constant-ratio solution; for definiteness, an np-control switch (i.e. a $u^* = -1$ to $u^* = +1$ switch) is assumed. Suppose the next control switch (or zero of η_2) occurs after a time t_1. The state (x_1^1, x_2^1) and adjoint $(\eta_1^1, 0)$ coordinates at this switch are readily calculated by integrating the state and adjoint equations, under the constant control $u^* = +1$, from $t = 0$ to $t = t_1$. For a constant-ratio solution these coordinates must satisfy†

$$(x_1^1, x_2^1) = (-\rho^2 x_1, -\rho x_2)$$

$$(\eta_1^1, 0) = (-\rho^3 \eta_1, 0)$$

In particular,

$$x_2^1 = x_2(t_1) = x_2 + t_1 = -\rho x_2 \tag{9.64a}$$

$$x_1^1 = x_1(t_1) = x_1 + x_2 t_1 + \tfrac{1}{2} t_1^2 = -\rho^2 x_1 \tag{9.64b}$$

$$\eta_1^1 = \eta_1(t_1) = \eta_1 + 2[x_1 t_1 + \tfrac{1}{2} x_2 t_1^2 + \tfrac{1}{6} t_1^3] = -\rho^3 \eta_1 \tag{9.64c}$$

$$\eta_2^1 = \eta_2(t_1) = -\eta_1 t_1 - 2[\tfrac{1}{2} x_1 t_1^2 + \tfrac{1}{6} x_2 t_1^3 + \tfrac{1}{24} t_1^4] = 0 \tag{9.64d}$$

Solving equations (9.64a, b, c) for x_1, x_2 and η_1 yields

$$\begin{aligned} x_2 &= -(1+\rho)^{-1} t_1 < 0 \\ x_1 &= \tfrac{1}{2}(1+\rho)^{-1}(1+\rho^2)^{-1}(1-\rho) t_1^2 > 0 \\ \eta_1 &= -\tfrac{1}{3}(1+\rho)^{-1}(1+\rho^2)^{-1}(1+\rho^3)^{-1}(1-2\rho-2\rho^2+\rho^3) t_1^3 \end{aligned} \tag{9.65}$$

and substituting in (9.64d) gives, after straightforward algebraic manipulation, a 6th order polynomial equation for ρ, viz.

$$\rho^6 - 3\rho^5 - 5\rho^4 + 5\rho^2 + 3\rho - 1 = 0 \tag{9.66}$$

which may be rewritten as

$$(\rho^2 - 1)(\rho^4 - 3\rho^3 - 4\rho^2 - 3\rho + 1) = 0$$

Now the roots $\rho = \pm 1$ are inadmissible as they do not correspond to constant-ratio trajectories which converge to the origin and hence

$$\rho^4 - 3\rho^3 - 4\rho^2 - 3\rho + 1 = 0 \tag{9.67}$$

Noting that the left-hand side of (9.67) is a reciprocal polynomial in ρ, the standard substitution

$$\sigma = \rho + \rho^{-1} \tag{9.68}$$

yields the quadratic equation

$$\sigma^2 - 3\sigma - 6 = 0 \tag{9.69}$$

† That $\eta_1^1 = -\rho^3 \eta_1$ is easily seen as follows. Evaluating the maximised Hamiltonian (identically zero) at $t = 0$ and $t = t_1$ gives
(i) $\eta_1 x_2 - (x_1^2) = 0 \Rightarrow \eta_1 = (x_1)^2/x_2$
(ii) $\eta_1^1 x_2^1 - (x_1^1)^2 = 0 \Rightarrow \eta_1^1 = (x_1^1)^2/x_2^1 = \rho^4 (x_1)^2/(-\rho x_2) = -\rho^3 (x_1)^2/x_2 = -\rho^3 \eta_1$

with roots

$$\sigma_1 = \tfrac{1}{2}(3 + \sqrt{33});\ \sigma_2 = \tfrac{1}{2}(3 - \sqrt{33})$$

Consequently, the four roots of (9.67) correspond, via (9.68), to the roots of the pair of quadratic equations: $\rho^2 - \sigma_1\rho + 1 = 0$ and $\rho^2 - \sigma_2\rho + 1 = 0$, viz.

$$\rho = \begin{cases} \tfrac{1}{4}[3 + \sqrt{33} + \sqrt{26 + 6\sqrt{33}}] \\ \tfrac{1}{4}[3 + \sqrt{33} - \sqrt{26 + 6\sqrt{33}}] \\ \tfrac{1}{4}[3 - \sqrt{33} + \sqrt{26 - 6\sqrt{33}}] \\ \tfrac{1}{4}[3 - \sqrt{33} - \sqrt{26 - 6\sqrt{33}}] \end{cases} \tag{9.70}$$

Now, roots 3 and 4 are complex and hence may be disregarded. Of the remaining two, root 1 gives $\rho > 1$ which corresponds to an inadmissible *divergent* constant-ratio trajectory. This leaves

$$\rho = \tfrac{1}{4}[3 + \sqrt{33} - \sqrt{26 + 6\sqrt{33}}] \simeq 0{\cdot}242\,121\,373 \tag{9.71}$$

as the only admissible root which corresponds to a convergent constant-ratio trajectory. Hence, a convergent constant-ratio solution of the state and adjoint equations exists. The set of constant-ratio switch points comprises the switching curve $\Gamma = \Gamma^+ \cup \Gamma^- \cup \{0\}$, where Γ^+ is the set of np-switch points and Γ^- is the set of pn-switch points. From (9.65), Γ^+ has the characterisation

$$\Gamma^+ = \left\{x\colon x_1 = \frac{1}{2}\left[\frac{1-\rho^2}{1+\rho^2}\right]x_2^2;\ x_2 < 0\right\} \tag{9.72}$$

and, by symmetry,

$$\Gamma^- = \left\{x\colon x_1 = -\frac{1}{2}\left[\frac{1-\rho^2}{1+\rho^2}\right]x_2^2;\ x_2 > 0\right\} \tag{9.73}$$

Combining (9.72) and (9.73) yields the overall switching curve

$$\Gamma = \Gamma^+ \cup \Gamma^- \cup \{0\} = \left\{x\colon \xi(x) = x_1 + \frac{1}{2}\left[\frac{1-\rho^2}{1+\rho^2}\right]x_2|x_2| = 0\right\} \tag{9.74}$$

and the optimal control is given by

$$u = u(x) = -\operatorname{sgn}(\Xi(x));\qquad \Xi(x) = \begin{cases} \xi(x); & \xi(x) \neq 0 \\ x_2; & \xi(x) = 0 \end{cases} \tag{9.75}$$

The optimal state portrait is depicted in Fig. 9.5.

That the origin is attained in finite time along a constant-ratio trajectory generated by (9.75) is easily verified as follows. Suppose that (x_1, x_2) are the state coordinates of the starting point, then the time t_1 to the first switch point on Γ is easily calculated as

$$t_1 = x_2 \operatorname{sgn}(\Xi(x)) + \sqrt{(1+\rho^2)[x_1 \operatorname{sgn}(\Xi(x)) + \tfrac{1}{2}x_2^2]} \tag{9.76}$$

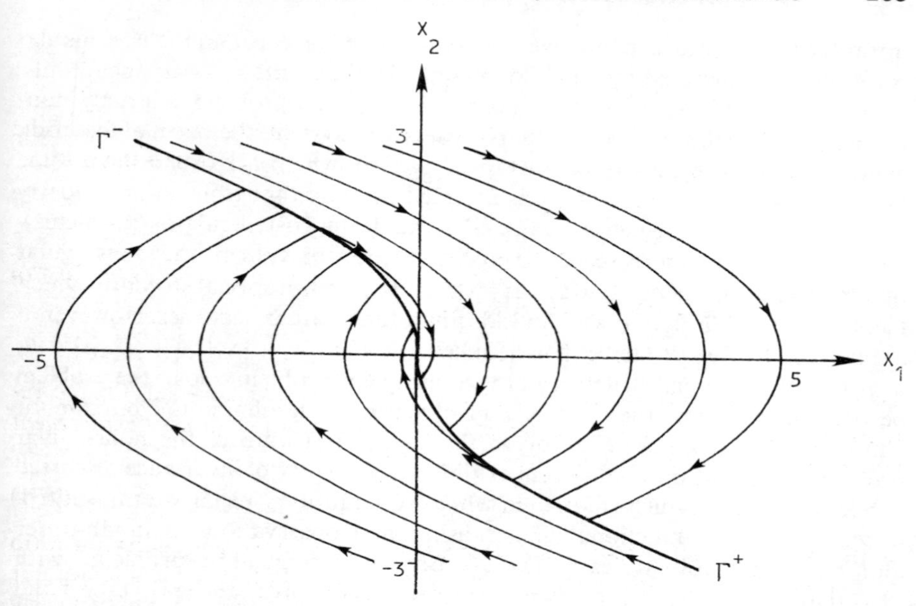

Fig. 9.5 *Example 2*
Optimal synthesis for $\mu = 0$ (nonsingular case)

The duration t_2 of the next constant-ratio control interval is given by

$$\begin{aligned} t_2 &= (1+\rho)x_2(t_1) \\ &= (1+\rho)\sqrt{(1+\rho^2)[x_1 \operatorname{sgn}(\Xi(x)) + \tfrac{1}{2}x_2^2]} \end{aligned} \tag{9.77}$$

Now, on a constant-ratio trajectory, if t_i is the duration of the ith interval of constant control, then $t_{i+1} = \rho t_i$ is the duration of the $(i+1)$th constant control interval. Hence, the total duration t_f of the trajectory from (x_1, x_2) is given by

$$\begin{aligned} t_f &= t_1 + t_2(1 + \rho + \rho^2 + \cdots) \\ &= t_1 + (1-\rho)^{-1}t_2 < \infty \end{aligned}$$

Specifically, from (9.76) and (9.77)

$$\begin{aligned} t_f = T(x) = x_2 \operatorname{sgn}(\Xi(x)) \\ + 2(1-\rho)^{-1}\sqrt{(1+\rho^2)[x_1 \operatorname{sgn}(\Xi(x)) + \tfrac{1}{2}x_2^2]} < \infty \end{aligned} \tag{9.78}$$

where ρ and Ξ are given by (9.71) and (9.75), respectively.

9.2.8 Discussion

The previous examples serve to illustrate the characteristic features of the optimal quadratic-cost regulator problem which, in general, admits singular solutions in a certain singular set. These singular solutions may be synthesised by linear feedback control acting over an *infinite* time horizon, and the nonsingular solutions are generated by bang-bang controls. When the x_1 component, only, of the state vector appears in the cost functional, the problem is strictly nonsingular and, in the case of the double integrator system, the optimal feedback control has been explicitly characterised and shown to act over a *finite* time horizon only, i.e. the state origin is attained in finite time (this latter property also holds for more general *nonsingular* quadratic-cost regulator problems). Hence, at least in the case of the double integrator system, the nonsingular quadratic-cost problem (i.e. $J(u) = \int_0^{t_f} x_1^2(t)\, dt$) exhibits a special structure which facilitates the analysis and which yields finite-time state trajectories. However, if a quadratic x_2 penalty term is introduced, (i.e. $J(u) = \int_0^{t_f} [x_1(t)^2 + \mu x_2(t)^2]\, dt$, $\mu > 0$) then the special structure of the nonsingular problem is lost; the problem becomes singular and the optimal control acts over an infinite time horizon for almost all initial states. The key to the special structure of the nonsingular problem is that the optimal system exhibits the property of invariance discussed in Section 3.4 and which is forfeited when a quadratic x_2 penalty term is introduced into the cost functional. Generalising these observations to an nth-order multiple integrator system, the quadratic-cost regulator problem, with $M = \text{diag}\,(\mu_1, \mu_2, \ldots, \mu_n)$ satisfying (9.4), exhibits the invariance property if and only if

$$\mu_1 > 0; \qquad \mu_i = 0 \qquad i = 2, 3, \ldots, n$$

i.e. $J(u) = \int_0^{t_f} \mu_1 x_1(t)^2\, dt$. This raises the question as to how the other components x_i ($i = 2, 3, \ldots, n$) may be incorporated in the cost functional without destroying the invariance property. Recalling relation (3.24) of Section 3.4, the answer is to adopt an integral cost

$$J(u) = \int_0^{t_f} L(x_1(t), x_2(t), \ldots, x_n(t))\, dt; \qquad t_f \text{ free}$$

with integrand (nonquadratic in general) which satisfies the homogeneity-type condition of the (special) invariance property, viz.

$$L(\kappa^n x_1, \kappa^{n-1} x_2, \ldots, \kappa x_n) = \alpha(\kappa) L(x_1, x_2, \ldots, x_n); \qquad \alpha(\kappa) > 0 \tag{9.79}$$

For example, the double integrator system ($n = 2$) with cost functional

$$J(u) = \int_0^{t_f} [\mu_1 x_1(t)^2 + \mu_2 x_2(t)^4]\, dt; \qquad t_f \text{ free}$$

where the component x_2 is now penalised to the fourth power, will exhibit the property of invariance for all $\mu_1 > 0$, $\mu_2 \geq 0$. The triple integrator system with cost functional

$$J(u) = \int_0^{t_f} [\mu_1 x_1(t)^2 + \mu_2 |x_2(t)|^3 + \mu_3 x_3(t)^6]\, dt; \qquad t_f \text{ free}$$

also exhibits the invariance property. More generally, an nth order multiple integrator system with cost functional

$$J(u) = \int_0^{t_f} f(x(t))\, dt$$

$$= \int_0^{t_f} \sum_{i=1}^{n} \mu_i |x_i(t)|^{n\nu/(n+1-i)}\, dt; \qquad \nu > 0; \qquad \mu_i \geq 0$$

exhibits the requisite invariance property.

The above considerations form the basis of the nonquadratic cost regulator problem studied in the remainder of the present chapter.

9.3 Nonquadratic-cost regulator problem

Again, the canonical system (9.2) is assumed but now associated with the *non-quadratic* cost functional

$$J(u) = \int_0^{t_f} \sum_{i=1}^{n} \mu_i |x_i(t)|^{n\nu/(n+1-i)}\, dt; \qquad t_f \text{ free}$$

with (9.80)

$$\nu \geq 1; \qquad \mu_1 > 0; \qquad \mu_i \geq 0 \qquad i = 2, 3, \ldots, n.$$

9.3.1 Application of the maximum principle

The Hamiltonian for the nonquadratic-cost problem is

$$H(x, u, \eta) = \langle \eta, Ax \rangle + \langle \eta, b \rangle u - \sum_{i=1}^{n} \mu_i |x_i|^{n\nu/(n+1-i)} \tag{9.81}$$

and, in accordance with the maximum principle (theorem 2.4.1), if $u^*(t)$, $0 \leq t \leq t_f$, is optimal then

$$H(x(t), u^*(t), \eta(t)) = \max_{|u| \leq 1} H(x(t), u, \eta(t))$$
$$= H^*(x(t), \eta(t)) \qquad \text{a.a.} \quad 0 \leq t \leq t_f \tag{9.82}$$

where $\eta(\cdot)$ satisfies the adjoint equations

$$\dot{\eta}_1 = -\frac{\partial H}{\partial x_1} = -a_1 \eta_n + \mu_1 \nu |x_1|^{\nu-1} \operatorname{sgn}(x_1)$$

$$\dot{\eta}_i = -\frac{\partial H}{\partial x_i} = -\eta_{i-1} - a_i \eta_n + \mu_i \frac{n\nu}{n+1-i} |x_i|^{\frac{n\nu}{(n+1-i)} - 1} \operatorname{sgn}(x_i)$$

$$i = 2, 3, \ldots, n \tag{9.83}$$

Moreover, as the problem is autonomous, H^* vanishes identically along an optimal trajectory (corollary 2.4.1), viz.

$$\begin{aligned} H^*(x(t), \eta(t)) &= \max_{|u| \leq 1} H(x(t), u, \eta(t)) \\ &= \langle \eta(t), Ax(t) \rangle + |\eta_n(t)| - \sum_{i=1}^{n} \mu_i |x_i(t)|^{n\nu/(n+1-i)} \\ &= 0; \qquad 0 \leq t \leq t_f \end{aligned} \tag{9.84}$$

If $\eta_n(t) \neq 0$ for almost $0 \leq t \leq t_f$, then the optimal control is nonsingular and satisfies

$$u^*(t) = \operatorname{sgn}(\eta_n(t)) \qquad \text{a.a.} \quad 0 \leq t \leq t_f \tag{9.85}$$

However, as in the quadratic cost problem, the singular condition may arise wherein $\eta_n(t)$ vanishes on some interval of positive measure. Due to the nonlinear character of the underlying differential equations, analysis of singular solutions is considerably more involved and many unresolved difficulties remain. For these reasons and to simplify the presentation, systems of first and second order only, will be investigated here.

9.3.2 First-order systems

In this case of $n = 1$, it is readily seen that the problem is nonsingular. The region of null controllability is given by

$$\mathscr{C} = \begin{cases} \mathbb{R}; & a_1 \ (= -\lambda) > 0 \\ \left\{ x_1 : |x_1| < \dfrac{1}{a|a_1|} \right\}; & a_1 \ (= -\lambda) < 0 \end{cases} \tag{9.86}$$

Within $\mathscr{C}$ the optimal control is identical to the time-optimal control (and the optimal quadratic-cost control) and is given by

$$u(x_1) = -\operatorname{sgn}(x_1); \qquad x_1 \neq 0 \tag{9.87}$$

with $u(0) = 0$.

9.3.3 Second-order systems

For notational convenience, a normalised gain parameter $a = 1$ will be as assumed throughout, corresponding to a rescaling of the problem, viz.

$$x_1 \mapsto \frac{x_1}{a}; \qquad x_2 \mapsto \frac{x_2}{a}; \qquad \mu \mapsto a^\nu \mu; \qquad J \mapsto a^\nu J \tag{9.88}$$

In this case the state equations are given by

$$\dot{x}_1 = x_2; \qquad \dot{x}_2 = a_1 x_1 + a_2 x_2 + u; \qquad |u| \leq 1 \tag{9.89}$$

The cost functional (9.80) may be expressed as

$$J(u) = \int_0^{t_f} [|x_1(t)|^\nu + \mu|x_2(t)|^{2\nu}]\, dt;$$

$$\nu \geq 1; \qquad \mu \geq 0; \qquad t_f \text{ free} \tag{9.90}$$

and the maximised Hamiltonian function is given by

$$H^*(x, \eta) = \max_{|u| \leq 1} H(x, u, \eta) = \eta_1 x_2 + \eta_2(a_1 x_1 + a_2 x_2) + |\eta_2| - |x_1|^\nu - \mu|x_2|^{2\nu} \equiv 0 \tag{9.91}$$

The adjoint equations are

$$\dot{\eta}_1 = -a_1 \eta_2 + \nu|x_1|^{\nu-1} \operatorname{sgn}(x_1) \tag{9.92a}$$

$$\dot{\eta}_2 = -\eta_1 - a_2 \eta_2 + 2\mu\nu|x_2|^{2\nu-1} \operatorname{sgn}(x_2) \tag{9.92b}$$

If $\eta_2(t) \neq 0$ for almost all $0 \leq t \leq t_f$, then the optimal control is nonsingular and satisfies (9.85) with $n = 2$. On the other hand, the singular case may arise and is considered below.

Singular controls: If the problem is singular, then $\eta_2(t)$ vanishes on some finite interval I, so that $\dot{\eta}_2 \equiv 0$ on the interior of this interval and, from (9.92b),

$$\eta_1 = 2\mu\nu|x_2|^{2\nu-1} \operatorname{sgn}(x_2) \tag{9.93}$$

with derivative

$$\begin{aligned} \dot{\eta}_1 &= 2\mu\nu(2\nu - 1)|x_2|^{2(\nu-1)}\dot{x}_2 \\ &= 2\mu\nu(2\nu - 1)|x_2|^{2(\nu-1)}(a_1 x_1 + a_2 x_2 + u) \end{aligned} \tag{9.94}$$

Hence, from (9.92a) and (9.94), a singular control u must satisfy

$$u = \frac{|x_1|^{\nu-1} \operatorname{sgn}(x_1)}{2\mu(2\nu - 1)|x_2|^{2(\nu-1)}} - a_1 x_1 - a_2 x_2 \tag{9.95}$$

on a singular subarc $x(t)$, $t \in I$.

Attention will now be restricted to the case of $a_1 = 0 = a_2$, i.e. the double integrator system. The treatment is later extended to the general case. A similar study can be found in Ryan (1979).

9.3.4 *Double integrator system*

For this case the system equations are

$$\dot{x}_1 = x_2; \qquad \dot{x}_2 = u; \qquad |u| \leq 1 \tag{9.96}$$

Singular subarcs: The singular condition (9.95) reduces to

$$u = \frac{|x_1|^{\nu-1} \operatorname{sgn}(x_1)}{2\mu(2\nu - 1)|x_2|^{2(\nu-1)}} \tag{9.97}$$

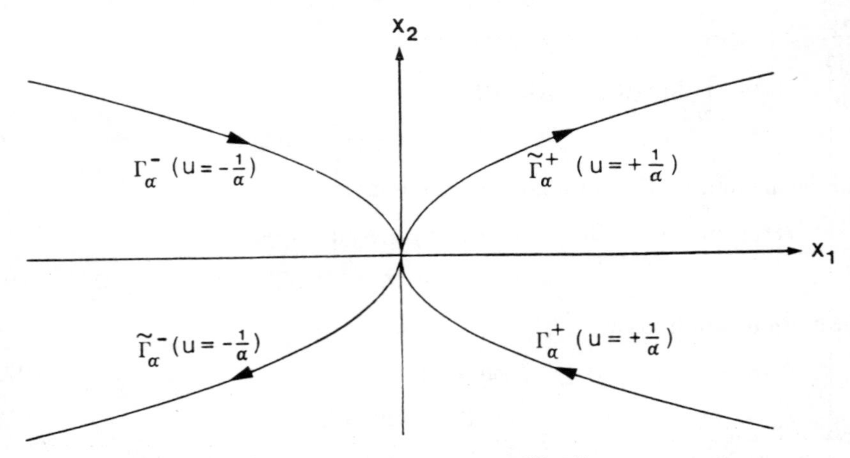

Fig. 9.6 *Convergent* $(\Gamma_\alpha^+, \Gamma_\alpha^-)$ *and divergent* $(\tilde{\Gamma}_\alpha^+, \tilde{\Gamma}_\alpha^-)$ *singular subarcs*

and admissible controls satisfying (9.97) are sought. Consider now state trajectories generated by a *constant* control

$$u = \pm\frac{1}{\alpha}; \qquad \alpha = \text{constant} > 0 \tag{9.98}$$

and which pass through the state origin. In the state plane these trajectories satisfy

$$x_1 = \pm\frac{\alpha}{2}\,x_2^2 \tag{9.99}$$

and have the general form shown in Fig. 9.6 consisting of the state origin and the four parabolic arcs: Γ_α^+, Γ_α^- (which converge to the origin) and $\tilde{\Gamma}_\alpha^+$, $\tilde{\Gamma}_\alpha^-$ (which diverge from the origin).

Setting

$$\alpha = 2[\mu(2\nu - 1)]^{1/\nu} \tag{9.100}$$

then, in view of (9.99), it is easily verified that the constant control (9.98) and (9.100) satisfies the singular condition (9.97). Thus, imposing the saturation constraint $|u| \leq 1$, it may be concluded that singular solutions exist for parameter values $(\mu, \nu) \in \mathscr{R}_s$, where $\mathscr{R}_s$ is the singular region of the parameter space defined by

$$\mathscr{R}_s = \{(\mu, \nu)\colon 2[\mu(2\nu - 1)]^{1/\nu} \geq 1;\ \mu \geq 0;\ \nu \geq 1\} \tag{9.101}$$

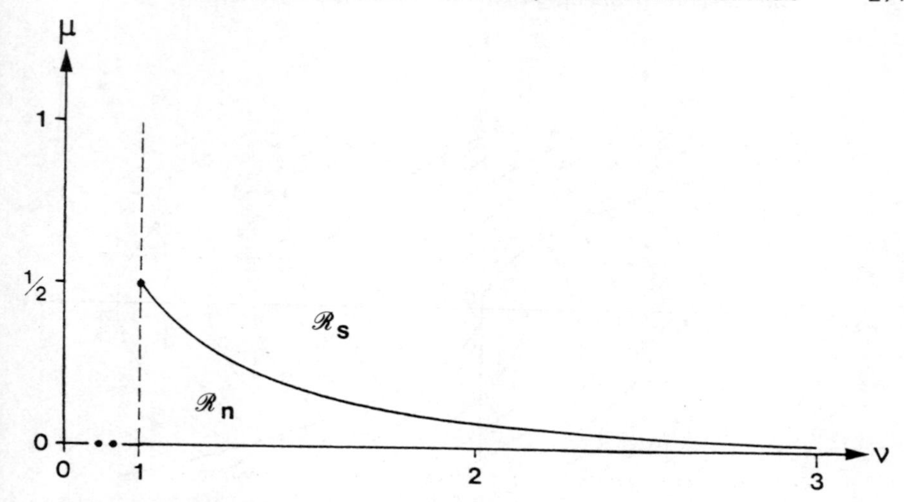

Fig. 9.7 *Singular* ($\mathscr{R}_s$) *and nonsingular* ($\mathscr{R}_n$) *regions in* (ν, μ) *parameter plane*

while, for $(\mu, \nu) \in \mathscr{R}_n$ the optimal control is strictly nonsingular, where $\mathscr{R}_n$ is the nonsingular region of the parameter space defined by

$$\mathscr{R}_n = \{(\mu, \nu)\colon 2[\mu(2\nu - 1)]^{1/\nu} < 1; \mu \geq 0; \nu \geq 1\} \tag{9.102}$$

The regions $\mathscr{R}_s$ and $\mathscr{R}_n$ are depicted in Fig. 9.7. Furthermore, for $(\mu, \nu) \in \mathscr{R}_s$, it is readily verified that (9.98) and (9.100) characterise *all* admissible controls which satisfy *both* conditions (9.91) and (9.97).

Referring to Fig. 9.6, as the subarcs $\tilde{\Gamma}_\alpha^+$, $\tilde{\Gamma}_\alpha^-$ diverge from the origin, it is not plausible that they could form part of an optimal trajectory.† Consequently the convergent subarcs Γ_α^+, Γ_α^- only are considered and, from (9.99)–(9.101), it may be concluded that the admissible set of singular subarcs has the characterisation

$$\Gamma_\alpha = \Gamma_\alpha^+ \cup \Gamma_\alpha^- = \{x\colon x_1 + \tfrac{1}{2}\alpha x_2|x_2| = 0; \alpha = 2[\mu(2\nu - 1)]^{1/\nu}; \\ (\mu, \nu) \in \mathscr{R}_s\} \tag{9.103}$$

Note that, for $x = x(0) \in \Gamma_\alpha$, the time t_f to the origin along a singular subarc is given by

$$t_f = \alpha|x_2| \tag{9.104}$$

Optimal control synthesis

(i) Singular case $(\mu, \nu) \in \mathscr{R}_s$: In the singular case, the complete optimal vector field (state portrait) is comprised of normal subarcs leading to the singular set Γ_α and may be constructed by integrating, in reverse time $\tau = -t$, the state and

† More rigorously, suppose that part of a divergent subarc is optimal. Then, by the invariance property, the entire singular trajectory from the origin is optimal. However, this trajectory goes to infinity and, hence, cannot be optimal; this contradiction invalidates the original supposition of optimality.

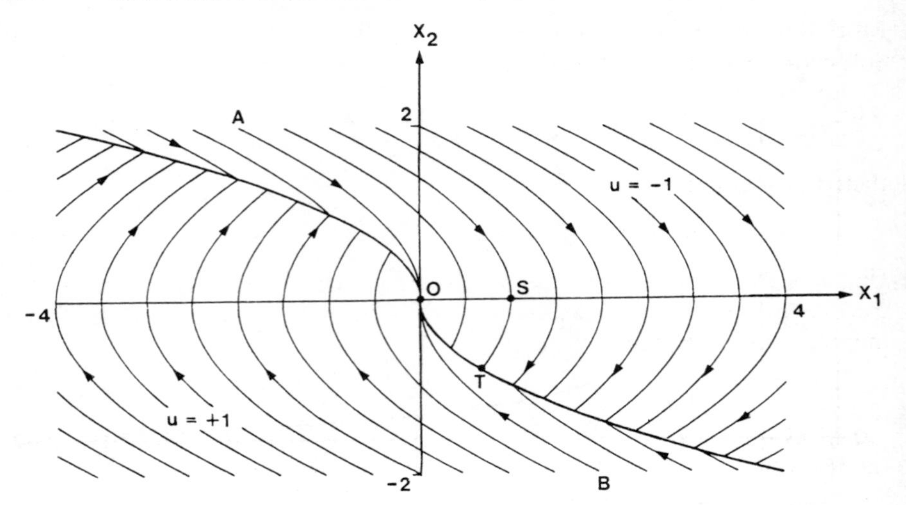

Fig. 9.8 *Optimal synthesis for cost parameters satisfying the relation*: $\mu(2\nu - 1) = 1$

adjoint equations under the normal control $u(\tau) = \text{sgn}\,(\eta_2(\tau))$, $\tau > 0$, with the initial (reverse time) conditions at $\tau = 0$ on Γ_α:

$$\begin{aligned} &x_2(0) = \gamma \text{ (non-zero constant)} \\ &x_1(0) = -\tfrac{1}{2}\alpha x_2(0)\,|x_2(0)| \qquad \text{(since } x(0) \in \Gamma_\alpha) \\ &\eta_2(0) = 0 \\ &\eta_1(0) = 2\mu\nu\,|x_2(0)|^{2\nu-1}\,\text{sgn}\,(x_2(0)) \text{ (from (9.93))} \\ &u(0) = \pm 1; \qquad u(\tau) = \text{sgn}\,(\eta_2(\tau)), \qquad \tau > 0 \end{aligned} \tag{9.105}$$

It may be verified that on such a reverse-time trajectory $|\eta_2(\tau)| > 0 \;\forall\; \tau > 0$, so that the normal subarcs do not contain any switches in control.

For an illustrative value of $\alpha = 2$, the complete state portrait is shown in Fig. 9.8 which, in view of (9.100), is valid for all admissible parameter pairs (μ, ν) satisfying $\mu(2\nu - 1) = 1$. Note that, with the exception of the normal subarcs AO, BO which lead directly to the state origin, all state trajectories terminate in a singular subarc converging to the origin which is reached in finite time in all cases. For example, trajectory STO corresponds to the unit step response (trajectory with initial state $x(0) = (1, 0)'$); from S the trajectory initially follows the normal subarc ST (with $u = -1$) of duration $t_1 = \sqrt{\tfrac{2}{3}}$, and subsequently follows the singular subarc TO (with $u = 1/\alpha = \tfrac{1}{2}$) of duration $t_2 = \alpha\,|x_2(t_1)| = 2\sqrt{\tfrac{2}{3}}$ directly to the state origin.

Clearly, the optimal normal subarcs ($x \notin \Gamma_\alpha$) can be synthesised by the feedback law (see e.g. Fig. 9.8)

$$u(x) = -\text{sgn}\,(\xi_\alpha(x)) = -\text{sgn}\,(x_1 + \tfrac{1}{2}\alpha x_2|x_2|) \tag{9.106}$$

Moreover, in view of the results of Zinober and Fuller (1973) as outlined in Subsection 5.2.3(i), control law (9.106) will also synthesise the singular subarcs.

Now, if

$$\alpha = 2[\mu(2\nu - 1)]^{1/\nu} = 1 \tag{9.107}$$

then the optimal feedback control is given by

$$u(x) = -\mathrm{sgn}\,(\xi_1(x)) = -\mathrm{sgn}\,(x_1 + \tfrac{1}{2}x_2|x_2|) \tag{9.108}$$

which is the same as the time-optimal feedback law derived in Subsection 5.2.1(i), i.e. the time-optimal control for the double integrator system (9.96) also minimises cost functional (9.90) if

$$\mu = [2^\nu(2\nu - 1)]^{-1} \tag{9.109}$$

(*ii*) *Nonsingular case:* $(\mu, \nu) \in \mathscr{R}_n$: In this case the optimal control satisfies $|u^*(t)| = 1$ almost everywhere on $[0, t_f]$ and is given by

$$u^*(t) = \mathrm{sgn}\,(\eta_2(t)); \qquad \text{a.a.} \quad 0 \le t \le t_f \tag{9.110}$$

Similar to lemma 9.2.2 (for the quadratic cost problem with $\mu = 0$), the following lemma holds for the nonquadratic cost problem with $(\mu, \nu) \in \mathscr{R}_n$. This lemma can be established via the arguments of Fuller (1960c) and the proof is omitted here.

Lemma 9.3.1: For $(\mu, \nu) \in \mathscr{R}_n$, every optimal trajectory is a constant-ratio trajectory generated by a nonsingular optimal control which contains a countable infinity of switches such that, if (x_1, x_2) are the state coordinates at one such switch, then $(-\rho^2 x_1, -\rho x_2)$ are the state coordinates at the next switch, where $\rho = \rho(\mu, \nu) \in (0, 1)$. Moreover, while the number of control switches is infinite, the state origin is always attained in *finite* time $t_f < \infty$. The optimal feedback synthesis is given by

$$u(x) = -\mathrm{sgn}\,(\xi_\rho(x)); \qquad \xi_\rho(x) = x_1 + \tfrac{1}{2}\left[\frac{1 - \rho^2}{1 + \rho^2}\right] x_2|x_2| \tag{9.111}$$

Lemma 9.3.1 implies that, for values of $0 \le \mu < [2^\nu(2\nu - 1)]^{-1}$, under control (9.111), damped oscillatory motion is encountered in the response $x_1(t)$, $0 \le t \le t_f$, such that the peak ratio of successive overshoots and undershoots is a constant $-\rho^2$, where $\rho \in (0,1)$. The value of the constant ratio parameter ρ depends on the cost functional parameters (μ, ν) and, in general, must be determined numerically (see also Fuller, 1960c). However, in certain cases, explicit solutions can be found as will now be illustrated.

Example 9.3.1 $\nu = 2$: In this case, the state and adjoint equations are

$$\dot{x}_1 = x_2; \qquad \dot{x}_2 = \mathrm{sgn}\,(\eta_2) \tag{9.112}$$

$$\dot{\eta}_1 = 2x_1; \; \dot{\eta}_2 = -\eta_1 + 4\mu x_2^3 \tag{9.113}$$

Adopting a similar approach to that employed in the proof of lemma 9.2.2, let (x_1, x_2) and $(\eta_1, 0)$ be the state and adjoint coordinates at an arbitrary switch point of the constant-ratio solution; for definiteness an *np*-switch is assumed. Suppose the next switch (i.e. zero of η_2) occurs after a time t_1, then the state (x_1^1, x_2^1) and adjoint $(\eta_1^1, 0)$ coordinates at this switch are readily calculated by integrating the state and adjoint equations, under the constant control $u^* \equiv 1$, from $t = 0$ to $t = t_1$. Specifically,

$$x_2^1 = x_2(t_1) = x_2 + t_1 = -\rho x_2 \tag{9.114a}$$

$$x_1^1 = x_1(t_1) = x_1 + x_2 t_1 + \frac{t_1^2}{2} = -\rho^2 x_1 \tag{9.114b}$$

$$\eta_1^1 = \eta_1(t_1) = \eta_1 + 2\left[x_1 t_1 + \tfrac{1}{2} x_2 t_1^2 + \frac{t_1^3}{6}\right] = -\rho^3 \eta_1 \tag{9.114c}$$

$$\begin{aligned}\eta_2^1 = \eta_2(t_1) &= \int_0^{t_1} [-\eta_1(t) + 4\mu[x_2(t)]^3]\, dt \\ &= (-\eta_1 + 4\mu x_2^3)t_1 - (x_1 - 6\mu x_2^2)t_1^2 \\ &\quad - (\tfrac{1}{3} - 4\mu)x_2 t_1^3 - (\tfrac{1}{12} - \mu)t_1^4 = 0\end{aligned} \tag{9.114d}$$

Solving (9.114a, b, c) for x_1, x_2 and η_1 yields

$$x_2 = -(1 + \rho)^{-1} t_1 \tag{9.115a}$$

$$x_1 = (1 + \rho)^{-1}(1 + \rho^2)^{-1}(1 - \rho)\frac{t_1^2}{2} \tag{9.115b}$$

$$\eta_1 = -(1 + \rho)^{-1}(1 + \rho^2)^{-1}(1 + \rho^3)^{-1}(1 - 2\rho - 2\rho^2 + \rho^3)\frac{t_1^3}{3} \tag{9.115c}$$

and substituting in (9.114d) gives, after straightforward algebraic manipulation, a sixth-order polynomial equation for ρ, viz.

$$\rho^6 - \rho^5 + \alpha\rho^4 + \beta\rho^3 + \alpha\rho^2 - \rho + 1 = 0 \tag{9.116a}$$

where

$$\alpha = \frac{9(4\mu + 1)}{12\mu - 1}; \qquad \beta = \frac{14 - 24\mu}{12\mu - 1} \tag{9.116b}$$

Noting that (9.116a) is a reciprocal polynomial, the substitution

$$\sigma = \rho + \rho^{-1} \tag{9.117}$$

yields the cubic equation

$$\sigma^3 - \sigma^2 + \gamma\sigma + \gamma = 0 \tag{9.118a}$$

where

$$\gamma = \frac{12}{12\mu - 1} \tag{9.118b}$$

which may be solved by standard methods giving, via (9.117), the required roots of (9.116), only one of which will lie in the admissible interval (0, 1) corresponding to the convergent constant-ratio trajectory (for the special case of $\mu = 0$, the result (9.71) is recovered). For example, if $\mu = \frac{1}{36}$, then (9.118) becomes

$$\sigma^3 - \sigma^2 - 18\sigma - 18 = 0 \tag{9.119a}$$

with roots

$$\sigma = \{-3, 2 \pm \sqrt{10}\} \tag{9.119b}$$

giving, via (9.117), the admissible root of (9.116) (i.e. the root satisfying $0 < \rho < 1$) as

$$\rho = \tfrac{1}{2}[2 + \sqrt{10} - \sqrt{10 + 4\sqrt{10}}] \tag{9.120}$$

so that the feedback control (9.111), with ρ given as above, minimises the cost functional

$$J(u) = \int_0^{t_f} [x_1^2 + \tfrac{1}{36}x_2^4]\, dt$$

Example 9.3.2 $\mu = 0$, $v \to \infty$: double integrator with minmax performance criterion : In this case, the cost functional is given by

$$J(u) = \int_0^{t_f} |x_1(t)|^v\, dt; \qquad v \to \infty; \qquad t_f \text{ free} \tag{9.121}$$

The associated optimal control problem has been studied by Fuller (1960c) and Johnson (1967) who have established the correspondence between this control problem and the minmax control problem, viz. a control u^* which minimises (9.121) as $v \to \infty$, also minimises the maximum value of the function $|x_1(t)|$, $0 \le t \le t_f$. To see this correspondence, first note that a control u^* which minimises $\int_0^{t_f} |x_1(t)|^v\, dt$ also minimises

$$\hat{J}(u) = \left[\int_0^{t_f} |x_1(t)|^v\, dt\right]^{1/v}$$

where $[\cdot]^{1/v}$ denotes the positive, real v-th root. In other words, the optimal control problem is equivalent to minimising the L^v norm of $x_1(\cdot)$; for $v = \infty$, the cost function $\hat{J}$ corresponds to the L^∞ norm of $x_1(\cdot)$, i.e. ess $\sup_{0 \le t \le t_f} |x_1(t)|$ (or $\max_{0 \le t \le t_f} |x_1(t)|$ since x_1 is continuous). Tailored to the control problem at hand, the latter observation can be illustrated (as in Johnson 1967) as follows. In particular, it will be shown that

$$\lim_{\nu\to\infty}\left[\int_0^{t_f} |x_1(t)|^\nu\, dt\right]^{1/\nu} = M = \max_{0\le t\le t_f} |x_1(t)| \tag{9.122}$$

For $\delta \in (0, M)$, define the function $y(t)$, $0 \le t \le t_f$, as follows

$$y(t) = \begin{cases} 0; & |x_1(t)| < M - \delta \\ M - \delta; & |x_1(t)| \ge M - \delta \end{cases}$$

so that

$$y(t) \le |x_1(t)| \le M, \qquad 0 \le t \le t_f$$

Hence

$$\left[\int_0^{t_f} |y(t)|^\nu\, dt\right]^{1/\nu} \le \left[\int_0^{t_f} |x_1(t)|^\nu\, dt\right]^{1/\nu} \le \left[\int_0^{t_f} M^\nu\, dt\right]^{1/\nu} = M(t_f)^{1/\nu}$$

Now

$$\left[\int_0^{t_f} |y(t)|^\nu\, dt\right]^{1/\nu} = (M - \delta)(K)^{1/\nu}$$

where $0 < K < t_f$ is the measure of the set of values of t for which $y(t) = M - \delta$. Consequently,

$$(M - \delta)(K)^{1/\nu} \le \left[\int_0^{t_f} |x_1(t)|^\nu\, dt\right]^{1/\nu} \le M(t_f)^{1/\nu}$$

On taking the limit as $\nu \to \infty$,

$$(M - \delta) \le \lim_{\nu\to\infty}\left[\int_0^{t_f} |x_1(t)|^\nu\, dt\right]^{1/\nu} \le M = \max_{0\le t\le t_f} |x_1(t)|$$

and, since the above relation holds for arbitrarily small δ, the required result (9.122) is obtained as $\delta \to 0^+$. Hence, a control u^* which minimises (9.121) is also a minmax control in the sense that it minimises the maximum value of the function $|x_1(t)|, 0 \le t \le t_f$.

A nonrigorous derivation of the optimal control will now be given. By lemma 9.3.1, an optimal solution which minimises $J(u) = \int_0^{t_f} |x_1(t)|^\nu\, dt$ is a constant-ratio solution and has the general form shown in Fig. 9.9. As ν increases monotonically, large values of $|x_1(t)|$ contribute progressively more to the cost of a trajectory $x(\cdot)$ with the result that the cost minimising control $u^*(\cdot)$ progressively reduces the set of times on which $|x_1(t)|$ takes 'large' values at the expense of increasing the set of times on which $|x_1(t)|$ takes 'small' values. This behaviour of the optimal solution is reflected in a monotonic increase in constant-ratio parameter $\rho = \rho(\nu) \in (0, 1)$. As $\nu \to \infty$, the limit to the reduction of the

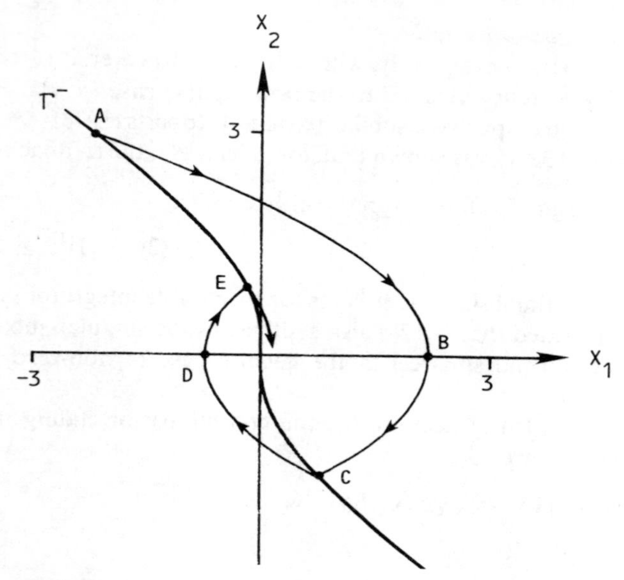

Fig. 9.9 *Example 9.3.2*
Minmax trajectory

effects of large values of $|x_1(t)|$, obtainable in this way, occurs when the first control switch (at $t = t_1$) is such that $|x_1(t_1)|$ is equal to the magnitude of the next peak overshoot/undershoot (at $t = t_p$) of the subsequent response $x_1(t)$, $t_1 \leq t \leq t_f$. Referring to Fig. 9.9,

$$|x_1^{\mathrm{C}}| = |x_1(t_1)| = |x_1^{\mathrm{D}}| = |x_1(t_p)| = \max_{t_1 \leq t \leq t_f} |x_1(t)| \tag{9.123}$$

Clearly, by the symmetry of the parabolic arcs BC and CD of Fig. 9.9,

$$|x_1^{\mathrm{B}}| = 3|x_1^{\mathrm{D}}| \tag{9.124}$$

Moreover, as the trajectory is a constant-ratio trajectory, it follows that

$$x_1^{\mathrm{D}} = -\rho^2 x_1^{\mathrm{B}} \tag{9.125}$$

Combining (9.124) and (9.125) it may be concluded that the constant-ratio parameter ρ is given by

$$\rho = \frac{1}{\sqrt{3}} \tag{9.126}$$

and, in view of (9.111), the optimal feedback control may be expressed as

$$u(x) = -\operatorname{sgn}\,[x_1 + \tfrac{1}{4}x_2|x_2|] \tag{9.127}$$

9.3.5 General second-order systems

The singular case $(\mu, \nu) \in \mathscr{R}_s$, only, will be treated. However, it is remarked that the investigation can be extended to the nonsingular case $(\mu, \nu) \in \mathscr{R}_n$ by adopting a computational approach similar to that of Roberts (1973).

In Subsection 9.3.4 it was shown that, for $(\mu, \nu) \in \mathscr{R}_s$, the feedback control

$$u(x) = -\text{sgn}\,(\xi_\alpha(x)) = -\text{sgn}\,(x_1 + \tfrac{1}{2}\alpha x_2|x_2|);$$
$$\alpha = 2[\mu(2\nu - 1)]^{1/\nu} \geq 1 \qquad (9.128)$$

synthesised the optimal singular subarcs for the double integrator system. It will now be demonstrated that (9.128) also synthesises the singular subarcs (but not necessarily the normal subarcs) in the general case of non-zero system parameters $a_1, a_2 \neq 0$.

Recalling eqn (3.15) of Section 3.3, the conditions for sliding on the curve $\Gamma_\alpha \doteq \{x: \xi_\alpha(x) = 0\}$ are

$$\text{(a)} \quad |\langle \nabla\xi_\alpha(x), Ax\rangle| < |\langle \nabla\xi_\alpha(x), b\rangle| \qquad (9.129a)$$

$$\text{(b)} \quad \langle \nabla\xi_\alpha(x), b\rangle > 0 \qquad (9.129b)$$

If conditions (9.129) are satisfied, then, under control (9.128), the state point is constrained to remain on the curve Γ_α and converges to the state origin in finite time t_f (given by (9.104)) under an effective input (see eqn (3.18) of Section 3.3)

$$u = -\frac{\langle \nabla\xi_\alpha(x), Ax\rangle}{\langle \nabla\xi_\alpha(x), b\rangle} \qquad (9.130)$$

Now, from (9.89) and (9.128), on the curve $\Gamma_\alpha = \{x: \xi_\alpha(x) = 0\}$,

$$\begin{aligned}\langle \nabla\xi_\alpha(x), Ax\rangle &= x_2 + \alpha a_1 x_1|x_2| + \alpha a_2 x_2|x_2| \\ &= x_2(1 + \alpha a_2|x_2| - \tfrac{1}{2}\alpha^2 a_1 x_2^2)\end{aligned} \qquad (9.131)$$

and

$$\langle \nabla\xi_\alpha(x), b\rangle = \alpha|x_2| \qquad (9.132)$$

so that (9.129b) is satisfied for all $x_2 \neq 0$. Condition (9.129a) becomes

$$|1 + \alpha a_2|x_2| - \tfrac{1}{2}\alpha^2 a_1 x_2^2| < \alpha \qquad (9.133)$$

and, from (9.130), the effective sliding control input satisfies

$$u = -\frac{1}{\alpha}\,\text{sgn}\,(x_2) - a_1 x_1 - a_2 x_2 \qquad (9.134)$$

Noting that, on the curve Γ_α,

$$\xi_\alpha(x) = 0 \Rightarrow |x_1|^{\nu-1}\,\text{sgn}\,(x_1) = -\frac{2}{\alpha}\left[\frac{\alpha}{2}\right]^\nu |x_2|^{2(\nu-1)}\,\text{sgn}\,(x_2) \qquad (9.135)$$

it can readily be seen that (9.134) also satisfies the singular condition (9.95), i.e. for $(\mu, \nu) \in \mathscr{R}_s$, the feedback control (9.128) will generate the singular subarcs under an effective singular control input (9.134), provided that the sliding condition (9.133) is satisfied on the curve Γ_α. Consequently, the set of singular subarcs $\Gamma_\alpha^s \subseteq \Gamma_\alpha$ for the general case has the characterisation†

$$\Gamma_\alpha^s = \{x\colon x_1 + \tfrac{1}{2}\alpha x_2|x_2| = 0;\ |1 + \alpha a_2|x_2| - \tfrac{1}{2}\alpha^2 a_1 x_2^2| < \alpha;$$
$$\alpha = 2[\mu(2\nu - 1)]^{1/\nu};\ (\mu, \nu) \in \mathscr{R}_s\} \subseteq \Gamma_\alpha \qquad (9.136)$$

However, although (9.128) synthesises the singular subarcs, it does not necessarily generate the normal subarcs (in contrast to the case of the double integrator system of Subsection 9.3.4) as is now illustrated for two specific systems, viz. the harmonic oscillator ($a_1 < 0,\ a_2 = 0$) and the integrator-plus-lag system ($a_1 = 0, a_2 < 0$).

Example 9.3.3 Harmonic oscillator: In this case, the state and adjoint equations are

$$\dot{x}_1 = x_2; \qquad \dot{x}_2 = a_1 x_1 + u; \qquad |u| \leq 1; \qquad a_1 < 0 \qquad (9.137a)$$
$$\dot{\eta}_1 = -a_1\eta_2 + \nu|x_1|^{\nu-1}\ \mathrm{sgn}\ (x_1)$$
$$\dot{\eta}_2 = -\eta_1 + 2\mu\nu|x_2|^{2\nu-1}\ \mathrm{sgn}\ (x_2) \qquad (9.137b)$$

The set of singular subarcs becomes

$$\Gamma_\alpha^s = \left\{x\colon x_1 + \tfrac{1}{2}\alpha x_2|x_2| = 0;\ |x_2| < \frac{1}{\alpha}\left[\frac{2(\alpha-1)}{|a_1|}\right]^{1/2};\right.$$
$$\left.\alpha = 2[\mu(2\nu-1)]^{1/\nu};(\mu, \nu) \in \mathscr{R}_s\right\} \qquad (9.138)$$

i.e. a bounded segment of the curve Γ_α (see Fig. 9.10). Hence, singular solutions exist for all values of $\alpha > 1$ (note that for $\alpha = 1$ the set Γ_α^s collapses to the state origin).

As before, the complete state portrait is comprised of the set Γ_α^s together with the normal subarcs leading to Γ_α^s and may be constructed by integrating the state and adjoint equations (9.137), in reverse time $\tau = -t$, from the singular set Γ_α^s with control $u(\tau) = \mathrm{sgn}\ (\eta_2(\tau))$, $\tau > 0$, and initial (reverse time) conditions (9.105), where, in view of (9.138), the constant γ of (9.105) is now restricted to satisfy

$$|\gamma| < \frac{1}{\alpha}\left[\frac{2(\alpha-1)}{|a_1|}\right]^{1/2}$$

† Note that, for $a_1 = 0 = a_2$ (the double integrator system), the sliding condition (9.133) is trivially satisfied for $(\mu, \nu) \in \mathscr{R}_s$, i.e. (9.136) and (9.103) are equivalent.

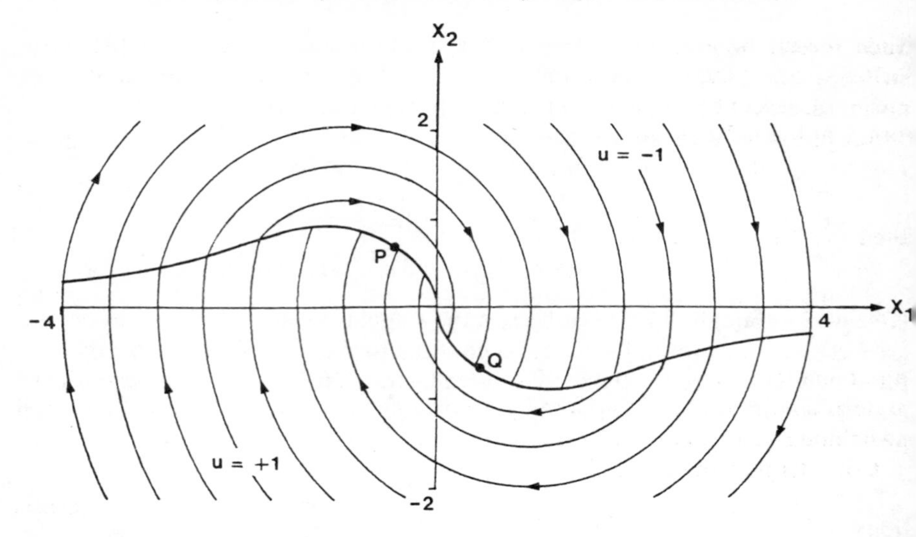

Fig. 9.10 *Harmonic oscillator* ($a_1 = -1$, $a_2 = 0$)
Optimal synthesis for cost parameter values $\nu = 2$, $\mu = \frac{1}{3}$

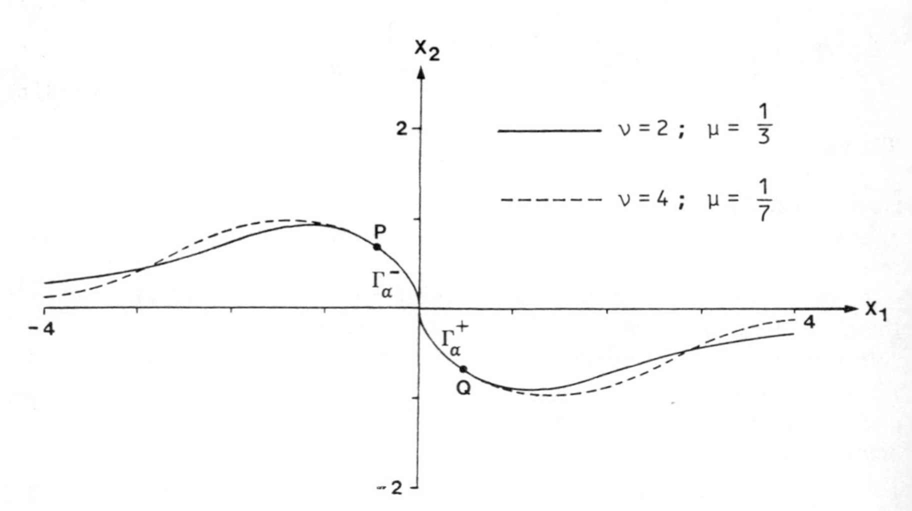

Fig. 9.11 *Harmonic oscillator* ($a_1 = -1$, $a_2 = 0$)
Optimal switching curves for cost parameter values
(i) $\nu = 2$, $\mu = \frac{1}{3}$ (solid line)
(ii) $\nu = 4$, $\mu = \frac{1}{7}$ (broken line)

Such reverse-time normal subarcs contain an unbounded number of control switches; the locus of the switch points in the (x_1, x_2) plane constitutes the normal segment Γ_α^n of the overall switching boundary.

Adopting the illustrative parameter values

$$a_1 = -1; \qquad v = 2; \qquad \mu = \tfrac{1}{3} \tag{9.139}$$

then $\alpha = 2$ and the singular set is given by

$$\Gamma_2^s = \left\{x\colon x_1 + x_2|x_2| = 0; |x_2| < \frac{1}{\sqrt{2}}\right\} \tag{9.140}$$

and is represented by segment PQ of Fig. 9.10. Reverse-time integration of the state and adjoint equations from Γ_2^s yields the normal segment Γ_2^n of the overall switching boundary and the state portrait of Fig. 9.10.

Consider now the case of

$$a_1 = -1; \qquad v = 4; \qquad \mu = \tfrac{1}{7} \tag{9.141}$$

It is easily verified that the singular set is again given by (9.140). However, the normal segment of the switching boundary differs significantly from that obtained for the (μ, v) parameter pair (9.139), as can be seen from Fig. 9.11.

Example 9.3.4 Integrator-plus-lag: In this case, the state and adjoint equations are

$$\dot{x}_1 = x_2; \qquad \dot{x}_2 = a_2 x_2 + u; \qquad |u| \leq 1; \qquad a_2 < 0 \tag{9.142a}$$

$$\dot{\eta}_1 = v|x_1|^{v-1} \operatorname{sgn}(x_1)$$

$$\dot{\eta}_2 = -\eta_1 - a_2\eta_2 + 2\mu v|x_2|^{2v-1}\operatorname{sgn}(x_2) \tag{9.142b}$$

The set of singular subarcs becomes

$$\Gamma_\alpha^s = \left\{x\colon x_1 + \tfrac{1}{2}\alpha x_2|x_2| = 0; |x_2| < \frac{\alpha+1}{\alpha|a_2|}; \right.$$

$$\left. \alpha = 2[\mu(2v-1)]^{1/v}; (\mu, v) \in \mathscr{R}_s\right\} \tag{9.143}$$

For example, adopting the illustrative parameter values

$$a_2 = -1; \qquad v = 2; \qquad \mu = \tfrac{1}{3} \tag{9.144}$$

then $\alpha = 2$ and the singular set is given by

$$\Gamma_2^s = \{x\colon x_1 + x_2|x_2| = 0; |x_2| < \tfrac{3}{2}\} \tag{9.145}$$

and corresponds to segment PQ of Fig. 9.12. As before, the normal segment of the overall switching boundary can be constructed by reverse-time integration of the state and adjoint equations (9.142) from Γ_2^s with initial (reverse-time)

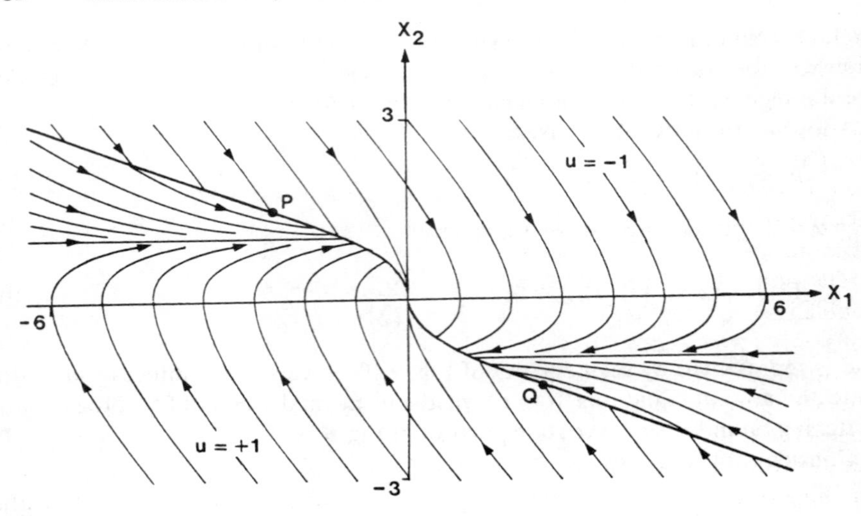

Fig. 9.12 *Integrator-plus-lag* ($a_1 = 0$, $a_2 = -1$)
Optimal synthesis for cost parameter values $\nu = 2$, $\mu = \frac{1}{3}$

conditions (9.105), where the constant γ of (9.105) is now restricted to satisfy

$$|\gamma| < \frac{(\alpha + 1)}{\alpha |a_2|}$$

In this case, the reverse-traced normal subarcs are found to contain at most one switch in control. Fig. 9.12 depicts the computed state portrait.

9.3.6 Discussion on nonquadratic-cost problem

A nonquadratic-cost regulator problem (with cost parameters μ, ν) has been investigated which, when specialised to multiple integrator systems, yields optimal solutions which exhibit the property of invariance (property 3.4.3) discussed earlier in Subsection 3.4.3. In the case of second-order systems, singular solutions have been shown to exist for all values of the cost parameters (μ, ν) in the region $\mathscr{R}_s$ defined by (9.101); the associated singular controls have been explicitly characterised. An extension of these results to the third-order case is contained in Ryan and Dorling (1981).

For the double integrator system, it has been found that the optimal control is identical to the time-optimal control on the specific locus (9.107) in the (μ, ν)-parameter plane which forms the boundary between the singular and nonsingular cases. The analysis of Fuller (1960c) has been extended to the latter (nonsingular) case of the double integrator system. A (nonsingular) minmax control problem for the double integrator system has also been investigated (Fuller, 1960c; Johnson, 1967); minmax problems have also been considered by Oldenburger (1966a), Bass and Webber (1966) and Fuller (1978) (see also bibliography therein). In the singular case, the convergent singular subarcs for the double integrator system constitute the (optimal) switching curve Γ_α which has been explicitly characterised. In the singular case for more general second-order systems, the convergent singular subarcs constitute only a bounded segment Γ_α^s of the (optimal) switching curve; the normal or nonsingular segments of the overall switching curve have been computed in two specific cases, viz. the harmonic oscillator and integrator-plus-lag plants.

The nonsingular case for *general* second-order systems (i.e. other than the double integrator system) has not been considered. In this case, numerical solutions only seem feasible. To this end, the numerical techniques of Roberts (1973) are applicable, wherein the analytical results for the double integrator system provide appropriate initial conditions for backtracking or reverse-time integration of the state and adjoint equations from a neighbourhood of the state origin (see also Brennan and Roberts 1962, Eggleston 1963). Special nonsingular cases for third-order systems have been studied by Grensted and Fuller (1965) and Ryan and Dorling (1981).

Chapter 10

Open-loop and quasioptimal feedback control techniques

10.1 Introduction

The previous nine chapters have been exclusively concerned with optimising (in various senses) saturating control system performance with particular emphasis on characterising optimal *feedback* structures. Explicit optimal feedback solutions have been obtained in a variety of cases of up to fourth order. However, these solutions frequently exhibit a high level of complexity which may prove unacceptable in many practical applications; this is especially evident in the time-optimal feedback control laws for the third and fourth-order systems of Chapters 6 and 7, many of which involve logarithmic and exponential functions of system state which may prove difficult to synthesise accurately. Furthermore, in these third- and fourth-order system studies, *linear*, *single-input* plants only were considered; the analytical difficulties presented by nonlinear, multi-input third and fourth-order plants are of such complexity that explicit feedback solutions are unavailable at present. For fifth- and higher-order systems, even in the case of linear single-input plants, the likelihood of obtaining explicit feedback solutions seems remote.† In addition, the explicit feedback solutions (when available) are derived on the basis of a plant model which is assumed to be precise; however, if the actual process parameter values differ from those of the model (as is almost certainly the case in practice) then some degradation in system performance is to be expected (as discussed in Sections 5.2.3 and 6.6).

† For example, extending the approach of Subsection 4.5.2 to the time-optimal synthesis problem for fifth-order, linear, single-input systems leads to the following observation:

In the case of the fourth-order systems of Chapter 7, solution of the synthesis problem crucially relied on the availability of closed-form expressions for the dependence on the subsystem initial state of the optimal switching and final times for the third-order subsystem; derivation of these expressions involved the general solution (by radicals) of the quartic equation (see Section 6.5). In a similar manner, solution of the time-optimal synthesis problem for fifth-order systems requires the availability of analogous closed-form expressions for the fourth-order subsystem; however, in this case, derivation of such expressions will require the general solution of the quintic equation—an impossible task!

Thus, a need exists for control techniques which, while being based on the mathematical theory of optimal control, nevertheless circumvent some of the practical difficulties alluded to above. These techniques should be readily implementable and should yield near-optimal performance. A selection of such techniques is presented in this chapter, largely in relation to the time-optimal control problem and appropriate to systems for which

(i) the exact optimal feedback solution is either unavailable or of impracticable complexity;

(ii) the underlying plant dynamics are imperfectly known.

In relation to case (i) above, open-loop computational techniques are briefly discussed in Section 10.2, predictive control techniques are analysed in Section 10.3 and suboptimal feedback control methods are investigated in Section 10.4. Finally, adaptive control techniques, appropriate to case (ii) above, are described in Section 10.5. Because of the scope and diversity of these approaches to quasi-optimal control, the treatment of this chapter is, of necessity, descriptive and somewhat lacking in rigour. The intention is to illustrate a selection of existing techniques; for deeper analyses, the reader is referred to the cited references.

10.2 Time-optimal open-loop control

As optimal feedback control is the main concern of this volume, the *open-loop* control techniques of this section are only briefly discussed. Attention is restricted to the time-optimal regulator problem for the linear, autonomous system

$$\dot{x} = Ax + Bu(t); \qquad x(0) = x^0; \qquad |u_j(t)| \le 1, j = 1, 2, \ldots, m \tag{10.1}$$

with the normality condition

$$\text{Rank } [b^j : Ab^j : A^2b^j : \cdots : A^{n-1}b^j] = n, \qquad j = 1, 2, \ldots, m \tag{10.2}$$

where b^j is the jth column vector of the $n \times m$ matrix B. From the theory of Section 2.6, the time-optimal control $u^*(t)$, $0 \le t \le t_f$ (when it exists), which steers x^0 to the origin in minimum time t_f, is uniquely determined (almost everywhere) by the relations

$$\begin{aligned} u_j^*(t) &= \text{sgn } \langle \eta(t), b^j \rangle = \text{sgn } \langle \exp(-A^T t)\eta^0, b^j \rangle \\ &= \text{sgn } \langle \eta^0, \exp(-At)b^j \rangle; \qquad j = 1, 2, \ldots, m \end{aligned} \tag{10.3}$$

for some initial adjoint vector η^0.

10.2.1 Computation of the vector η^0

For the appropriate choice of η^0, the open-loop control system is depicted in Fig. 10.1. Clearly, η^0 depends on the initial state x^0, but the form of this (nonlinear) dependence is, in general, unknown. Many iterative techniques have been

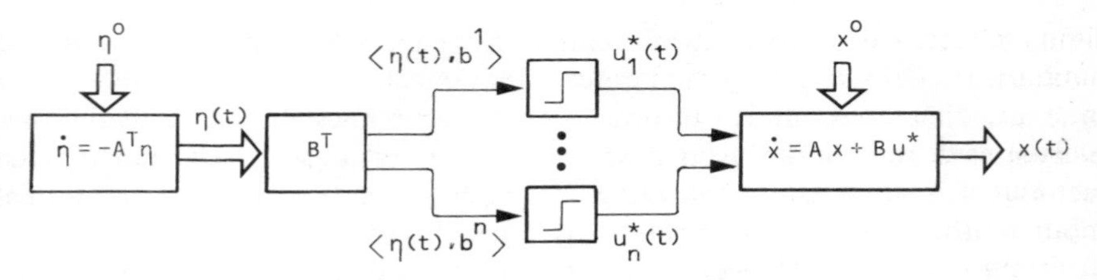

Fig. 10.1 *Time-optimal open-loop control*

proposed for computing the vector η^0. Generally speaking, these techniques generate (in different ways) a sequence of vectors $\{\eta^k\}$ which converge to the desired vector η^0; the various methods by which this sequence is determined are largely based on the geometry of the optimal solution and, in particular, on the convexity and compactness of the attainable set (as discussed in Section 2.5). Among these techniques are those of Neustadt (1960), Eaton (1962), Fadden and Gilbert (1964), Knudsen (1964), Fujisawa and Yasuda (1967) (see also Neustadt and Paiewonsky, 1963; Paiewonsky, 1963; Boltyanskii, 1971; and Gorlov, 1975; for related discussions), some of which are also applicable to time-varying systems and non-normal systems.

10.2.2 Nonoscillatory systems: calculation of the switching times

In cases for which the eigenvalues of the A matrix of system (10.1) are real, it follows from theorem 4.1.1 that each component $u_j^*(\cdot)$ of the optimal bang-bang control function $u^*(t)$, $0 \leq t \leq t_f$ can contain at most $(n-1)$ discontinuities on $(0, t_f)$. This extra structure is exploited in a second category of open-loop techniques which are based on direct computation of the $(n-1)$ switching times and final time t_f. Included in this category are the techniques of Smith (1961), Yastreboff (1969) (see Elkin and Daly (1973) for a quantitative assessment of the latter technique), Vena and Bershad (1971), Wolek (1971) and Farlow (1970, 1973); related approaches to the control of distributed parameter systems are contained in Goldwyn, Sriram and Graham (1967) and Babary and Pelczewski (1977), and time-varying nonlinear systems are considered by Davidson and Munro (1971).

The above techniques (and those of Subsection 10.2.1) suffer from the well known drawbacks of open-loop control; in particular, it is necessary to compute the optimal control for every new initial state. In contrast, the suboptimal control methods of the ensuing subsections all have a basic feedback structure.

10.3 Quasi-time-optimal predictive control

The general principle of predictive control can loosely be summarised as follows. The control logic incorporates a model of the plant to be controlled simulated on a fast time scale. The overall controller operates with sampled data, the actual plant state being measured at the start of each sampled-data interval; this measurement is used to set the initial conditions of the fast model. During each sampled-data interval (real time) the response (responses) of the fast model to some control schedule (schedules) is (are) observed; the control input to the actual plant during the next sampled-data interval is logically determined on the basis of the observed fast model response (responses). This principle of prediction forms the basis of the various control techniques described and analysed in Coales and Noton (1956), Chestnut, Sollecito and Troutman (1961), Adey, Coales and Stiles (1963), Gulko (1963), Gulko and Kogan (1963), Gulko *et al.* (1964), Kaufman and de Russo (1964, 1966), Mikhailov and Novoseltseva (1964), Adey, Billingsley and Coales (1966), Billingsley and Coales (1968), Fuller (1971; 1973a, b, c) and Ryan (1976a).

The strategy of Coales and Noton (1956) and its generalisation by Gulko (1963), Gulko and Kogan (1963) and Gulko *et al.* (1964) was discussed in Section 4.5. The time-optimality of the idealised strategy (Fuller 1973b) was demonstrated in the case of the serially decomposed plant of Fig. 4.3; this, in turn, facilitated the derivation of time-optimal feedback control solutions for a variety of plants of up to fourth order in Chapters 5, 6 and 7. This strategy (for single-input plants), in its original context as a predictive control technique, will now be discussed together with a practical suboptimal modification developed by Billingsley and Coales (1968).

Specifically, the general principle of predictive control may be interpreted, in the case of the strategy of Gulko *et al.* when applied to the serially decomposed single-input plant of Fig. 4.3, as follows. During each sampled-data interval, $(n-1)$ subsystem states of the overall nth-order fast model are driven time-optimally to the subspace origin; the polarity of the remaining state variable (x_1) is then inspected and the control input to the actual plant is set to the opposite polarity during the next sampled-data interval. The $(n-1)$ subsystem states may, in turn, be controlled by a complete predictive controller which uses a second 'nested' model, of order $(n-1)$, operating on a faster time scale than that of the original nth-order fast model. This procedure may be continued yielding a set of $(n-1)$ nested models, each successive model being reduced in order by unity and operating on a progressively faster time scale, and culminating in an innermost model of first order. In principle, if the nested model responses can be generated instantaneously, then the above strategy regulates the actual plant time-optimally. In practice the strategy is suboptimal (due to the finite times required to generate the model responses) and, moreover, is restricted to plants of low-order as the iterative speeds of the inner models are difficult to achieve. A suboptimal modification (Billingsley and Coales 1968) to the strategy, which circumvents the latter restriction, is described below. This modification uses a single nth-order fast model which is not controlled but merely observed, thereby dispensing with the inner lower-order nested models.

10.3.1 Predictive strategy of Billingsley and Coales

As outlined above, the difficulty of generating the optimal subsystem control input to the fast model, required by the time-optimal strategy of Gulko *et al.*, approaches that of controlling the plant itself if the latter is of high order. Billingsley and Coales (1968) have proposed a simple predictive control technique in which the fast model is not controlled but instead two fast model test responses to constant inputs are observed and the control input to the actual plant is determined on the basis of these observations. Successful *empirical* applications of the strategy to give quasi-time-optimal control for single input plants up to sixth-order have been reported by Billingsley and Coales (1968), who also suggest that the controller is insensitive to variations in plant dynamics. However, there is as yet no *theoretical* justification for the strategy applied to plants of fourth and higher order. In the case of second-order plants, the strategy is essentially that of Gulko *et al.* and hence is time-optimal. Detailed analysis of the strategy applied to a triple integrator plant has been supplied by Fuller (1971, 1973a). This analysis is outlined below (the reader is referred to Fuller (1971, 1973a) for details) to illustrate the principles of the predictive strategy and to lend some plausibility to its effectiveness in more general cases.

10.3.2 Quasi-time-optimal predictive control of a triple integrator plant

The plant to be controlled is the following

$$\dot{x}_1 = x_2; \qquad \dot{x}_2 = x_3; \qquad \dot{x}_3 = \frac{u}{a}; \qquad |u| \leq 1; \qquad a > 0 \tag{10.4}$$

The predictive controller incorporates a model of (10.4) operating on a fast time scale.† At the start of each sampled-data interval the model state is set equal to the current plant state. Two fast model test responses $x_1^+(\cdot)$ and $x_1^-(\cdot)$ are then generated, the first corresponding to a constant positive (fast model) control $u \equiv +1$ and the second corresponding to a constant negative (fast model) control $u \equiv -1$. One of the two test responses $x_1^+(\cdot)$ and $x_1^-(\cdot)$ is then selected (by a method to be described) and its sign at the *final* stationary point (i.e. the final point at which its derivative is zero) is observed; the control input to the actual plant is set to the opposite sign and the next sampled-data interval is initiated. The method of selecting the appropriate response from the two fast model test responses will now be discussed.‡

In what Billingsley and Coales (1968) have termed the modified strategy, the test response with the greater number of stationary points is first identified. If the final stationary point of this response is such that the response subsequently

† To avoid proliferation of symbols, the notation of (10.4) is used for both the actual plant and the fast model; specifically, t will denote both the actual and fast time scales, the correct interpretation being obvious from the context.

‡ In fact, Billingsley and Coales (1968) proposed two methods of selection in what they refer to as the basic and modified strategies; their modified strategy only is considered here. Analysis of both the basic and modified strategies is contained in Fuller (1971, 1973a).

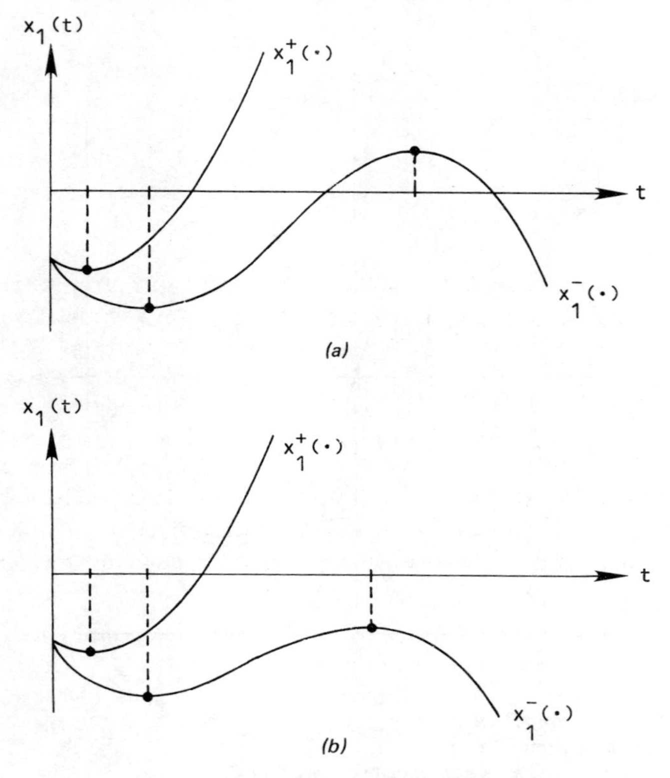

Fig. 10.2 *Test responses*

changes sign (as on response $x_1^-(\cdot)$ of Fig. 10.2*a*), then this test response is selected by the strategy; otherwise the other test response (as in the case of Fig. 10.2*b*) is selected. The sign of the value of the selected response at the final stationary point is determined and the actual control is set to the opposite sign; if the selected response has no stationary point, the actual control is simply set to the opposite sign of the initial value of the selected response.

Analysis in (Θ, Φ)*-space:* In analysing the above strategy, the two-dimensional (Θ, Φ)-space of Section 3.5 will be employed. Furthermore, the strategy is assumed to be idealised in the sense that the model test responses $x_1^+(\cdot)$ and $x_1^-(\cdot)$ can be generated instantaneously. These test responses correspond to the first components of the test trajectories $(x_1^+(\cdot), x_2^+(\cdot), x_3^+(\cdot))'$ and $(x_1^-(\cdot), x_2^-(\cdot), x_3^-(\cdot))'$ which, in turn, map to a p-path of Fig. 3.9 and an n-path of Fig. 3.10. Stationary points of the test responses $x_1^+(\cdot)$ and $x_1^-(\cdot)$ correspond, in (Θ, Φ) space, to crossings of the Φ axis (where $\Theta = 0 \Rightarrow x_2 = \dot{x}_1 = 0$) or to crossings of the (equivalent) edges $\Theta = \pm\pi$ ($\Rightarrow x_2 = \dot{x}_1 = 0$) by the model test

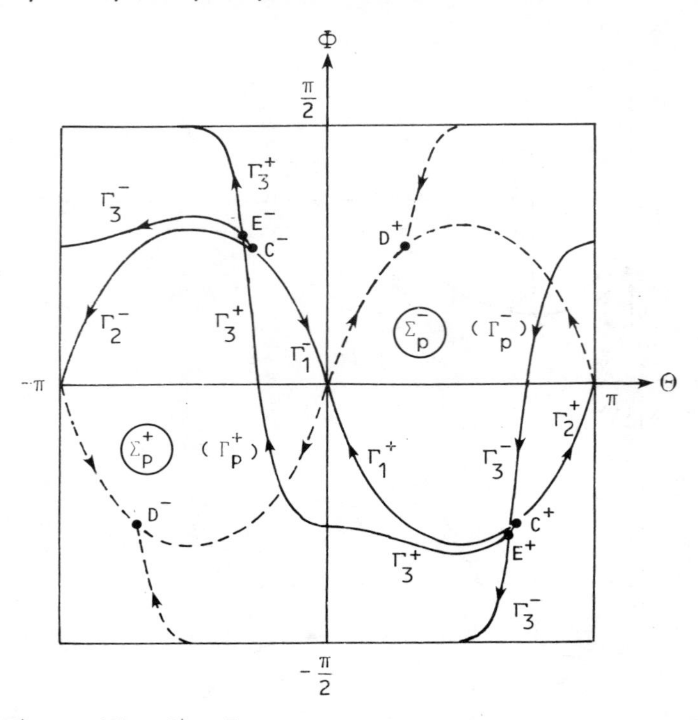

Fig.10.3 Γ_p^+ : p-*path from* E^+ *to* E^-
Γ_p^- : n-*path from* E^- *to* E^+
Σ_p^- : *region lying above composite curve* $\Gamma_p^+ \cup \Gamma_p^-$
Σ_p^+ : *region lying below composite curve* $\Gamma_p^+ \cup \Gamma_p^-$

trajectories. Hence, the portraits of p-paths and n-paths of Figs. 3.9 and 3.10 play a central role in the analysis. Referring to Fig. 10.3, the following notation is introduced: Γ_1^+ and Γ_1^- denote the p- and n-paths leading, from the singular points C^+ and C^-, to the origin in (Θ, Φ)-space (i.e. to the positive x_1 semi-axis in state space); Γ_2^+ and Γ_2^- denote the p- and n-paths leading, from the singular points C^+ and C^-, to the (equivalent) points $(\pi, 0)$ and $(-\pi, 0)$, respectively, (i.e. to the negative x_1 semi-axis in state space); Γ_3^+ denotes the p-path, from C^+, leading to the upper edge $\Phi = \pi/2$ (i.e. to the positive x_3 semi-axis in state space) and Γ_3^- denotes the n-path, from C^-, leading to the lower edge $\Phi = -\pi/2$ (i.e. to the negative x_3 semi-axis in state space).† The latter paths (Γ_3^+ and Γ_3^-) intersect at the points E^+ and E^-. The segment of Γ_3^+ joining E^+ to E^- is denoted by Γ_p^+ and the segment of Γ_3^- joining E^- to E^+ is denoted by Γ_p^-. It will be shown that the composite curve $\Gamma_p = \Gamma_p^+ \cup \Gamma_p^-$ (i.e. the closed path $E^+E^-E^+$) constitutes an effective nonlinear switching boundary which partitions (Θ, Φ)- space into

† The broken curves of Fig. 10.3 represent the continuation of the paths Γ_i^+ and Γ_i^- $(i = 1, 2, 3)$, terminating at the singular points D^+ and D^-.

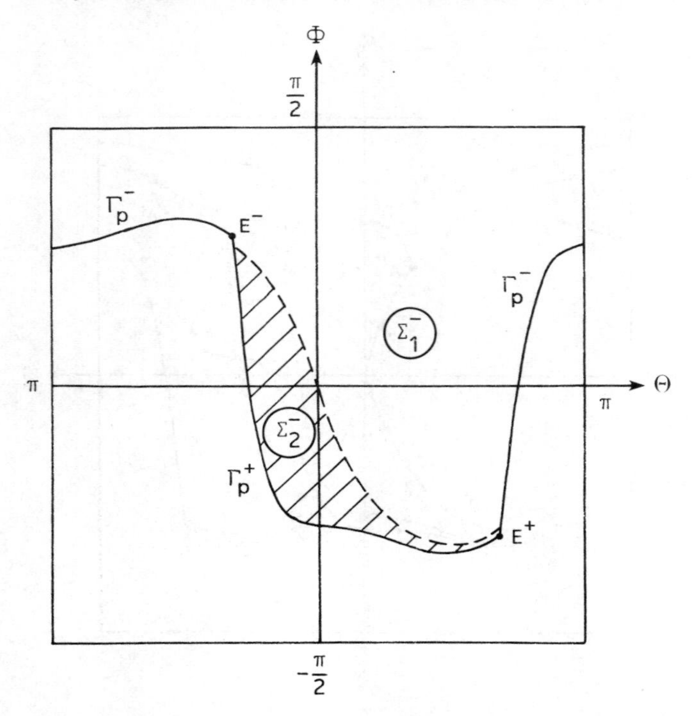

Fig. 10.4 *Decomposition of Σ_p^- (region lying above $\Gamma_1^+ \cup \Gamma_p^-$) into subregions Σ_1^- and Σ_2^-*

two mutually exclusive regions Σ_p^- (lying above Γ_p) and Σ_p^+ (lying below Γ_p) wherein the predictive control strategy defines the actual control input to the plant as $u = -1$ in Σ_p^- and $u = +1$ in Σ_p^+. For definiteness, it will be shown that the strategy sets the plant control to $u = -1$ in Σ_p^- (the result $u = +1$ in Σ_p^+ follows from symmetry). To this end, the region Σ_p^- $(= \Sigma_1^- \cup \Sigma_2^-)$ is decomposed into two subregions Σ_1^- and Σ_2^- as shown in Fig. 10.4, where Σ_2^- is defined to be the *interior* of the shaded area, i.e. the area lying above the curve Γ_p and below the composite curve $\Gamma_2^- \cup C^- \cup \Gamma_1^- \cup \Gamma_1^+ \cup C^+ \cup \Gamma_2^+$. Starting points in each of these subregions will be considered separately.

Case (i): Starting points in Σ_1^-: Suppose initially that the starting point (initial plant state) lies in the subregion Σ_1^-. It may be verified that, in all such cases, the fast model test response $x_1^-(\cdot)$ has the greater number of stationary points. For example, from point A $\in \Sigma_1^-$ of Fig. 10.5, the fast model test responses $x_1^+(\cdot)$ and $x_1^-(\cdot)$ correspond to the x_1 components along the p- and n-paths γ_A^+ and γ_A^-, respectively; γ_A^+ does not cross the Φ axis $(\Theta = 0)$ or the edges $\Theta = \pm\pi$ and, consequently, the test response $x_1^+(\cdot)$ has *no* stationary points. On the other hand, γ_A^- intersects the Φ axis *once* corresponding to the single

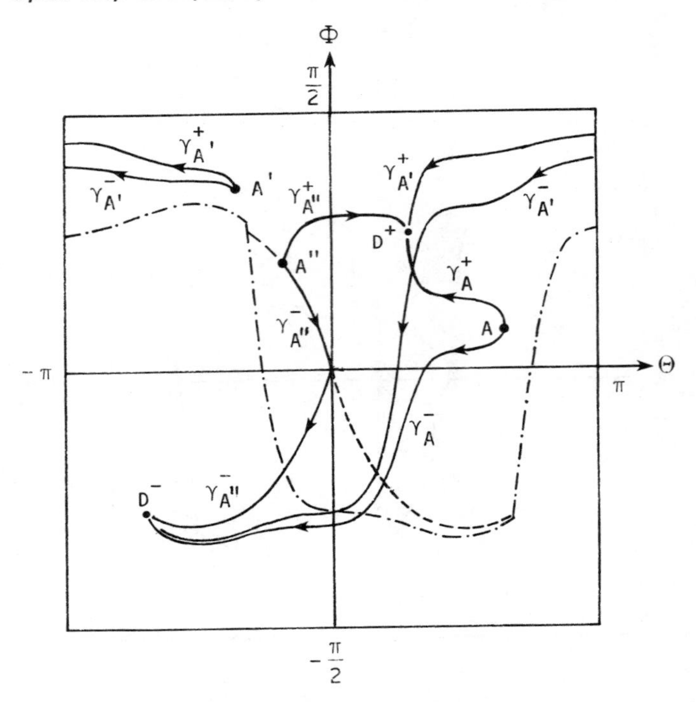

Fig. 10.5 *Typical test trajectories from starting points in* Σ_1^-

stationary point of the test response $x_1^-(\cdot)$. As a second possibility (corresponding to a test response of the form shown in Fig. 10.2*a*), consider the starting point A′ of Fig. 10.5. The model test trajectories from this starting point are $\gamma_{A'}^+$, which intersects the (equivalent) edges $\Theta = \pm\pi$ *once*, and $\gamma_{A'}^-$, which intersects the (equivalent) edges $\Theta = \pm\pi$ *once* and which also intersects the Φ axis *once*, so that it may again be concluded that the test response $x_1^-(\cdot)$ has the greater number of stationary points. Finally, consider the starting point $A'' \in \Gamma_1^- \subset \Sigma_1^-$; in this (nongeneric) case, the model test trajectories are $\gamma_{A''}^+$, which intersects the Φ axis *once*, and $\gamma_{A''}^-$, which passes through the origin in (Θ, Φ)-space and which corresponds to the confluence of *two* stationary points of the test response $x_1^-(\cdot)$. If confluent stationary points are counted with full multiplicity, $x_1^-(\cdot)$ again has the greater number of stationary points. A general starting point in the subregion Σ_1^- will exhibit one of the above three patterns of behaviour. Hence, at each starting point in the subregion Σ_1^-, the test response $x_1^-(\cdot)$ has the greater number of stationary points; in each case, the final stationary point corresponds to a point on the Φ axis so that $\Theta = 0 \Rightarrow x_1 > 0$ at the final stationary point of the test response $x_1^-(\cdot)$. As a second possibility (corre-
subregion Σ_1^-, the test response $x_1^-(\cdot)$ tends to the singular point D^- (at which point $x_1 < 0$) so that $x_1^-(\cdot)$ must change sign subsequent to its final (positive-

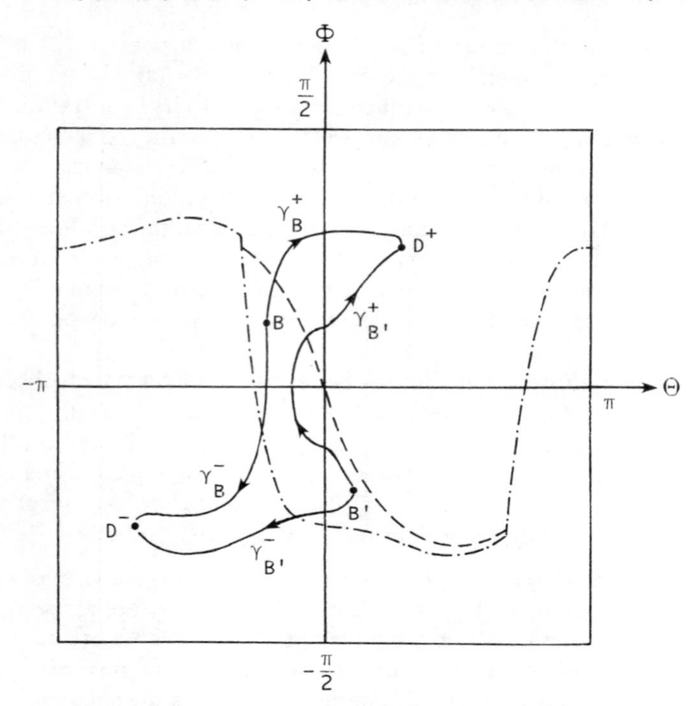

Fig. 10.6 *Typical test trajectories from starting points in* Σ_2^-

valued) stationary point, thereby fulfilling the second condition for selection by the strategy. Therefore, in accordance with the strategy, the control input to the actual plant is set to the value $u = -1$ (i.e. the negative of the sign of the selected test response at its final stationary point). This establishes the result that, at all actual plant states in the subregion Σ_1^-, the idealised strategy sets the actual plant control intput to the value $u = -1$.

Case (ii) : Starting points in Σ_2^- *:* Now suppose that the starting point lies in the subregion Σ_2^-. It may be verified in this case that the test response $x_1^+(\cdot)$ has the greater number of stationary points. For example, referring to Fig. 10.6, from the point $\mathrm{B} \in \Sigma_2^-$, the model test trajectories are γ_{B}^+, which intersects the Φ axis *once*, and γ_{B}^-, which does not intersect the Φ axis or the edges $\Theta = \pm\pi$ and consequently has *no* stationary points. As a second possibility, consider the starting point $\mathrm{B}' \in \Sigma_2^-$ of Fig. 10.6. From this starting point, the model test trajectories are $\gamma_{\mathrm{B}'}^+$, which intersects the Φ axis *twice*, and $\gamma_{\mathrm{B}'}^-$, which intersects the Φ axis *once*. All other starting points in Σ_2^- will give rise to model test trajectories akin to one or other of the above two possibilities. Consequently, for starting points in the subregion Σ_2^-, the test response $x_1^+(\cdot)$ has the greater number of stationary points. However, in each case, the final stationary point corresponds to an intersection of the Φ axis, at which point $x_1^+(\cdot)$ is positive-valued; moreover, $x_1^+(\cdot)$ tends to the singular point D^+ (at which point $x_1 > 0$) and, hence, must remain positive-valued subsequent to its final stationary point. The latter behaviour of the test response $x_1^+(\cdot)$ violates the second condition for selection by the strategy with the result that the other test response $x_1^-(\cdot)$ is again selected. In accordance with the strategy, it then follows that at all points in the subregion Σ_2^- the actual plant control input is set to the value -1 (i.e. the negative of the sign of $x_1^-(\cdot)$ at its only stationary point in the case of a test trajectory of the form $\gamma_{B'}^-$ of Fig. 10.6, or the negative of the sign of $x_1^-(\cdot)$ at its starting point in the case of a test trajectory with no stationary points, as in γ_B^- of Fig. 10.6).

Combining the above results, it may be seen that, under the idealised predictive strategy, the actual plant control takes the value -1 at all points of the region $\Sigma_p^- = \Sigma_1^- \cup \Sigma_2^-$ lying above Γ_p. By symmetry, it follows that the actual plant control takes the value $+1$ at all points of the region Σ_p^+ lying below the curve Γ_p. On the curve $\Gamma_p = \Gamma_p^+ \cup \Gamma_p^-$ itself, the appropriate interpretation of the idealised strategy is that $u = +1$ on Γ_p^+ and $u = -1$ on Γ_p^-.

Equivalent feedback control: The curve Γ_p in (Θ, Φ)-space corresponds to a surface Γ_p in x-state which plays the role of a switching surface associated with a feedback control structure which is equivalent to the idealised predictive control system. To illustrate this equivalent feedback form, the surface Γ_p and the regions Σ_p^+ and Σ_p^- may be explicitly characterised in x-state space as

$$\Gamma_p = \{x\colon \xi_p(x) = 0\} \tag{10.5a}$$

$$\Sigma_p^+ = \{x\colon \xi_p(x) < 0\} \tag{10.5b}$$

$$\Sigma_p^- = \{x\colon \xi_p(x) > 0\} \tag{10.5c}$$

where

$$\xi_p(x) = x_1 + ax_2x_3 \operatorname{sgn}(\Xi_p^s) + \frac{a^2x_3^3}{3} + a^{-1}\frac{2\sqrt{2}}{3} \times [ax_2 \operatorname{sgn}(\Xi_p^s) + \tfrac{1}{2}(ax_3)^2]^{3/2} \operatorname{sgn}(\Xi_p^s) \tag{10.5d}$$

with

$$\Xi_p^s = \Xi_p^s(x_2, x_3) = \begin{cases} \xi_p^s(x_2, x_3) = x_2 + \dfrac{\sqrt{3}}{4}ax_3|x_3|; & \xi_p^s(x_2, x_3) \neq 0 \\ x_3; & \xi_p^s(x_2, x_3) = 0 \end{cases} \tag{10.5e}$$

Derivation of the above expressions is straightforward but lengthy and is omitted here, the reader is referred to Fuller (1973a) for details. Defining the (suboptimal) switching function $\Xi_p\colon \mathbb{R}^3 \to \mathbb{R}$ as

$$\Xi_p(x) = \begin{cases} \xi_p(x); & \xi_p(x) \neq 0 \\ \Xi_p^s(x_2, x_3); & \xi_p(x) = 0 \end{cases} \tag{10.6a}$$

the equivalent feedback control may be expressed as

$$u = u(x) = -\operatorname{sgn}(\Xi_p(x)) \tag{10.6b}$$

By way of comparison, referring to the results of Section 6.4, the precise time-optimal switching surface is given by

$$\Gamma = \{x: \xi(x) = 0\} \tag{10.7a}$$

where

$$\xi(x) = x_1 + ax_2x_3 \operatorname{sgn}(\Xi^s) + \frac{a^2x_3^3}{3} + a^{-1}[ax_2 \operatorname{sgn}(\Xi^s) + \tfrac{1}{2}(ax_3)^2]^{3/2} \operatorname{sgn}(\Xi^s) \tag{10.7b}$$

with

$$\Xi^s = \Xi^s(x_2, x_3) = \begin{cases} \xi^s(x_2, x_3) = x_2 + \tfrac{1}{2}ax_3|x_3|; & \xi^s(x_2, x_3) \neq 0 \\ x_3; & \xi^s(x_2, x_3) = 0 \end{cases} \tag{10.7c}$$

Inspection of (10.5) and (10.7) reveals that the switching surface Γ_p, implicit in the idealised predictive strategy, is similar to the time-optimal switching surface Γ; this similarity is further emphasised by Fig. 10.7 which depicts the surfaces Γ and Γ_p in (Θ, Φ)-space and which are seen to be in close agreement.

Finally, as the surface Γ_p in (Θ, Φ)-space is composed of the p-path from E^+ to E^- and the n-path from E^- to E^+, it follows that, on reaching Γ_p, the state point will subsequently trace a limit cycle $E^+E^-E^+$ in (Θ, Φ)-space which, recalling the results of Subsection 6.6.2, corresponds to a convergent *constant-ratio trajectory* to the origin in x-state space such that the plant response $x_1(\cdot)$ exhibits damped oscillatory motion† with the ratio of successive peak overshoots and undershoots equal to a constant $-\rho^3$, where the constant-ratio parameter ρ is easily calculated (Fuller 1973a) as

$$\rho = 2 - \sqrt{3} \simeq 0{\cdot}2679 \tag{10.8}$$

Estimate of suboptimality of the predictive strategy: As an indication of the suboptimality of the predictive strategy, the time t_f to the origin on a trajectory $x(\cdot)$ from an initial state $x(0) = (x_1, 0, 0)'$, $x_1 \neq 0$, on the x_1 state axis can be calculated (see Fuller (1973a) for details) as

$$t_f = 2[3(1 + \sqrt{2})a|x_1|]^{1/3} \tag{10.9}$$

† To eliminate such oscillatory motion, Billingsley and Coales (1968) (see also Fuller 1971, 1973a) introduced a further refinement (termed slugging) of their strategy which has the effect of distorting the surface Γ_p so as to enforce nonoscillatory sliding motion to the origin; slugging has the additional beneficial effect of reducing sensitivity to plant uncertainties and disturbances. This refinement will not be analysed here.

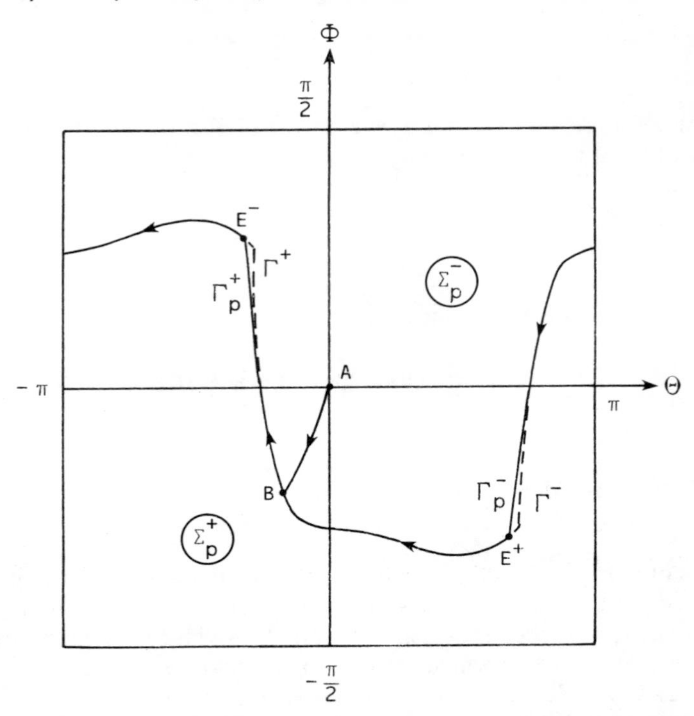

Fig. 10.7 *Equivalent switching surface* (Γ_p) *of predictive strategy and time-optimal switching surface* (Γ) *in* (Θ, Φ)*-space*

For example, with $x_1 > 0$, all such trajectories map to the path $ABE^-E^+E^-$ of Fig. 10.7. From the starting point A (at time $t = 0$), the n-path AB is followed until the point $B \in \Gamma_p$ is reached after a time t_1, where

$$t_1 = [3(3 - 2\sqrt{2})a\,|x_1|]^{1/3} \tag{10.10}$$

At $B \in \Gamma_p$ a control switch occurs and the p-path BE^- (in Γ_p) is followed to the point E^-; the duration t_2 of the path BE^- is given by

$$t_2 = [1 + 2\sqrt{2 - \sqrt{3}}]t_1 \tag{10.11}$$

From E^-, the limit cycle $E^-E^+E^-$ is traced in (Θ, Φ)-space, corresponding to a constant-ratio trajectory (with parameter (10.8)) in state space which converges in finite time to the origin. Again, recalling the results of Subsection 6.6.2, the duration t_{cr} of the constant-ratio trajectory from E^- is given by

$$t_{cr} = (1 + \rho)(1 - \rho)^{-1}a\,|x_3^{E^-}| = \sqrt{3}\,a\,|x_3^{E^-}|$$

where $x_3^{E^-}$ is the x_3 state coordinate at the point E^-, i.e. at the second switch point of the overall trajectory. Specifically,

$$x_3^{E-} = x_3(t_1 + t_2) = x_3(t_1) + \frac{t_2}{a}$$

$$= x_3(0) - \frac{t_1}{a} + \frac{t_2}{a} = \frac{1}{a}(t_2 - t_1)$$

$$= \frac{1}{a}[2\sqrt{2 - \sqrt{3}}]t_1$$

and hence,

$$t_{cr} = 2\sqrt{3}[2 - \sqrt{3}]^{1/2} t_1 \tag{10.12}$$

Adding (10.10), (10.11) and (10.12) gives the total time to the origin $t_f = t_1 + t_2 + t_{cr}$ as in (10.9). In Subsection 6.5.3, the minimum time t_f^* from $x(0) = (x_1, 0, 0)'$ to the origin was calculated as

$$t_f^* = (32a|x_1|)^{1/3} = 2(4a|x_1|)^{1/3} \tag{10.13}$$

Combining (10.9) and (10.13) gives the fractional increase in settling time under the (idealised) predictive strategy as

$$\Delta = \frac{t_f - t_f^*}{t_f^*} = [\tfrac{3}{4}(1 + \sqrt{2})]^{1/3} - 1 \simeq 0{\cdot}219 \tag{10.14}$$

Hence, for initial states on the x_1 axis, the idealised predictive strategy yields times to the origin which exceed the corresponding minimum times by approximately 22%.

10.4 Suboptimal feedback control techniques

Another approach to suboptimal control, appropriate to systems for which precise optimal feedback solutions are either unobtainable or of impracticable complexity, is to impose on the system a readily synthesised feedback structure predetermined up to a set of design parameters. The latter parameters are then adjusted so as to 'best' approximate the exact but infeasible optimal feedback solution. In other words, the original optimal feedback control problem is replaced by one of parametric optimisation on a prescribed feedback structure. In earlier chapters, it was seen that, for many optimal relay and saturating control problems, the feedback solutions (when available) involve highly nonlinear feedback functions (switching functions) associated with nonlinear switching hypersurfaces in state space. The philosophy underlying the suboptimal techniques described below is to adopt simpler feedback functions at the expense of some (but acceptable) degradation in system performance. Regulating control problems are assumed throughout, i.e. the target set is assumed to be the origin in state space.

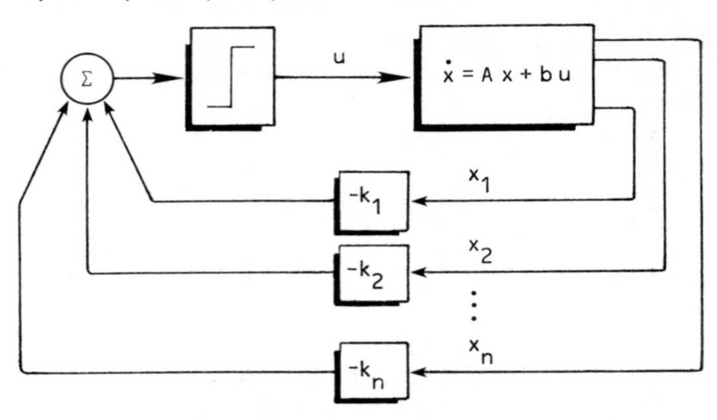

Fig. 10.8 *Relay control system with linear feedback*

10.4.1 Linear feedback control of relay systems

Because of their simplicity, relay control systems with linear feedback have received considerable attention in the literature. For such systems, the inputs to the relays are simply linear combinations of the components of the current state vector; the single input case is depicted in Fig. 10.8. Phase-plane studies of such control systems go back to Léauté (1885, 1891); for discussion of these and other early studies of relay systems under linear feedback see Fuller (1967a). More recent investigations of such systems include Flügge-Lotz (1968), Roberts (1970, 1973) and Coupé (1976). In Subsection 9.2.6, a second-order example of an optimal quadratic-cost saturating control system was presented for which the optimal solution is synthesized in the form of a relay controller with linear feedback. However, such cases are unusual and relay control systems (of second and higher order) under linear feedback are, for the most part, suboptimal with respect to the most commonly adopted measures of system performance (e.g. minimum-time, minimum-fuel, integral-square-error etc.). The stability of third and higher-order relay control systems under linear feedback has been investigated by Michaels and Frederick (1968) and Fuller (1969). In cases where the plant to be controlled is a multiple integrator plant of order n, Fuller (1969) has shown that, for $n \geq 3$, the feedback system is unstable-in-the-large (i.e. there exist trajectories which do *not* go to the origin and hence are not even approximately optimal). This latter result serves to illustrate the limitations of suboptimal linear feedback control for relay systems and argues the case for the *nonlinear* feedback techniques considered below. Adaptive linear feedback control of relay systems will be considered in Section 10.5.

10.4.2 Quasi-time-optimal nonlinear feedback control of relay and saturating systems

Attention will be restricted to single-input systems of the form

$$\dot{x} = Ax + bu; \qquad x(t) \in \mathbb{R}^n; \qquad |u(t)| \leq 1 \tag{10.15a}$$

The pair (A, b) is assumed controllable, i.e. rank $[b : Ab : A^2b : \cdots : A^{n-1}b] = n$. Hence, no loss in generality is incurred in assuming A and b to be of the canonical forms

$$A = \begin{bmatrix} 0 & 1 & 0 & \cdots & 0 \\ 0 & 0 & 1 & \cdots & 0 \\ \vdots & \vdots & \vdots & \ddots & \vdots \\ 0 & 0 & 0 & \cdots & 1 \\ a_1 & a_2 & a_3 & \cdots & a_n \end{bmatrix}; \quad b = \begin{bmatrix} 0 \\ 0 \\ \vdots \\ 0 \\ \frac{1}{a} \end{bmatrix} \quad (a > 0) \tag{10.15b}$$

so that the state vector $x = (x_1, x_2, \ldots, x_n)' \in \mathbb{R}^n$ is composed of the *phase variables*

$$x_1, x_2 = \frac{dx_1}{dt}, \ldots, x_n = \frac{d^{n-1}x_1}{dt^{n-1}}$$

It is required to determine an 'easily synthesised' feedback control $u = F(x)$ which 'quasi-optimally' *regulates* the system (i.e. such that $x(t_f) = 0$ for some finite time $t_f < \infty$, or such that $x(t) \to 0$ as $t \to \infty$). While the terms 'easily synthesised' and 'quasioptimal' cannot be readily quantified, the underlying notion is to strike a balance between the (reduction in) difficulty of realising a (simplified) feedback control and the resulting (degradation in) system performance.

Quasi-time-optimal control techniques are the main concern of this section, in which case the controller design problem is that of characterising an easily synthesised feedback law such that the system is driven from a (null controllable) state $x(0) = x^0$ to the state origin $x(t_f) = 0$ in a time $t_f \geq t_f^*$ which is acceptably close to the theoretical minimum time t_f^*. As pointed out by Fuller (1967a), naïve application of the minimum-time criterion casts doubts on the usefulness of many suboptimal techniques; this is due to the fact that the minimum-time criterion penalises all non-zero states equally, and only when the state *exactly* attains the origin is there no further contribution to the cost. In practice, small errors are frequently acceptable; consequently, in assessing the effectiveness of quasi-time-optimal techniques the time t_f required to attain (and subsequently remain in) a sufficiently small neighbourhood of the origin (rather than the origin itself) would appear to be a more appropriate measure of performance. To this end, the suboptimal response time (or the cost of a suboptimal trajectory) is defined as the first time t_f at which the following inequality holds,

$$\|x(t_f)\| \leq \epsilon \|x(0)\|; \qquad \epsilon \geq 0 \tag{10.16}$$

i.e. t_f is the time taken to reduce the state vector norm to within ϵ times its initial value (for example, $\epsilon = 10^{-6}$ is adopted in Subsection 10.4.4).

For convenience, the matrix A is assumed to be stable† (i.e. all eigenvalues have nonpositive real parts) so that, by theorem 4.3.1, a time-optimal path to the origin exists from every point of the state space $\mathbb{R}^n$; moreover, by theorem 2.6.2, the associated time-optimal control is of bang-bang form. Most quasi-time-optimal techniques also impose this bang-bang structure on the system so that the feedback control is assumed to be of the form

$$u = -\operatorname{sgn}(\Xi(x)) \tag{10.17}$$

for some switching function $\Xi: \mathbb{R}^n \to \mathbb{R}$. In the case of linear feedback, Ξ is of the linear form $\Xi(x) = \langle k, x \rangle = k_1 x_1 + k_2 x_2 + \cdots + k_n x_n$ where $k = (k_1, k_2, \ldots, k_n)'$ is the (constant) vector of design parameters to be selected in some appropriate manner. However, as mentioned in the previous section, Fuller (1969) has shown that, for $n \geq 3$, this feedback control is unstable-in-the-large when applied to multiple integrator systems. Frederick and Franklin (1967) have proposed the use of piecewise linear switching functions with design parameters defining the associated slopes and break-points; the parameter values are selected so as to minimise a weighted sum of suboptimal settling times from a representative set of initial states. Smith (1966) has investigated piecewise linear switching functions with parameters determined such that the associated switching surfaces give a least-squares approximation to the (computed) time-optimal switching surface. In the case of a third-order system, typical response times which exceed the corresponding minimum times by up to a factor of 2 are reported. DeRooy (1970) also proposed a least squares fitting technique to approximating time-optimal switching hypersurfaces by suboptimal hypersurfaces defined via polynomial forms.

The above techniques rest largely on a heuristic basis and, in practice, are frequently restricted to a bounded region of the state space; moreover, while a computed set of switching function parameter values may yield acceptable performance for a plant of given structure, these values must be recomputed whenever the plant characteristics are altered. The following sections (Subsections 10.4.3 and 10.4.4) consider nonlinear feedback controllers which are based on properties of the optimal system and which circumvent the aforementioned restrictions. Specifically, multiple integrator systems are considered in Subsection 10.4.3 and a (simplified) nonlinear switching function structure is specified such that the suboptimal system exhibits the invariance property of the optimal system. In Subsection 10.4.4 a linear state transformation is proposed which enables an optimal or suboptimal controller, designed for a multiple integrator system, to be directly applied to more general systems of the same order.

† In principle, many of the ensuing suboptimal techniques are applicable in the case of an unstable matrix A; clearly, in this case, attention is restricted to the domain of null controllability $\mathscr{C}$. However, under suboptimal feedback control, it frequently turns out that trajectories to the origin may not exist from all points of $\mathscr{C}$; in effect, the suboptimal feedback system may define a smaller suboptimal set of null controllable states within $\mathscr{C}$. These aspects will not be pursued here.

10.4.3 Quasi-time-optimal nonlinear feedback control of multiple integrator systems

As previously outlined in Section 3.4, Persson (1963) proposed the adoption of a specific nonlinear feedback structure (determined up to a set of design parameters) on the intuitive basis that system response should be 'independent of signal amplitude'. In contrast to the methods outlined in Subsections 10.4.1 and 10.4.2, this structure implies that, if the design parameter values are calculated to give acceptable performance for 'small' initial conditions, then the same parameter values also yield acceptable performance for 'large' initial conditions, i.e. system response should be invariant under signal and time scale changes. Subsequently, (as described in Section 3.4) Fuller (1970, 1971) clarified the theoretical basis of Persson's approach and rigorously established such invariance properties (properties 3.4.1, 3.4.2 and 3.4.3) in the case of multiple integrator systems of the form

$$\begin{aligned} \dot{x}_i &= x_{i+1}; \qquad i = 1, 2, \ldots, n-1 \\ \dot{x}_n &= \frac{u}{a}; \qquad a > 0; \qquad |u(t)| \leq 1 \end{aligned} \tag{10.18}$$

i.e. (10.15) with $a_1 = a_2 = \cdots = a_n = 0$. In the context of quasi-time-optimal feedback control of (10.18), it is the property of invariance of the feedback system (property 3.4.2) which is required. Restricting attention to the case of special invariance, it may be concluded, from properties 3.4.2, 3.4.3 together with equations (3.34) and (3.35), that if

$$(x_1(t), x_2(t), \ldots, x_n(t))', \qquad 0 \leq t \leq t_f \tag{10.19a}$$

is a trajectory of system (10.18) under the feedback control (10.17), then

$$(\kappa^{-n}x_1(\kappa t), \kappa^{-(n-1)}x_2(\kappa t), \ldots, \kappa^{-1}x_n(\kappa t))', \qquad 0 \leq t \leq \kappa^{-1}t_f \tag{10.19b}$$

is also a trajectory of (10.18) for all $\kappa > 0$ provided that the switching function $\Xi: \mathbb{R}^n \to \mathbb{R}$ of (10.17) satisfies

$$\Xi(x) = \Xi((x_i)) = \beta(\kappa)\Xi((X_i^{-1}(x_i))) \tag{10.20}$$

for all $x = (x_i)$ and for some $\beta = \beta(\kappa) > 0$ where

$$X_i^{-1}(x_i) = \kappa^{-(n+1-i)}x_i; \qquad i = 1, 2, \ldots, n \tag{10.21}$$

Hence, for the property of invariance to hold, the switching function $\Xi: \mathbb{R}^n \to \mathbb{R}$ should be such that (10.20) is satisfied. Note initially, that a linear switching function $\Xi(x) = \langle k, x \rangle = k_1 x_1 + k_2 x_2 + \cdots + k_n x_n$ fails to satisfy (10.20). A second practicable form for the switching function is a linear combination of functions of single state variables, viz.

$$\Xi(x) = f_1(x_1) + f_2(x_2) + \cdots + f_n(x_n) \tag{10.22}$$

Persson proposed the adoption of the following functions f_i (see also Fuller 1971)

$$f_i(x_i) = k_i\,|\,ax_i\,|^{H/(n+1-i)}\ \mathrm{sgn}\ (x_i); \quad k_i > 0, \qquad i = 1, 2, \ldots, n \qquad (10.23a)$$

which, when substituted in (10.22), satisfy the invariance condition (10.20) for $\beta(\kappa) = \kappa^{-H} > 0$. Moreover, to further simplify the controller design, Persson suggested that H should be chosen so as to make one of the functions f_i linear, i.e. H is restricted to be an integer in the range

$$1 \leq H \leq n \qquad (10.23b)$$

For example, in the second-order case (i.e. $n = 2$), selecting the value $H = 2$ yields a switching function

$$\Xi(x) = k_1\, ax_1 + k_2\, ax_2\,|\,ax_2\,| \qquad (10.24a)$$

which can be made time-optimal† (see eqn. (5.15)) by setting

$$k_1 = \frac{1}{a}; \qquad k_2 = \frac{1}{2a} \qquad (10.24b)$$

In other words, the suboptimal approach can yield time-optimal performance in the second-order case. For the triple integrator system ($n = 3$), the suboptimal switching function becomes

$$\Xi(x) = k_1(a\,|\,x_1\,|)^{H/3}\ \mathrm{sgn}\ (x_1) + k_2(a\,|\,x_2\,|)^{H/2}\ \mathrm{sgn}\ (x_2) + k_3(a\,|\,x_3\,|)^{H}\ \mathrm{sgn}\ (x_3) \qquad (10.25)$$

which is considerably easier to synthesise than the exact time-optimal switching function (6.15) derived in Section 6.4. As (10.25) satisfies the invariance condition (10.20), the dynamic behaviour of the suboptimal feedback system can be analysed in (Θ, Φ)-space (such an analysis is carried out in Fuller (1971)). To illustrate the performance of the third-order feedback system, the following heuristic method is adopted for selecting the controller parameter values H, k_1, k_2, k_3. First recall that the reasonable assumption of *negative feedback* has been made which restricts the parameters k_i to be positive-valued. Now, as the *sign* of the switching function determines the control u, k_1 can be set to unity without loss of generality, i.e.

$$k_1 = 1 \qquad (10.26)$$

For each admissible integer value of H (= 1, 2, 3), the function (of the remaining parameters k_2, k_3)

$$L(k_2, k_3) = \Delta_1(k_2, k_3) + \Delta_2(k_2, k_3) \qquad (10.27)$$

† With this switching function, appropriate interpretation (details omitted) of the associated feedback law $u = -\mathrm{sgn}\ (\Xi(x))$ must be made when $\Xi(x) = 0$.

is minimised numerically. Here, Δ_1 and Δ_2 are the computed fractional increases in settling time from initial states $x^1 = (x_1, 0, 0)'$ and $x^2 = (0, x_2, 0)'$ on the x_1 and x_2 state axes, respectively. Note that, as the feedback system exhibits the property of invariance, the value of Δ_1 (or Δ_2) will be the same (for fixed values of k_2 and k_3) for all non-zero initial states x^1 (or x^2) on the x_1 (or x_2) state axis. Hence, without loss in generality, the initial states $x^1 = (1/a, 0, 0)'$ and $x^2 = (0, 1/a, 0)'$ may be assumed. Now, by the results of Section 6.5.3, the exact minimum times to the origin from x^1 and x^2 are, respectively, given by

$$t^*_{f_1} = 2(4)^{1/3} \quad \text{and} \quad t^*_{f_2} = 2(2+\sqrt{5})^{1/2} \tag{10.28a}$$

Hence, L may be expressed as

$$L(k_2, k_3) = \frac{t_{f_1}(k_2, k_3)}{2(4)^{1/3}} + \frac{t_{f_2}(k_2, k_3)}{2(2+\sqrt{5})^{1/2}} - 2 \tag{10.28b}$$

where $t_{f_1}(k_2, k_3)$ and $t_{f_2}(k_2, k_3)$ are the suboptimal times to the origin from x^1 and x^2. The latter times must be determined numerically by digital simulation of the feedback system.† Adopting the suboptimal trajectory termination condition (10.16) with $\varepsilon = 10^{-4}$ (i.e. the origin was deemed to be attained when the norm of the state vector was reduced by a factor of 10^{-4} times its initial value), the following controller parameter values are found to minimise (at least locally) the function L:

$$H = 2; \quad k_1 = 1; \quad k_2 = 1{\cdot}41; \quad k_3 = 0{\cdot}85 \tag{10.29}$$

These compare favourably with the parameter values obtained by Fuller (1971) using a different approach. The final suboptimal controller design now becomes

$$u = -\operatorname{sgn}(\Xi(x)) \tag{10.30a}$$

with

$$\Xi(x) = |ax_1|^{2/3} \operatorname{sgn}(x_1) + 1{\cdot}41ax_2 + 0{\cdot}85a^2x_3|x_3| \tag{10.30b}$$

The associated switching surface $\Gamma_s = \{x: \Xi(x) = 0\}$ is depicted in (Θ, Φ)-space in Fig. 10.9, together with the exact time-optimal switching surface Γ. The suboptimal feedback system turns out to be globally stable and yields suboptimal response times which typically are within 23% of the corresponding minimum values. The computed suboptimality figures for an ensemble of initial states are contained in Table 10.1.

In the case of a fourth-order integrator system, a similar numerical procedure to the above yields the following suboptimal feedback design

$$u = -\operatorname{sgn}(\Xi(x)) \tag{10.31}$$

† Digital simulation techniques are described in Fuller (1971) and Ryan (1974, 1975).

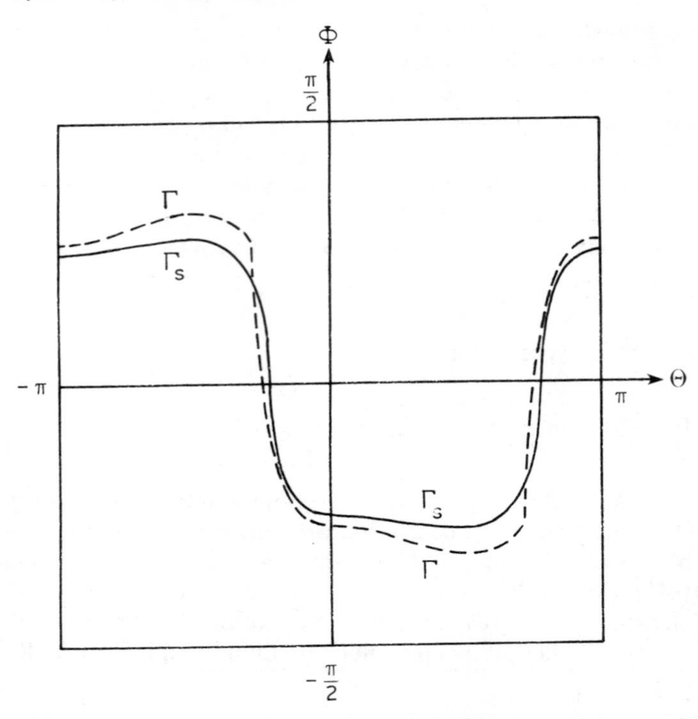

Fig. 10.9 *Time-optimal* (Γ) *and suboptimal* (Γ_s) *switching surfaces in* (Θ, Φ)*-space*

where the suboptimal switching function $\Xi\colon \mathbb{R}^4 \to \mathbb{R}$ is given by

$$\Xi(x) = (a|x_1|)^{3/4}\ \mathrm{sgn}\ (x_1) + 2{\cdot}29ax_2 + 3{\cdot}02(a|x_3|)^{3/2}\ \mathrm{sgn}\ (x_3) + 3{\cdot}21(ax_4)^3 \qquad (10.32)$$

Under this control, suboptimal response times which exceed the corresponding

Table 10.1

Initial state x_1	x_2	x_3	Fractional increase in response time $\Delta = (t_f - t_f^*)/t_f^*$
C	0	0	0·2161
0	C	0	0·1702
0	0	C	0·1380
C^3	$C\|C\|$	0	0·1814
C^3	$C\|C\|$	C	0·1576
(arbitrary $C \neq 0$)			

Table 10.2

Initial state x_1	x_2	x_3	x_4	Fractional increase in response time $\Delta = (t_f - t_f^*)/t_f^*$
C	0	0	0	0·7263
0	C	0	0	0·7438
0	0	C	0	0·7096
0	0	0	C	0·6085
$C^3\|C\|$	C^3	0	0	0·7581
$C^3\|C\|$	C^3	$C\|C\|$	0	0·7129
$C^3\|C\|$	C^3	$C\|C\|$	C	0·6545

(arbitrary $C \neq 0$)

minimum times by approximately 75% are to be expected, as indicated by Table 10.2. These suboptimality figures are not surprising when the simplicity of (10.32) is contrasted with the high level of complexity of the exact time-optimal switching function (7.15) derived in Section 7.4. The figures of Table 10.2 do, however, seem to suggest that for higher-order systems the suboptimal switching functions of the simple form (10.22) may prove inadequate. Quasi-time-optimal controllers of increased complexity for fourth- and fifth-order integrator systems are described and analysed in Ryan (1976a, 1978a); these controllers again satisfy the invariance condition (10.20) and may be interpreted (and indeed implemented) as a modification of the time-optimal predictive strategy of Gulko (1963), Gulko and Kogan (1963) and Gulko *et al.* (1964).

A linear state transformation, whereby optimal and quasioptimal feedback controls designed for multiple integrator plants may be applied to more general plants, will now be described.

10.4.4 Application of multiple integrator system control to general plants with real eigenvalues

Suppose now that the eigenvalues λ_i, $i = 1, 2, \ldots, n$, of the canonical system matrix A of (10.15) are real; moreover, it is assumed that $\lambda_i \leq 0$, i.e. the eigenvalues are non-positive-valued (extension to the unstable case is possible but, for brevity, is not considered here). It may be verified that the system representation is transformed from the 'phase variable' form (10.15) to the *serially decomposed* form

$$\left.\begin{aligned} \dot{v}_i &= \lambda_i v_i + v_{i+1}; \qquad i = 1, 2, \ldots, n-1 \\ \dot{v}_n &= \lambda_n v_n + \frac{u}{a}; \qquad a > 0; \qquad |u| \leq 1 \end{aligned}\right\} \tag{10.33a}$$

or

$$\dot{v} = Cv + bu; \qquad C = \begin{bmatrix} \lambda_1 & 1 & 0 & \cdots & 0 & 0 \\ 0 & \lambda_2 & 1 & \cdots & 0 & 0 \\ \vdots & & & & & \vdots \\ 0 & 0 & 0 & \cdots & \lambda_{n-1} & 1 \\ 0 & 0 & 0 & \cdots & 0 & \lambda_n \end{bmatrix} \tag{10.33b}$$

by the introduction of the linear state transformation

$$v = Px \qquad (\text{with } C = PAP^{-1}) \tag{10.34a}$$

where $P = [p_{i,j}]$ is a nonsingular lower triangular matrix with elements

$$\left.\begin{aligned} p_{i,i} &= 1 \\ p_{i,j} &= 0, \qquad i < j \leq n \end{aligned}\right\} \quad i = 1, 2, \ldots, n \tag{10.34b}$$

$$\left.\begin{aligned} p_{i,1} &= \prod_{k=1}^{i-1} (-\lambda_k) \\ p_{i,j} &= -\lambda_{i-1} p_{i-1,j} + p_{i-1,j-1}; \qquad 1 < j < i \end{aligned}\right\} \quad i = 2, 3, \ldots, n \tag{10.34c}$$

For example, in the second-order case,

$$n = 2: \qquad P = \begin{bmatrix} 1 & 0 \\ -\lambda_1 & 1 \end{bmatrix} \tag{10.35}$$

and, in the third-order case,

$$n = 3: \qquad P = \begin{bmatrix} 1 & 0 & 0 \\ -\lambda_1 & 1 & 0 \\ \lambda_1 \lambda_2 & -(\lambda_1 + \lambda_2) & 1 \end{bmatrix} \tag{10.36}$$

Consider initially the special case in which the eigenvalues λ_i are in the simple ratio $n : n-1 : \cdots : 2 : 1$, viz.

$$\lambda_i = (n + 1 - i)\lambda, \qquad i = 1, 2, \ldots, n; \qquad \lambda < 0 \tag{10.37}$$

In this case, the system equivalence property of Section 3.6 implies that if a feedback control $u = -\text{sgn}\,(\Xi(x))$, *designed for the multiple integrator system* (10.18), gives an acceptable fractional increase in response time $\Delta_0 = (t_f - t_f^*)/t_f^*$ for an initial state x^0 of (10.18), then the feedback control

$$u = -\text{sgn}\,(\Xi(v)) \tag{10.38}$$

when applied to system (10.33) and (10.37) will yield a fractional increase in response time Δ_λ, from the same initial state $v(0) = x^0$, of

$$\Delta_\lambda = \frac{\ln(1 - \lambda t_f) - \ln(1 - \lambda t_f^*)}{\ln(1 - \lambda t_f^*)} \qquad (\text{using transformation (3.52)})$$

or equivalently,

$$\Delta_\lambda = \frac{\ln(1 - \lambda(1 + \Delta_0)t_f^*) - \ln(1 - \lambda t_f^*)}{\ln(1 - \lambda t_f^*)} \tag{10.39a}$$

Assuming $\Delta_0 \neq 0$ small, then for large values of $-\lambda t_f^* \gg 0$,

$$\Delta_\lambda \simeq \frac{\ln(1 + \Delta_0)}{\ln(-\lambda t_f^*)} \simeq \frac{\Delta_0}{\ln(-\lambda t_f^*)} \ll \Delta_0 \qquad (\text{for } -\lambda t_f^* \gg 0) \tag{10.39b}$$

while

$$\Delta_\lambda \uparrow \Delta_0 \quad \text{as} \quad -\lambda t_f^* \downarrow 0 \tag{10.39c}$$

Hence, for system (10.33) and (10.37) with initial state x^0, the (suboptimal) feedback control (10.38) yields a fractional increase in response time Δ_λ which is less than the corresponding value Δ_0 for the multiple integrator system (with the same initial state) for which (10.38) was originally designed. Assuming that the value Δ_0 is an acceptable suboptimality figure then it is reasonable to assume that the value $\Delta_\lambda < \Delta_0$ is acceptable for system (10.33) and (10.37) with eigenvalues in simple ratio. Note that (10.39b, c) imply that, near the state origin (i.e. for $-\lambda t_f^* \downarrow 0$), the performance figure for system (10.33) and (10.37) approaches that of the multiple integrator system, while for initial states remote from the state origin the performance approaches true optimality. In the above discussion it was implicitly assumed that (10.38) is suboptimal, in cases where $\Xi: \mathbb{R}^n \rightarrow \mathbb{R}$ is the true time-optimal switching function for the multiple integrator system ($\Delta_0 = 0$), then (10.38) also gives precise time-optimal control ($\Delta_\lambda = \Delta_0 = 0$) for system (10.33)–(10.37).

In terms of the original system representation (10.15), the above analysis implies that, if a feedback control $u = -\operatorname{sgn}(\Xi(x))$ is designed for the multiple integrator system (i.e. when $a_1 = a_2 = \cdots = a_n = 0$ in (10.15)), then the same feedback structure, applied via the linear transformation (10.34), viz.

$$u = -\operatorname{sgn}(\Xi(Px)) \tag{10.40}$$

is an appropriate design for system (10.15) if the eigenvalues of the system matrix A are in simple ratio. However, in general the system eigenvalues will not be in the simple ratio of (10.37). In such cases, the following heuristic technique proves effective (Ryan 1975, 1976a, 1978a). The underlying approach is again to carry over the multiple integrator feedback control to the general case via a linear state transformation (as in (10.40)) with the important distinction that the matrix P is replaced by a matrix $\hat{P}$ associated with a nominal plant with nominal eigenvalues $\hat{\lambda}_i$ in simple ratio. In particular,

(*a*) a nominal plant is defined with nominal eigenvalues $\hat{\lambda}_i$ in simple ratio

$$\hat{\lambda}_i = (n + 1 - i)\hat{\lambda}, \qquad i = 1, 2, \ldots, n; \qquad \hat{\lambda} < 0 \tag{10.41}$$

and such that the trace of the nominal system matrix is equal to the trace of the actual system matrix A or equivalently, the sum of the nominal eigenvalues is equal to the sum of the actual eigenvalues, i.e.

$$\sum_{i=1}^{n} \hat{\lambda}_i = \sum_{i=1}^{n} \lambda_i \tag{10.42}$$

which, when combined with (10.41), defines the nominal plant parameter $\hat{\lambda}$ as

$$\hat{\lambda} = \frac{\sum_{i=1}^{n} \lambda_i}{\sum_{i=1}^{n} (n+1-i)} = \frac{\text{trace A}}{\sum_{i=1}^{n} (n+1-i)} = \frac{\text{trace A}}{\frac{1}{2}n(n+1)} \tag{10.43}$$

(*b*) The multiple integrator feedback control is then applied assuming that the actual plant behaves like the nominal plant with eigenvalues in simple ratio. In other words, the feedback control

$$u = -\text{sgn}\,(\Xi(\hat{P}x)) \tag{10.44}$$

is implemented on the actual system, where $\hat{P}$ is the state transformation (10.34) with the nominal eigenvalues $\hat{\lambda}_i$ replacing the actual eigenvalues λ_i.

For example, in the general second-order case with real eigenvalues λ_1, λ_2,

$$n = 2: \qquad \hat{P} = \begin{bmatrix} 1 & 0 \\ -\hat{\lambda}_1 & 1 \end{bmatrix} = \begin{bmatrix} 1 & 0 \\ -\frac{2}{3}(\lambda_1 + \lambda_2) & 1 \end{bmatrix} \tag{10.45}$$

and, in the third-order case,

$$n = 3: \qquad \hat{P} = \begin{bmatrix} 1 & 0 & 0 \\ -\hat{\lambda}_1 & 1 & 0 \\ \hat{\lambda}_1 \hat{\lambda}_2 & -(\hat{\lambda}_1 + \hat{\lambda}_2) & 1 \end{bmatrix}$$
$$= \begin{bmatrix} 1 & 0 & 0 \\ -\frac{1}{2}(\lambda_1 + \lambda_2 + \lambda_3) & 1 & 0 \\ \frac{1}{6}(\lambda_1 + \lambda_2 + \lambda_3)^2 & -\frac{5}{6}(\lambda_1 + \lambda_2 + \lambda_3) & 1 \end{bmatrix} \tag{10.46}$$

Note that, if the actual eigenvalues are in simple ratio (i.e. satisfying (10.37)), then $\hat{\lambda}_i = \lambda_i$ and $\hat{P} = P$ so that the previously discussed control (10.40) is recovered. Also, to apply the technique, only the trace of the actual system matrix A need be known; precise knowledge of the eigenvalues is not a prerequisite.

Evidence to support the effectiveness of the above suboptimal technique is largely empirical; successful application of the technique to the control of a wide range of plants of up to fifth-order is demonstrated by the simulation studies of Ryan (1975, 1976a, 1978a). To illustrate the principles of the technique, some second-order examples will now be considered (for a more detailed analysis of the second-order case see Ryan (1975)).

Second-order systems: Attention is now restricted to systems of the form

$$\dot{x} = \begin{bmatrix} \dot{x}_1 \\ \dot{x}_2 \end{bmatrix} = \begin{bmatrix} 0 & 1 \\ a_1 & a_2 \end{bmatrix} \begin{bmatrix} x_1 \\ x_2 \end{bmatrix} + \begin{bmatrix} 0 \\ \frac{1}{a} \end{bmatrix} u = Ax + bu; \qquad |u| \leq 1 \qquad (10.47)$$

where the eigenvalues λ_1, λ_2 of A are real, nonpositive and ordered $\lambda_1 \leq \lambda_2 \leq 0$ so that $\alpha = \lambda_1/\lambda_2 \geq 1$. In the case of the double integrator ($a_1 = 0 = a_2$), the time-optimal feedback control is

$$u = -\operatorname{sgn}(\Xi(x)); \qquad (10.48a)$$

where $\Xi\colon \mathbb{R}^2 \to \mathbb{R}$ is given by

$$\Xi(x) = \begin{cases} \xi(x) = x_1 + \frac{1}{2}ax_2|x_2|; & \xi(x) \neq 0 \\ x_2; & \xi(x) = 0 \end{cases} \qquad (10.48b)$$

involving easily synthesised linear and quadratic functions of state variables. In contrast, the time-optimal feedback solutions for other second-order systems (see Chapter 5) frequently involve logarithmic, exponential and other 'difficult-to-synthesise' functions of state variables; moreover, such feedback systems were shown in Subsection 5.2.3(ii) to be highly sensitive to inaccuracies in control realisation. In effect, (10.48) may be regarded as the simplest of all second-order time-optimal feedback solutions; its adoption by other second-order systems is clearly advantageous. The state transformation technique outlined above facilitates the application of control (10.48) to general second-order systems of the form (10.47); this application is achieved via (10.44), which for the case at hand, may be expressed as

$$u = -\operatorname{sgn}(\Xi(\hat{P}x)) \qquad (10.49a)$$

where, from (10.45)

$$\hat{P}x = \begin{bmatrix} x_1 \\ -\frac{2}{3}(\lambda_1 + \lambda_2)x_1 + x_2 \end{bmatrix} \qquad (10.49b)$$

In other words, the linear transformation $x_2 \to -\frac{2}{3}(\lambda_1 + \lambda_2)x_1 + x_2$ is first effected and control law (10.48) is then implemented. If $\alpha = \lambda_1/\lambda_2 = 2$ (i.e. when the actual and nominal eigenvalues are in the same simple ratio 2 : 1) then, in accordance with the system equivalence property of Section 3.6, (10.49) gives precise time-optimal control. It would appear reasonable to expect quasi-time-optimal performance when the ratio $\alpha \geq 1$ of actual eigenvalues is close to the nominal ratio, i.e. for $\alpha \simeq 2$. The two extreme cases of

(i) $\alpha = \infty$ (integrator-plus-lag plant)
(ii) $\alpha = 1$ (confluent eigenvalues)

will now be investigated.

(*i*) *Integrator-plus-lag plant :* Initially, the integrator-plus-lag plant is studied, with eigenvalues $\lambda_1 = 0$, $\lambda_2 = -1$ corresponding to (10.47) with $a_1 = 0$, $a_2 = -1$ and, for simplicity, a normalised gain parameter of $a = 1$ is assumed. In this case, the (primary) suboptimal switching function $\hat{\xi}: \mathbb{R}^2 \to \mathbb{R}$ becomes

$$\hat{\xi}(x) = \xi(\hat{P}x) = x_1 + \tfrac{1}{2}(\tfrac{2}{3}x_1 + x_2)|\tfrac{2}{3}x_1 + x_2| \tag{10.51}$$

In contrast to the simple form of (10.51), the exact time-optimal primary switching function (see Subsection 5.2.1(ii)) is given by

$$\xi^*(x) = x_1 + x_2 - \operatorname{sgn}(x_2) \ln(1 + |x_2|) \tag{10.52}$$

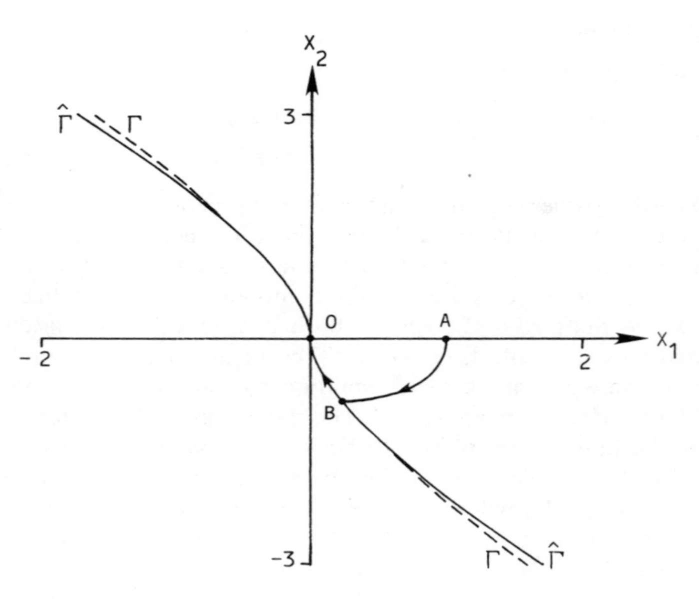

Fig. 10.10 *Integrator-plus-lag plant*
Optimal (Γ) and quasioptimal ($\hat{\Gamma}$) switching curves

involving a logarithmic function. The optimal $\Gamma = \{x: \xi^*(x) = 0\}$ and suboptimal $\hat{\Gamma} = \{x: \hat{\xi}(x) = 0\}$ switching curves are depicted in Fig. 10.10 and are seen to be in close agreement. As an indication of the suboptimality of the method, for all initial states on the x_1-axis, the fractional increase in response time $\Delta = (t_f - t_f^*)/t_f^*$ was found (by digital simulation) to satisfy

$$0 < \Delta < 0{\cdot}00082 \tag{10.53}$$

representing a maximum deterioration from time-optimal performance of less than 0·082%, for initial states $x^0 = (x_1, 0)'$ on the x_1 state axis. For example, for an initial state $x^0 = (1, 0)'$ the optimal t_f^* and suboptimal t_f times are calculated† as

$$t_f^* = 2{\cdot}170077 \qquad \text{and} \qquad t_f = 2{\cdot}171842$$

corresponding to a fractional increase in settling time of

$$\Delta = \frac{t_f - t_f^*}{t_f^*} = 0{\cdot}000813$$

In Fig. 10.10 the corresponding trajectories (ABO) are virtually indistinguishable.

(*ii*) *Repeated eigenvalues* ($\alpha = 1$): The case of confluent eigenvalues $\lambda_1 = -1 = \lambda_2$ will now be considered, corresponding to (10.47) with $a_1 = -1$ and $a_2 = -2$, and again a normalised gain parameter $a = 1$ is assumed. In this case the (primary) suboptimal switching function $\hat{\xi}\colon \mathbb{R}^2 \to \mathbb{R}$ becomes

$$\hat{\xi}(x) = \xi(\hat{P}x) = x_1 + \tfrac{1}{2}(\tfrac{4}{3}x_1 + x_2)|\tfrac{4}{3}x_1 + x_2| \qquad (10.54)$$

In contrast to this simple structure, the exact time-optimal primary switching function is given by‡

$$\xi^*(x) = \operatorname{sgn}(x_1 + x_2)(1 + |x_1 + x_2|) \ln(1 + |x_1 + x_2|) - x_2 \qquad (10.55)$$

again involving a logarithmic nonlinearity. The optimal $\Gamma = \{x\colon \xi^*(x) = 0\}$ and suboptimal $\hat{\Gamma} = \{x\colon \hat{\xi}(x) = 0\}$ switching curves are depicted in Fig. 10.11 and as in the previous example, are seen to be in close agreement. For all initial states $x^0 = (x_1, 0)'$ on the x_1 state axis, the fractional increase in response time $\Delta = (t_f - t_f^*)/t_f^*$ was found (by digital simulation) to satisfy

$$0 < \Delta < 0{\cdot}024 \qquad (10.56)$$

representing a maximum deterioration from time-optimal performance of less than 2·4% for initial states on the x_1 axis. For example, for an initial state $x^0 = (1, 0)'$, the computed optimal and suboptimal times are

$$t_f^* = 1{\cdot}917951 \qquad \text{and} \qquad t_f = 1{\cdot}960628$$

corresponding to a fractional increase in settling time of $\Delta = 0{\cdot}022251$. Note that the upper bound on Δ (for initial states on the x_1 axis) in (10.56), although remarkably small, is considerably greater than the corresponding figure in (10.53). This is due to the fact that, in the case of the integrator-plus-lag plant,

† As in all other cited cases, in the calculation of these times the optimal and suboptimal trajectories were generated by digital simulation (see Ryan (1975) for details). In each case the origin was deemed to be attained when

$$\|x(t_f^*)\| \le \epsilon\|x(0)\| \qquad \text{and} \qquad \|x(t_f)\| \le \epsilon\|x(0)\|$$

with a norm reduction factor of $\epsilon = 10^{-6}$.

‡ Note that the form of (10.55) differs from the expression obtained in Subsection 5.2.1(ix); this is due to the fact that the x_i state variables of the present section differ from those of Subsection 5.2.1(ix).

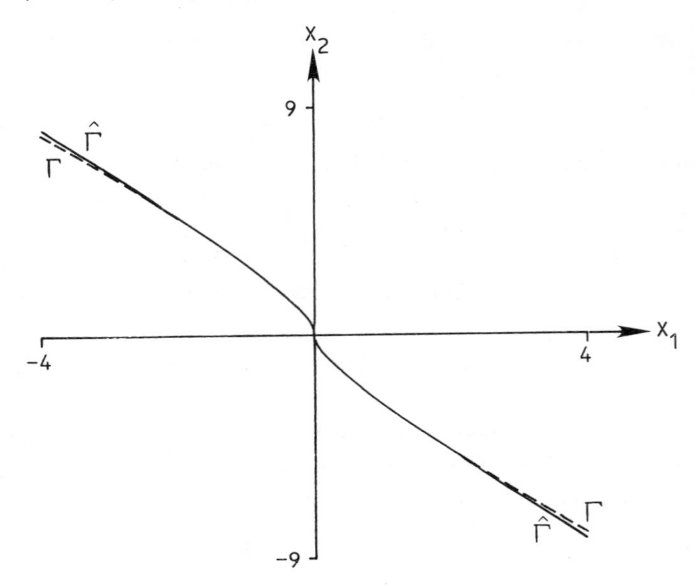

Fig. 10.11 *Plant with repeated eigenvalues* ($\alpha = 1$)
Optimal (Γ) and quasioptimal ($\hat{\Gamma}$) switching curves

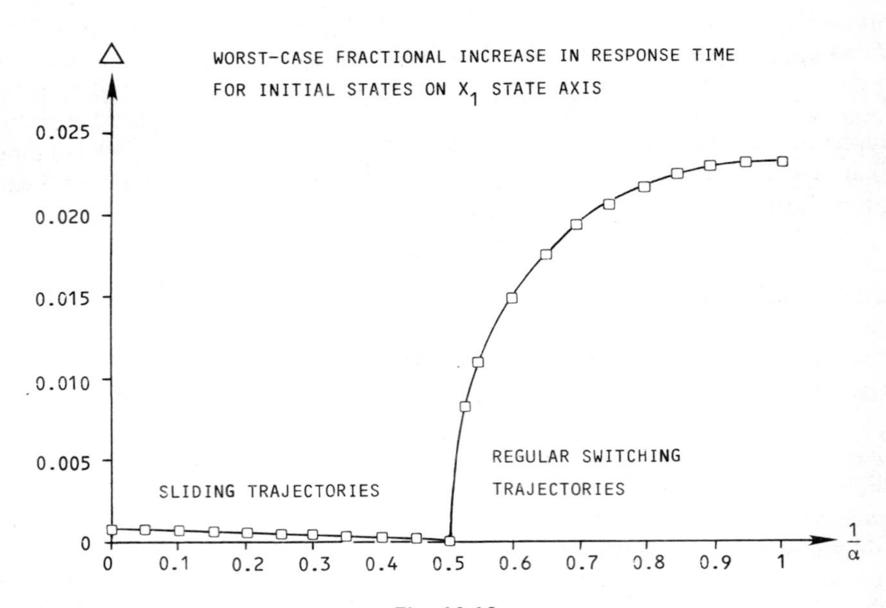

Fig. 10.12

the suboptimal switching curve lies below the optimal curve Γ in the second quadrant and above Γ in the fourth quadrant, whereas the converse is true in the case of the plant with repeated eigenvalues. As a result, in the first case (integrator-plus-lag) the suboptimal control switches earlier than the optimal control and terminates in an efficient sliding path to the origin; in the second case (repeated eigenvalues) the suboptimal control switches late, initially overshoots the origin and subsequently follows a less efficient (but highly damped) regular switching path which spirals in to the origin with switches occurring alternately in the second and fourth quadrants (these properties are rigorously established in Ryan (1975)).

(*iii*) *Intermediate cases*: $1 < \alpha < \infty$: For other eigenvalue ratio values satisfying $1 < \alpha < \infty$, it has been seen that the feedback system is precisely optimal for $\alpha = 2$. On the other hand, (*a*) when $\alpha > 2$, the suboptimal feedback system behaves in a manner akin to the integrator-plus-lag plant with trajectories terminating in efficient sliding paths which lead directly to the origin; whereas (*b*) for values of $1 < \alpha < 2$, the suboptimal system behaves in a manner akin to the plant with repeated eigenvalues with less efficient regular switching trajectories spiralling in to the origin. This is evident from Fig. 10.12 which depicts the 'worst-case' fractional increase in response time for initial states on the x_1 axis computed, as in (10.53) and (10.56), for a range of intermediate values of α. In summary, it may be concluded that, at least for initial states on the x_1 axis, the readily implemented feedback control (10.48)–(10.49) yields suboptimal response times t_f which are within 2·4% of the corresponding minimum values for *all* second-order plants of the form (10.47) with real nonpositive eigenvalues.

Although attention here was largely restricted to second-order systems, it is emphasised that the state transformation approach carries over to higher-order systems, enabling optimal and suboptimal feedback controls designed for multiple integrator systems to be adopted by more general systems of corresponding order. Relevant fourth- and fifth-order simulation studies are contained in Ryan (1975, 1976a, 1978a).

10.5 Adaptive control

The dynamic characteristics of most physical processes are subject to fluctuations, disturbances and environmental effects of various kinds which, in some cases, cause unacceptable degradation in system performance if a controller of fixed structure (designed on the basis of a nominal process model) is employed. For such cases, the notion of adaptive control is well established; the underlying concept is to up-date (in a variety of ways) the controller parameters to match the prevailing process characteristics. Loosely speaking, adaptive control techniques divide into two categories: (i) direct techniques in which the process dynamics are *explicitly* identified and the control determined on the basis of the up-dated accurate model; and (ii) indirect techniques in which controller adjustments are made on the basis of system monitoring which contains *implicit* information on the process dynamics. In other words, the direct approach decomposes the overall problem into two subproblems of identification and control (e.g. Hsu, Bacher and Kaufman, 1972), whereas the indirect approach circumvents the identification aspect (e.g. Kurokawa and Tamura, 1972). The indirect approach only will be discussed here.

In the context of relay and saturating control systems, many (indirect) adaptive control techniques exploit the possibility of enforcing sliding motion on switching hypersurfaces by an appropriate choice of controller structure. In this way, the dynamic behaviour of an imperfectly known process can be rendered insensitive to internal variations and external disturbances; in effect, if the state flow (over the range of process fluctuations) is directed towards the switching hypersurface from each side (as in Fig. 3.5) then the process can be forced to follow a sliding path (with desirable dynamic characteristics) which is unaffected by the process fluctuations. Variable structure systems with sliding modes fall within the scope of these techniques and are admirably described by Utkin (1977, 1978) (see also Drazenović, 1969; Babary and Pelczewski, 1975; Young, Kokotovic and Utkin, 1977). For single-input saturating control systems, Zinober (1975, 1977, 1979, 1980) has developed an efficient adaptive control technique for quasioptimal regulation which is based on similar principles; Zinober's technique neatly complements some of the theoretical results of earlier chapters and is described in some detail below.

10.5.1 Adaptive hyperplanes

To illustrate the underlying principles of the approach, consider the system

$$\left.\begin{aligned} &\dot{x}_i = x_{i+1}; \qquad i = 1, 2, \ldots, n-1 \\ &\dot{x}_n = f(x) + u/a; \qquad a > 0; \qquad f(0) = 0; \qquad |u| \leq 1 \end{aligned}\right\} \tag{10.57}$$

with $x(t) = (x_1(t), x_2(t), \ldots, x_n(t))' \in \mathbb{R}^n$ and where the (continuously differentiable) function f is not necessarily known precisely; the gain parameter $a > 0$ may also be imprecisely known. Suppose the control u is determined via a linear switching function ξ_k: $\mathbb{R}^n \to \mathbb{R}$, viz.

$$u = u(x) = -\operatorname{sgn}(\xi_k(x)) \tag{10.58a}$$

where

$$\xi_k(x) = \langle k, x \rangle = k_1 x_1 + k_2 x_2 + \cdots + k_{n-1} x_{n-1} + k_n x_n \tag{10.58b}$$

The components k_i of the vector k are assumed to be positive (implying negative feedback) and the nth component k_n is set to unity without loss of generality, i.e.

$$\left.\begin{aligned} &k_i > 0; \qquad i = 1, 2, \ldots, n-1 \\ &k_n = 1 \end{aligned}\right\} \tag{10.58c}$$

Hence, the set

$$\Pi_k = \{x: \xi_k(x) = \langle k, x \rangle = 0\} \tag{10.59}$$

constitutes a switching hyperplane which partitions the state space into two mutually exclusive regions

$$\{x: \xi_k(x) = \langle k, x \rangle < 0\} \tag{10.60a}$$

and

$$\{x: \xi_k(x) = \langle k, x \rangle > 0\} \tag{10.60b}$$

Straightforward application of the theory of Section 3.2 reveals that sliding motion occurs at a point $x \in \Pi_k$ if the following condition is satisfied:

$$\text{sliding condition:} \qquad |k_1 x_2 + k_2 x_3 + \cdots + f(x)| < \frac{1}{a} \tag{10.61}$$

Under appropriate conditions (e.g. for sufficiently small x) this condition can be satisfied, in which case the feedback system will follow a sliding path in Π_k; such a sliding path is a solution of the linear system with characteristic equation

$$\lambda^{n-1} + k_{n-1}\lambda^{n-2} + \cdots + k_2\lambda + k_1 = 0 \tag{10.62}$$

Clearly, the switching function parameters $k_i > 0$ can be chosen such that the roots of (10.62) have negative real parts so that associated sliding paths in Π_k are asymptotically stable. One such choice for the parameters k_i is the following

$$k_i = \frac{(n-1)!}{(n-i)!\,(i-1)!}\, q^{n-i}; \qquad i = 1, 2, \ldots, n; \qquad q > 0 \tag{10.63}$$

i.e. binomial coefficients, in which case (10.62) becomes

$$(\lambda + q)^{n-1} = 0 \tag{10.64}$$

Consequently, sliding paths in Π_k will correspond to asymptotically stable solutions of a linear homogeneous system with $n-1$ confluent negative real eigenvalues $\lambda_i = -q$. With the controller parameters k_i defined by (10.63), the original switching function ξ_k (with $n-1$ parameters) is reduced to a one-parameter switching function $\xi_q\colon \mathbb{R}^n \to \mathbb{R}$ which, in turn, defines the one-parameter switching hypersurface Π_q, viz.

$$\Pi_q = \{x: \xi_q(x) = 0\} \tag{10.65a}$$

where

$$\xi_q(x) = \sum_{i=1}^{n} \frac{(n-1)!}{(n-i)!\,(i-1)!}\, q^{n-i} x_i; \qquad q > 0 \tag{10.65b}$$

The basic principle underlying Zinober's adaptive control technique is to attempt to 'optimise' the sliding motion on Π_q by replacing fixed q in (10.65) by variable $q(t)$, the value of which is increased (as far as possible) so as to yield rapid decay to the state origin. Specifically, on detection of sliding motion on Π_q at some time t_1, the parameter value $q(t_1)$ is increased by a small amount, i.e. $q(t_1^+) = (1+\epsilon)q(t_1)$; $\epsilon > 0$. If, on reaching the new hyperplane $\Pi_{(1+\epsilon)q}$ at some time $t_2 > t_1$, the state continues to slide, then the parameter value is again increased, i.e. $q(t_2^+) = (1+\epsilon)q(t_2) = (1+\epsilon)q(t_1^+) = (1+\epsilon)^2 q(t_1)$. This procedure is repeated until the sliding condition fails to hold on the switching hyperplane. If the above strategy is idealised, i.e. $\epsilon \to 0^+$ and $q(t)$ is adjusted continuously in time, then, in many cases (as will be seen), the resulting time-varying linear switching function effectively defines a time-invariant nonlinear switching function with an associated nonlinear switching hypersurface which corresponds to the boundary of the set of sliding points of the feedback system. It is emphasised that the strategy identifies the latter sliding boundary (along which, in many cases, the state point travels, tracing a trajectory which terminates at the origin) without explicit knowledge of the process dynamics. Owing to the nonlinear nature of the overall adaptive feedback system, analysis of the general case is highly technical and many unresolved difficulties remain. Consequently, the strategy largely rests on a heuristic basis which is, however, well supported by simulation studies (as reported in Zinober 1975, 1977, 1979, 1980). The adaptive strategy applied to a double integrator system with parameter uncertainty will now be considered.

10.5.2 Adaptive control of a double integrator system with gain parameter uncertainty

In this case, the system equations are

$$\dot{x}_1 = x_2; \qquad \dot{x}_2 = \frac{u}{a}; \qquad |u| \le 1 \tag{10.66}$$

and it is supposed that the gain parameter a is a positive but otherwise unknown constant. In this case (10.65) becomes

$$\Pi_q = \{x\colon \xi_q(x) = qx_1 + x_2 = 0\} \tag{10.67}$$

and noting that $k_1 = q$, it may be concluded from (10.61), that the set

$$\Pi_q^s = \left\{x\colon qx_1 + x_2 = 0;\ |x_2| < \frac{1}{qa}\right\} \tag{10.68}$$

is the set of sliding points of the feedback system for *fixed* q. The points with coordinates

$$\left(-\frac{1}{q^2 a}, \frac{1}{qa}\right) \quad \text{and} \quad \left(\frac{1}{q^2 a}, -\frac{1}{qa}\right)$$

in the second and fourth quadrants of the state plane are the boundary points of the set Π_q^s. Now, if q is allowed to range over the positive reals then these boundary points generate loci Γ^+ and Γ^- in the second and fourth quadrants given by

$$\Gamma^+ = \{x: x_1 = -ax_2^2;\ x_2 > 0\} \tag{10.69a}$$

$$\Gamma^- = \{x: x_1 = ax_2^2;\ x_2 < 0\} \tag{10.69b}$$

The composition of these loci (and the state origin) yields the overall sliding boundary Γ, viz.

$$\Gamma = \Gamma^+ \cup \Gamma^- \cup \{0\} = \{x: \xi(x) = 0\} \tag{10.70a}$$

where

$$\xi(x) = x_1 + ax_2|x_2| \tag{10.70b}$$

The basic adaptive strategy identifies this boundary without knowledge of the parameter a as will now be illustrated. Referring to Fig. 10.13, suppose x^A is the initial state of the system. Noting, from (10.15), that the sliding segment Π_q^s of the switching line Π_q can be made arbitrarily large by taking q sufficiently small, the adaptive strategy initially sets q to such a value q_1 so that, on first reaching Π_{q_1} at the point B, sliding motion occurs, i.e. a sliding path BB′ in Π_{q_1} is followed. On detection of this sliding motion, in accordance with the strategy, the parameter q is updated to a value q_2 where

$$q_2 = (1 + \epsilon)q_1, \qquad \epsilon > 0$$

This corresponds to a clockwise rotation of the switching line as depicted in Fig. 10.13 so that the state follows the n-path B′C until the new switching line

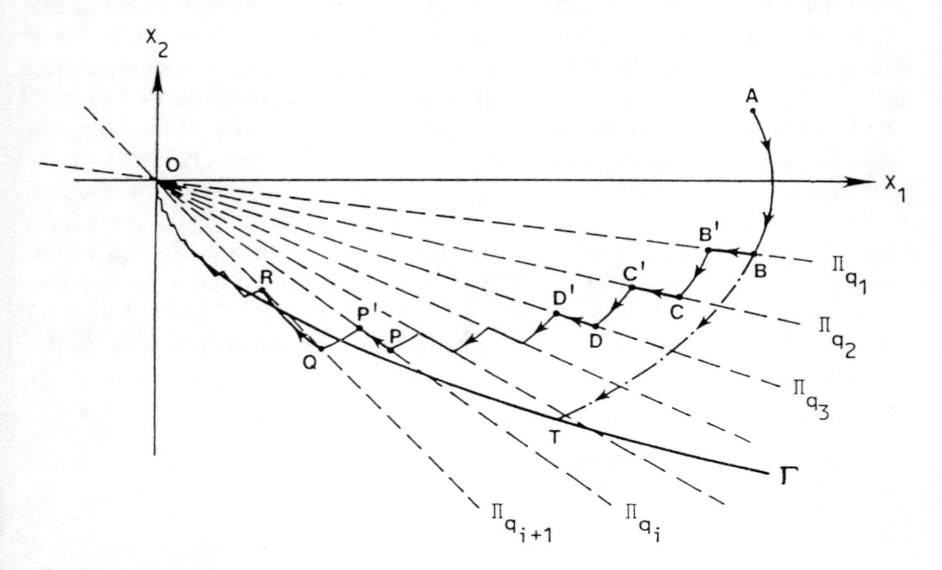

Fig. 10.13 *Dynamic behaviour of adaptive strategy*

Π_{q_2} is attained. Sliding motion is subsequently detected along the path CC′ in Π_{q_2} so that the parameter is again undated to a value $q_3 = (1+\epsilon)q_2$. Continuing this procedure, the subsequent trajectory C′DD′ ⋯ PP′ is generated. On detection of the sliding path PP′ in Π_{q_i} the parameter q is again updated to the value $q_{i+1} = (1+\epsilon)q_i$ and the n-path P′Q is followed to $\Pi_{q_{i+1}}$. However, this latter path crosses the sliding boundary Γ and leaves the sliding region so that Q is a regular switch point of the feedback system in which case the parameter q is held at the value q_{i+1} until the state returns to the line $\Pi_{q_{i+1}}$ via the p-path QR which recrosses the sliding boundary Γ. At R, sliding motion is again detected and the strategy proceeds as before, subsequently generating a path which 'hemstitches' along Γ to the origin. In the idealised case, for which $\epsilon \to 0^+$ and sliding motion can be detected instantaneously so that the above adjustments of the parameter q are carried out continuously in time, the path ATO is followed directly to the origin which is attained in finite time $t_f = t_{AT} + t_{TO}$, where

$$t_{AT} = a|x_2^A| + \sqrt{\tfrac{2}{3}[a|x_1^A| + \tfrac{1}{2}(ax_2^A)^2]} \tag{10.71a}$$

is the duration of the n-path AT and

$$t_{TO} = 2\sqrt{\tfrac{2}{3}[a|x_1^A| + \tfrac{1}{2}(ax_2^A)^2]} \tag{10.71b}$$

is the duration of the path along the sliding boundary Γ, from T to the origin.

Consider now an initial state x^A at the point A of Fig. 10.14. Under the (idealised) adaptive strategy, the path ABCO is followed to the origin. In this case, the state trajectory does not lead directly to the sliding boundary but first follows the p-path AB until the initialised switching line Π_{q_1} (q_1 small) is reached at B, from which point the strategy behaves as before and generates path BCO to the origin. If, instead, the state could be forced to follow the more direct path AB′O of Fig. 10.14, then the system response would be more efficient in the sense that the origin would be attained in less time. Zinober (1975) has introduced a modification to the basic strategy which yields this 'more efficient' dynamic behaviour. The logical details of this modification are omitted; the reader is referred to Zinober (1975) for a detailed description. Suffice it to remark here that, under the modified strategy, the state moves directly to the sliding boundary (and thence to the origin) from all initial states (e.g. paths ATO and AB′O of Figs. 10.13 and 10.14). Thus, the sliding boundary Γ effectively acts as a *nonlinear* switching curve and the (idealised) modified strategy is equivalent to the feedback control

$$u = u(x) = -\operatorname{sgn}(\xi(x)) \tag{10.72a}$$

where

$$\xi(x) = x_1 + ax_2|x_2| \tag{10.72b}$$

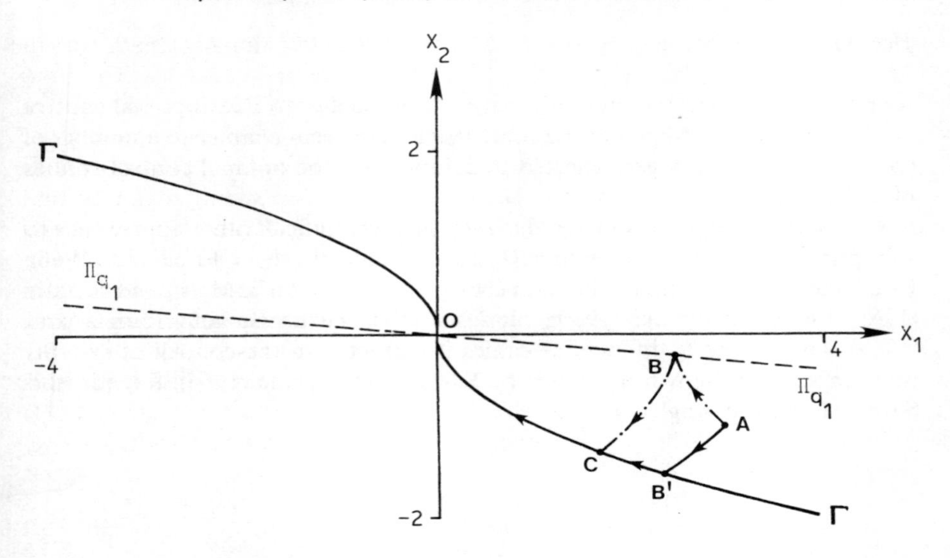

Fig. 10.14 *Typical trajectories generated by the basic* (ABCO) *and modified* (AB′O) *adaptive strategies*

Again, it is emphasised that, while the modified strategy behaves essentially as the feedback law (10.72), a priori knowledge of the gain parameter $a > 0$ is not required.

Although the basic and modified strategies were originally proposed as quasi-time-optimal control techniques, comparing the equivalent feedback law (10.72) with equations (9.88), (9.100) and (9.106) of Subsection 9.3.4, it may be concluded that the idealised modified strategy, applied to a double integrator plant, precisely minimises the nonquadratic cost functional

$$J(u) = \int_0^{t_f} [|x_1(t)|^\nu + a^\nu(2\nu - 1)^{-1}|x_2(t)|^{2\nu}]\, dt; \qquad t_f \text{ free} \tag{10.73}$$

for all $\nu \geq 1$.

The basic and modified strategies, as outlined above, carry over to general second-order systems of the form (10.57) for which neither the precise form of the function f nor the gain parameter value a need be known. For such second-order examples, see Zinober (1975). The adaptive strategy applied to a triple integrator plant is analysed in considerable detail in Zinober (1977); computational results on the application of the strategy to a fourth-order system is contained in Zinober (1979) and the strategy applied to the control of diffusion equations is described in Zinober (1980).

10.6 Discussion

As remarked earlier, the great diversity of approaches to quasioptimal control necessitated the restriction of the contents of the present chapter to a number of specific techniques, largely selected to complement the optimal control studies of earlier chapters.

Although no attempt is made to survey the overall field, other approaches to suboptimal control of saturating systems include those of Friedland (1966), Friedland and Sarachik (1966), Gunchev (1971), Lee (1967), Melsa and Schultz (1965, 1967), Mitsumaki (1960), Notley (1971), Yuvarajan and Ramaswami (1969). The theory of differential games is exploited in the control of systems with imperfectly known dynamics by Sarma and Ragade (1966), Ragade and Sarma (1967) and Singh (1977).

Chapter 11

Concluding remarks

While an attempt has been made to present a self-contained account of optimal relay and saturating control system synthesis in this book, many facets of the problem fall outside its scope.

Firstly, although some nonlinear processes have been investigated, optimal nonlinear feedback control of linear processes has been the main concern. Numerous theoretical and practical problems associated with optimal nonlinear control of nonlinear processes remain. Investigations into the latter area include (in addition to many of the references cited in earlier chapters and others too numerous to mention here) those of Arimoto and Gaafar (1977), Brockett (1973), Javid (1978), Krener (1973, 1974), Mohler (1973), Mohler and Rink (1969), Mohler and Ruberti (1972) and Sussmann (1972); the strongest results relate to bilinear processes in which the (saturating) control enters in a multiplicative manner.

Secondly, finite dimensional (lumped parameter) systems on $\mathbb{R}^n$, only, have been studied. In practice, many processes are of a 'distributed parameter' nature, involving delays (e.g. time lags in economic models, etc.), and/or spatially dependent state variables (e.g. temperature distributions in thermal processes, displacements and velocities in vibrating structures, etc.). Frequently, distributed parameter systems can be modelled with sufficient accuracy, by approximate finite dimensional models. In other applications, the distributed character of the process is an intrinsic feature of the control problem, in which case the state evolution of the system may be modelled by a delay differential equation or partial differential equation defined on some appropriate infinite dimensional function space. Investigations into problems of infinite dimensional saturating control include (among numerous others) Balakrishnan (1976), Butkovskii (1969), Falb (1964), Fattorini (1964), Fuller (1968), Hermes (1967), Knowles (1976, 1979), Koppel and Latour (1965), Lions (1971), McCausland (1965), Prabhu and McCausland (1970), Quinn (1971), Reeve (1970), Rogak *et al.* (1970) and Wang (1975).

Thirdly, a deterministic hypothesis has been adopted throughout. Stochastic systems have not been investigated. When stochastic effects are included in the formulation, the optimal saturating control synthesis problem becomes disproportionately more difficult. Explicit solutions are, as yet, available only in the simplest of first and second-order systems. Related studies of stochastic relay and saturating control systems include Akizuki and Tada (1977), Barrett (1960), Dorato *et al.* (1967), Fuller (1960d; 1961a, b; 1974c; 1975; 1980a, b, c), Fuller's collection of key papers (Fuller 1970b), Lee and Luecke (1974), Merklinger (1963), Robinson and Öner Yurtseven (1969), Wells and Kashiwagi (1969). It is stressed that these references (and also those cited in the previous two paragraphs) by no means constitute an exhaustive bibliography; they are intended only as a representative selection from which the interested reader may progress.

Notwithstanding the above aspects, it is hoped that this volume comprises a reasonably complete and up-to-date compilation of the theory and techniques of optimal feedback synthesis for finite-dimensional, deterministic, relay and saturating control systems.

References

ADEY, A. J., COALES, J. F., and STILES, J. A. (1963): 'Predictive control of an on-off system with two control variables', *Proc. 2nd IFAC Congress, Basle, Theory*, p. 41. (London, Butterworths/Munich, Oldenbourg, 1964)

ADEY, A. J., BILLINGSLEY, J., and COALES, J. F. (1966): 'Predictive control of higher order systems', *Proc. 3rd IFAC Congress, London*, paper 40f

AKIZUKI, K., and TADA, F. (1977): 'The analysis of relay control systems with coloured noise inputs', *Int. J. Control*, **25,** p. 949

AKULENKO, L. D. (1978): 'Time-optimal stabilization of a perturbed system with invariant norm', *J. Appl. Math. Mech.*, **42,** p. 646

ALMUZARA, J. L., and FLÜGGE-LOTZ, I. (1968): 'Minimum-time control of a nonlinear system', *J. Diff. Eqns.*, **4,** p. 12

ALY, G. M. (1978): 'The computation of optimal singular control', *Int. J. Control*, **28,** p. 681

ANDRÉ, J., and SEIBERT, P. (1956): Über stückweise lineare Differentialgleichungen, die bie Regelungsproblemen auftreten', *Arch. Math.*, **7,** p. 148

ANDRÉ, J., and SEIBERT, P. (1960): 'After end-point motions of general discontinuous control systems and their stability properties', *Proc. 1st IFAC Congress, Moscow*, p. 919, (London, Butterworths, 1961)

ARIMOTO, S., and GAAFAR, Y. (1977): 'On the structure of optimal control for certain systems linear in control', *Int. J. Control*, **25,** p. 545

ATANACKOVIC, T. M. (1978): 'Optimal control of dynamically controllable motions: an example', *J. Inst. Maths. Applics.*, **21,** p. 293

ATHANASSIADES, M. (1963): 'Optimal control for linear time-invariant plants with time-, fuel-, and energy constraints', *IEEE Trans. Appl. Ind.*, **81,** p. 321

ATHANASSIADES, M., and SMITH, O. J. M. (1961): 'Theory and design of high order bang-bang control systems', *IRE Trans. autom. Control*, **6,** p. 125

ATHANS, M. (1964): 'Minimum fuel control of second order systems with real poles', *IEEE Trans. Appl. Ind.*, **83,** p. 148

ATHANS, M. (1966): 'On the uniqueness of the extremal controls for a class of minimum fuel problems', *IEEE Trans. autom. Control*, **AC-11,** p. 660

ATHANS, M., and CANON, M. D. (1964): 'On the fuel-optimal singular control of nonlinear second-order systems', *IEEE Trans. autom. Control*, **AC-9,** p. 360

ATHANS, M., and FALB, P. L. (1966): 'Optimal Control: An Introduction to the Theory and its Applications', (New York, McGraw-Hill)

ATHANS, M., FALB, P. L., and LACOSS, R. T. (1963): 'Time-, fuel-, and energy-optimal control of nonlinear norm-invariant systems', *IEEE Trans. autom. Control*, **AC-8,** p. 196

ATHANS, M., FALB, P. L., and LACOSS, R. T. (1964): 'On optimal control of self-adjoint systems', *IEEE Trans. Appl. Ind.*, **83,** p. 161

BABARY, J.-P., and PELCZEWSKI, W. (1975): 'Application de la théorie des systèmes à structure variable à la commande de systèmes à paramètres répartis', *R.I.A.R.O. Automatique*, **9,** p. 43

BABARY, J.-P., and PELCZEWSKI, W. (1977): 'Commande en temps minimal d'un processus thermique par application de la méthod Pelczewski-Lawrynowicz', *ibid.*, **11,** p. 269

BALAKRISHNAN, A. V. (1976): 'Applied Functional Analysis' (Springer Verlag)

BARMISH, B. R., and SCHMITENDORF, W. E. (1980): 'A necessary and sufficient condition for local constrained controllability of a linear system', *IEEE Trans. autom. Control*, **AC-25,** p. 97

BARRETT, J. F. (1960): 'Application of Kolmogorov's equations to randomly disturbed automatic control systems', *Proc. 1st IFAC Congress, Moscow*, p. 724, (London, Butterworths, 1961)

BASS, R. W., and WEBBER, R. F. (1966): 'Optimal nonlinear feedback control derived from quartic and higher-order performance criteria', *IEEE Trans. autom. Control*, **AC-11,** p. 448

BECKER, N. (1980): 'A note on performance index sensitivity of time-optimal control systems', *IEEE Trans. autom. Control*, **AC-25,** p. 819

BELL, D. J. (1975): 'Singular problems in optimal control—a survey', *Int. J. Control*, **21,** p. 319

BELL, D. J. (1978): 'A note on junction conditions for partially singular trajectories', *Int. J. Control*, **28,** p. 67

BELL, D. J., and BOISSARD, M. (1979): 'Necessary conditions at the junction of singular and non-singular control subarcs', *Int. J. Control*, **29,** p. 981

BELL, D. J., and JACOBSON, D. H. (1975): 'Singular optimal control problems' (New York, Academic Press)

BELLMAN, R. (1957): 'Dynamic programming' (Princeton University Press)

BELLMAN, R., GLICKSBERG, I., and GROSS, O. (1956): 'On the bang-bang control problem', *Quart. Appl. Math.*, **14,** p. 11

BERKOVITZ, L. D. (1974): 'Optimal control theory', Applied Mathematical Sciences, vol. 12 (Springer-Verlag)

BERSHCHANSKII, Ya. M. (1976): 'Some problems of design of linear systems with a quadratic performance criterion', *Automation & Remote Control*, **37,** p. 315

BERSHCHANSKII, Ya. M. (1979): 'Fusing of singular and non-singular parts of optimal control', *Automation & Remote Control*, **40,** p. 325

BILLINGSLEY, J., and COALES, J. F. (1968): 'Simple predictive controller for high-order systems', *Proc. IEE*, **115,** p. 1568

BOGNER, I., and KAZDA, L. F. (1954): 'An investigation of the switching criteria for higher order servomechanisms', *AIEE Trans.*, pt. II, **73,** p. 118

BOLTYANSKII, V. G. (1958): 'A maximum principle in the theory of optimal processes', *Dokl. Akad. nauk. SSSR*, **119,** p. 1070

BOLTYANSKII, V. G. (1966): 'Sufficient conditions for optimality and justifiation of the dynamic programming method', *SIAM J. Control*, **4,** p. 326

BOLTYANSKII, V. G. (1971): 'Mathematical methods of optimal control' (New York, Holt, Rinehart & Winston)

BOLTYANSKII, V. G., GAMKRELIDZE, R. V., and PONTRYAGIN, L. S. (1956): 'A note on the theory of optimal processes', *Dokl. Akad. nauk. SSSR*, **110,** p. 7

BOLTYANSKII, V. G., GAMKRELIDZE, R. V., MISHCHENKO, E. F., and PONTRYAGIN, L. S. (1960): 'The maximum principle in the theory of optimal processes of control', *Proc. 1st IFAC Congress, Moscow*, p. 454, (London, Butterworths, 1961)

BRENNAN, P. J., and ROBERTS, A. P. (1962): 'Use of an analogue computer in the application of Pontryagin's maximum principle to the design of control systems with optimum transient response', *J. Electron. Control*, **12,** p. 345

BROCKETT, R. W. (1970): 'Finite dimensional linear systems' (Wiley)

BROCKETT, R. W. (1973): 'Lie algebras and Lie groups in control theory', *in* Mayne, D. Q., and Brockett, R. W. (Eds.): 'Geometric methods in system theory' (Dordrecht Holland: Reidel)

BRUNOVSKÝ, P. (1974): 'The closed-loop time optimal control I: optimality', *SIAM J. Control & Optim.*, **12**, p. 624

BRUNOVSKÝ, P. (1976): 'The closed-loop time optimal control II: stability', *SIAM J. Control & Optim.*, **14**, p. 156

BUELENS, P. F., VAN ROMPAY, P. V., and HELLINCKX, L. J. (1978): 'Bang-bang control with quadratic performance index', *Int. J. Control*, **27**, p. 525

BURMEISTER, H. L. (1961): 'Zeitoptimale Ubergangsvorgange mit beschrandter n-ter Ableitung', *z.f. Messen-Steuren-Regeln.*, **4**, p. 407

BUSHAW, D. W. (1953): 'Differential equations with a discontinuous forcing term'. Rep. No. 469, Experimental towing tank, Stevens Institute of Technology, Holboken, NJ

BUSHAW, D. W. (1958): 'Optimal discontinuous forcing terms' *in* Lefschetz, S. (Ed.): 'Contributions to the theory of nonlinear oscillations', **4**, p. 29

BUTKOVSKII, A. G. (1969): 'Theory of optimal control of distributed parameter systems' (American Elsevier)

CHANG, JEN-WEI (1961): 'A problem in the synthesis of optimal systems using maximum principle', *Automation & Remote Control*, **22**, p. 1170

CHAUDHURI, A. K., and CHAUDHURY, A. K. (1964a): 'On the optimum switching function of a certain class of third-order contactor servomechanism I', *J. Electron. Control*, **16**, p. 451

CHAUDHURI, A. K., and CHAUDHURY, A. K. (1964b): 'On the optimum switching function of a certain class of third-order contactor servomechanism II', *J. Electron. Control*, **17**, p. 465

CHESTNUT, H., SOLLECITO, W. E., and TROUTMAN, P. H. (1961): 'Predictive control system application', *AIEE Trans.*, pt. II, **80**, p. 128

COALES, J. F., and NOTON, A. R. M. (1956): 'An on-off servomechanism with predictive changeover', *Proc. IEE*, **103B**, p. 449

COUPÉ, G. M. (1976): 'Design of near-time-optimum rotation control procedures for a spinning symmetric satellite', *IEEE Trans. autom. Control*, **AC-21**, p. 288

CURTAIN, R. F., and PRITCHARD, A. J. (1977): 'Functional analysis in modern applied mathematics' (Academic Press)

CURTAIN, R. F., and PRITCHARD, A. J. (1978): 'Infinite dimensional linear systems theory'. Lecture notes in Control & Information Sciences, **8**, Springer-Verlag

DAVIES, M. J. (1970): 'A property of the switching curve for certain systems', *Int. J. Control*, **12**, p. 457

DAVIES, M. J. (1971): 'Discontinuities of switching curves', *Int. J. Control*, **14**, p. 175

DAVIES, M. J. (1972): 'Domains of controllability', *in* Bell, D. J. (Ed.): 'Recent mathematical developments in control' (Academic Press)

DAVISON, E. J., and MUNRO, D. M. (1971): 'A computational technique for finding 'bang-bang' controls of non-linear time-varying systems', *Automatica*, **7**, p. 255

DE ROOY, J. J. (1970): 'A method of realizing quasi-time-optimal control by means of an approximate switching surface', *Int. J. Control*, **11**, p. 255

DESOER, C. A. (1961): 'Pontryagin's maximum principle and the principle of optimality', *J. Franklin Inst.*, **271**, p. 361

DORATO, P., CHANG-MING HSIEH, and ROBINSON, P. N. (1967): 'Optimal bang-bang control of linear stochastic systems with small noise parameter', *IEEE Trans. autom. Control*, **AC-12**, p. 682

DRAŽENOVIĆ, B. (1969): 'The invariance conditions in variable structure systems', *Automatica*, **5**, p. 287

EATON, J. H. (1962): 'An iterative solution to time optimal control', *J. Math. Anal. Appl.*, **5**, p. 329

EDGAR, T. F., and LAPIDUS, L. (1972): 'The computation of optimal singular bang-bang control, Part I: linear systems; Part II: nonlinear systems', *Amer. Inst. Chem. Eng. J.*, **18**, p. 774

ELKIN, D. V. I., and DALY, K. C. (1973): 'Application of a time-optimal control algorithm to the design of approximately time-optimal systems'. IEE Conference on Computer Aided Control System Design, Cambridge, (London, IEE)

EGGLESTON, D. M. (1963): 'On the application of the Pontryagin maximum principle using reverse time trajectories', *Trans. ASME, J. bas. Engng,* **85D,** p. 478

EGGLESTON, H. G. (1958): 'Convexity'. Cambridge tracts in mathematics and mathematical physics, No. 47, (Cambridge University Press)

FADDEN, E. J., and GILBERT, E. G. (1964): 'Computational aspects of the time-optimal control problem', *in* Balakrishnan, A. V., and Neustadt, L. W. (Eds.): 'Computing methods in optimization problems' (New York, Academic Press)

FALB, P. L. (1964): 'Infinite dimensional control problems I: On the closure of the set of attainable states for linear systems', *J. Math. Anal. Appl.,* **9,** p. 12

FARLOW, S. J. (1970): 'A note on switching times for time-optimal control systems', *IEEE Trans. autom. Control,* **AC-15,** p. 118

FARLOW, S. J. (1973): 'On finding switching times in time-optimal control systems', *Int. J. Control,* **17,** p. 855

FATTORINI, H. O. (1964): 'Time-optimal control of solutions of operational differential equations', *SIAM J. Control,* **2,** p. 54

FELDBAUM, A. A. (1955): 'On the synthesis of optimal systems with the aid of phase space', *Avto. Telemekh.,* **16,** p. 129

FILIPPOV, A. F. (1960a): 'Differential equations with discontinuous right-hand sides', *Matemat. Sbornik.,* **51,** p. 99. English trans.: *Am. Math. Soc. Trans.,* **42,** p. 199 (1964)

FILIPPOV, A. F. (1960b): 'Application of the theory of differential equations with discontinuous right-hand sides to non-linear problems in automatic control', *Proc. 1st IFAC Congress, Moscow,* p. 923, (London, Butterworths, 1961)

FILIPPOV, A. F. (1962): 'On certain questions in the theory of optimal control', *SIAM J. Control,* **1,** p. 76

FILIPPOV, A. F. (1967): 'Classical solutions of differential equations with multivalued right-hand sides', *ibid.,* **5,** p. 609

FLAHERTY, J. E., and O'MALLEY, R. E. (1977): 'On the computation of singular controls', *IEEE Trans. autom. Control,* **AC-22,** p. 640

FLEMING, W. H., and RISHEL, R. W. (1975): 'Deterministic and stochastic optimal control', Applications of mathematics 1 (Springer-Verlag)

FLÜGGE-LOTZ, I. (1953): 'Discontinuous automatic control' (Princeton University Press)

FLÜGGE-LOTZ, I. (1968): 'Discontinuous and optimal control' (McGraw-Hill)

FLÜGGE-LOTZ, I., and MARBACH, H. (1963): 'The optimal control of some attitude control systems for different performance criteria', *Trans. ASME, J. bas. Engng,* **85D,** p. 165

FLÜGGE-LOTZ, I., and MIH YIN (1961): 'The optimum response of second-order, velocity-controlled systems with contactor control', *Trans. ASME, J. bas. Engng,* **83D,** p. 59

FLÜGGE-LOTZ, I., and TITUS, H. A. (1962): 'The optimum response of full third-order systems with contactor control', *Trans. ASME, J. bas. Engng,* **84D,** p. 554

FLÜGGE-LOTZ, I., and TITUS, H. A. (1963): 'Optimum and quasi-optimum control of third and fourth-order systems', *Proc. 2nd IFAC Congress, Basle, Theory,* p. 363, (London, Butterworths/Munich, Oldenbourg, 1964)

FOY, W. H. (1963): 'Fuel minimization in flight vehicle attitude control', *IEEE Trans. autom. Control,* **AC-8,** p. 84

FRANKENA, J. F., and SIVAN, R. (1979): 'A non-linear optimal control law for linear systems', *Int. J. Control,* **30,** p. 159

FREDERICK, D. K., and FRANKLIN, G. F. (1967): 'Design of piecewise-linear switching functions for relay control systems', *IEEE Trans. autom. Control,* **AC-12,** p. 380

FRIEDLAND, B. (1966): 'A technique of quasi-optimum control', *Trans. ASME, J. bas. Engng,* **88D,** p. 437

FRIEDLAND, B., and SARACHIK, P. E. (1966): 'A unified approach to sub-optimum control', *Proc. 3rd IFAC Congress, London,* paper 13A

FUJISAWA, T., and YASUDA, Y. (1967): 'An iterative procedure for solving the time-optimal regulator problem', *SIAM J. Control,* **5,** p. 501

FULLER, A. T. (1960a): 'On phase space in the theory of optimum control', *J. Electron. Control*, **8,** p. 381

FULLER, A. T. (1960b): 'Optimization of non-linear control systems with transient inputs', *ibid.*, **8,** p. 465

FULLER, A. T. (1960c): 'Relay control systems optimized for various performance criteria', *Proc. 1st IFAC Congress, Moscow*, **1,** p. 510 (London, Butterworths, 1961)

FULLER, A. T. (1960d): 'Optimization of non-linear control systems with random inputs', *J. Electron. Control*, **9,** p. 65

FULLER, A. T. (1961a): 'Optimization of a non-linear control system with a random telegraph signal input', *ibid.*, **10,** p. 61

FULLER, A. T. (1961b): 'Optimization of a saturating control system with Brownian motion input', *ibid.*, **10,** p. 157

FULLER, A. T. (1962): 'Bibliography of optimum non-linear control of determinate and stochastic-definite systems', *ibid.*, **13,** p. 589

FULLER, A. T. (1963a): 'Study of an optimum nonlinear control system', *ibid.*, **15,** p. 63

FULLER, A. T. (1963b): 'Bibliography of Pontryagin's maximum principle', *ibid.*, **15,** p. 513

FULLER, A. T. (1964a): 'Further study of an optimum non-linear control system', *ibid.*, **17,** p. 283

FULLER, A. T. (1964b): 'The absolute optimality of a non-linear control system with integral-square-error criterion', *ibid.*, **17,** p. 301

FULLER, A. T. (1966): 'Optimization of some non-linear control systems by means of Bellman's equation and dimensional analysis', *Int. J. Control*, **3,** p. 359

FULLER, A. T. (1967a): 'Linear control of non-linear systems', *ibid.*, **5,** p. 197

FULLER, A. T. (1967b): 'The replacement of saturation constraints by energy constraints in control optimization theory', *ibid.*, **6,** p. 201

FULLER, A. T. (1968): 'Optimal nonlinear control of systems with pure delay', *ibid.*, **8,** p. 145

FULLER, A. T. (1969): 'In-the-large stability of relay and saturating control systems with linear controllers', *ibid.*, **10,** p. 457

FULLER, A. T. (1970a): 'Dimensional properties of optimal and sub-optimal nonlinear control systems', *J. Franklin Inst.*, **289,** p. 379

FULLER, A. T. (editor) (1970b): 'Nonlinear stochastic control systems', (London, Taylor and Francis)

FULLER, A. T. (1971): 'Sub-optimal nonlinear controllers for relay and saturating control systems', *Int. J. Control*, **13,** p. 401

FULLER, A. T. (1973a): 'Notes on some predictive control strategies', *ibid.*, **18,** p. 637

FULLER, A. T. (1973b): 'Proof of the time-optimality of a predictive control strategy for systems of higher order', *ibid.*, **18,** p. 1121

FULLER, A. T. (1973c): 'Predictive control of a plant with complex poles', *ibid.*, **18,** p. 1129

FULLER, A. T. (1973d): 'Simplified time-optimal switching functions for plants containing lags', *ibid.*, **18,** p. 1141

FULLER, A. T. (1974a): 'Optimal and sub-optimal saturating control of plants containing lags', *J. Franklin Inst.*, **297,** p. 7

FULLER, A. T. (1974b): 'Simplification of some time-optimal switching functions', *IEEE Trans. autom. Control*, **AC-19,** p. 65

FULLER, A. T. (1974c): 'Analysis and partial optimization of a non-linear stochastic control system', *Int. J. Control*, **19,** p. 81

FULLER, A. T. (1974d): 'Time-optimal control in regions where all state coordinates have the same sign', *ibid.*, **20,** p. 705

FULLER, A. T. (1975): 'Dynamic programming applied to some non-linear stochastic control systems', *Proc. Roy. Soc. Edinburgh*, **74a,** p. 175

FULLER, A. T. (1978): 'Controllers which minimize the maximum deviation', *Int. J. Control*, **27,** p. 165

FULLER, A. T. (1980a): 'Exact analysis of a first-order relay control system with a white noise disturbance', *ibid.*, **31,** p. 841

FULLER, A. T. (1980b): 'Approximate analysis of stochastic relay control systems—I', *ibid.*, **31,** p. 1171

FULLER, A. T. (1980c): 'Approximate analysis of stochastic relay control systems—II', *ibid.*, **32,** p. 625

FULLER, A. T. (1980d): 'Time-optimal control on state axes', *ibid.*, **32,** p. 771

FULLER, A. T., and ZINOBER, A. S. I. (1977): 'On the existence of constant-ratio trajectories in nominally time-optimal control systems subject to parameter variation', *J. Franklin Inst.*, **303,** p. 359

GABASOV, R., and KIRILLOVA, F. M. (1978): 'Singular Optimal Controls', Mathematical concepts & methods in science & engineering, **10,** (Plenum Press)

GAMKRELIDZE, R. V. (1978): 'Principles of optimal control theory', (New York, Plenum Press)

GLEN, J. A. (1973): 'Time-optimal control and indifference curves of certain second-order systems', *Int. J. Control*, **18,** p. 161

GOLDWYN, R. M., SRIRAM, K. P. and GRAHAM, M. (1967): 'Time-optimal control of a linear diffusion process', *SIAM J. Control*, **5,** p. 295

GORLOV, V. M. (1975): 'The construction of attainable sets for a minimal-time problem', *USSR Comp. Math. & Math. Phys.*, **15,** 212

GRENSTED, P. E. W., and FULLER, A. T. (1965): 'Minimization of integral-square-error for non-linear control systems of third and higher order', *Int. J. Control*, **2,** p. 33

GRIMMELL, W. C. (1967): 'The existence of piecewise continuous fuel optimal controls', *SIAM J. Control*, **5,** p. 515

GULKO, F. B. (1963): 'A special feature of the structure of optimal processes', *Ivz. Akad. Nauk SSSR, OTN, Tekh. Kibernetika*, **1,** p. 91

GULKO, F. B., and KOGAN, B. YA. (1963): 'A method of optimal control prediction'. *Proc. 2nd IFAC Congress, Basle, Theory*, p. 63 (London, Butterworths/Munich, Oldenbourg, 1964)

GULKO, F. B., KOGAN, B. YA., LERNER, A. YA., MIKHAILOV, N. N., and NOVOSELTSEVA, Z. A. (1964): 'A prediction method using high-speed analog computers and its applications', *Avtomatika Telemekh.*, **25,** p. 896 (Translation: *Automation and Remote Control*, **25,** p. 803)

GUNCHEV, L. AT. (1971): 'On the construction of the equivalent structure of a near time-optimal controller', *Automatica*, **7,** p. 51

HAAS, V. B. (1978): 'The Clebsch and Jacobi conditions for singular extremals', *Int. J. Control*, **27,** p. 557

HÁJEK, O. (1971): 'Geometric theory of time-optimal control', *SIAM* J. *Control*, **9,** p. 339

HÁJEK, O. (1973): 'Terminal manifolds and switching locus', *Math. Systems Theory*, **6,** p. 289

HÁJEK, O., (1979a): 'Discontinuous differential equations, I', *J. Diff. Eqns*, **32,** p. 149

HÁJEK, O., (1979b): 'Discontinuous differential equations, II', *J. Diff. Eqns*, **32,** p. 171

HARVEY, C. A., and LEE, E. B. (1962): 'On the uniqueness of time-optimal control for linear processes', *J. Math. Anal. Appl.*, **5,** p. 258

HAZEN, H. L. (1934): 'Theory of servo-mechanisms', *J. Franklin Inst.*, **218,** p. 279

HERMES, H. (1967): 'On the closure and convexity of attainable sets in finite and infinite dimensions', *SIAM J. Control*, **5,** p. 409

HERMES, H., and LASALLE, J. P. (1969): 'Functional analysis and time-optimal control' (New York, Academic Press)

HOPKIN, A. M. (1951): 'A phase plane approach to the design of saturating servomechanisms', *Trans. AIEE*, **70,** p. 631

HSU, E. H., BACHER, S., and KAUFMAN, A. (1972): 'A self-adaptive time-optimal control algorithm for second-order processes', *Amer. Inst. Chem. Eng. J.*, **18,** p. 1133

JAVID, S. H. (1978): 'The time-optimal control of a class of non-linear singularly perturbed systems', *Int. J. Control*, **27,** p. 831

JOHNSON, C. D. (1965): *in* LEONDES, C. T. (Ed.): 'Advances in control systems' **2,** p. 209 (New York, Academic Press)

JOHNSON, C. D. (1967): 'Optimal control with Chebyshev minimax performance index', *Trans. ASME, J. Bas. Engng,* **89D,** p. 251

JOHNSON, C. D., and GIBSON, J. E. (1963): 'Singular solutions in problems of optimal control', *IEEE Trans. Autom. Control,* **AC-8,** p. 4

JOHNSON, C. D., and WONHAM, W. M. (1965): 'On a problem of Letov in optimal control', *Trans. ASME, J. bas. Engng,* **87D,** p. 81

KAUFMAN, H., and DERUSSO, P. M. (1964): 'An adaptive predictive control system for random signals', *IEEE Trans. Autom. Control,* **AC-9,** p. 540

KAUFMAN, H., and DERUSSO, P. M. (1966): 'Stability analysis of predictive control systems', *ibid.,* **AC-11,** p. 455

KISHI, F. H. (1963): 'The existence of optimal controls for a class of optimization problems', *ibid.,* **AC-8,** p. 173

KOCHENBURGER, R. J. (1950): 'A frequency response method for analysing and synthesizing contactor servomechanisms', *Trans. AIEE,* **69**(pt. 1), p. 270

KNOWLES, G. (1976): 'Time-optimal control in infinite-dimensional spaces', *SIAM J. Control & Optim.,* **14,** p. 919

KNOWLES, G. (1979): 'Some problems in the control of distributed systems, and their numerical solution', *ibid.,* **17,** p. 5

KOPPEL, L. B., and LATOUR, P. R. (1965): 'Time-optimal control of second-order overdamped systems with transportation lag', *Ind. Eng. Chem. Fundamentals,* **4,** p. 463

KNUDSEN, H. K. (1964): 'An iterative procedure for computing time-optimal controls', *IEEE Trans. Autom. Control,* **AC-9,** p. 23

KRASOVSKII, N. N. (1958): 'On the theory of optimal regulation', *Autom. and Remote Control,* **18,** p. 1005

KRASOVSKII, N. N. (1959): 'On the theory of optimum control', *Appl. Math. Mech.,* **23,** p. 899

KREINDLER, E. (1963): 'Contributions to the theory of time-optimal control', *J. Franklin Inst.,* **275,** p. 314

KREINDLER, E. (1969): 'On the sensitivity of time-optimal systems', *IEEE Trans. Autom. Control,* **AC-14,** p. 578

KREINDLER, E. (1972): 'On performance sensitivity of optimal control systems', *Int. J. Control,* **15,** p. 481

KRENER, A. J. (1973): 'The high order maximal principle' *in* Mayne, D. Q., and Brockett, R. W. (Eds.): 'Geometric methods in systems theory', (Dordrecht Holland, Reidel)

KRENER, A. J. (1974): 'A generalization of Chow's theorem and the bang-bang theorem to nonlinear control problems', *SIAM J. Control,* **12,** p. 43

KUROKAWA, T., and TAMURA, H. (1972): 'A self-organizing time-optimal controller', *Int. J. Control,* **16,** p. 225

LASALLE, J. P. (1953): 'Study of the basic principle underlying the bang-bang servo'. Goodyear Aircraft Corp. Report GER-5518; Abstract 247t., *Bull. Amer. Math. Soc.,* **60,** p. 154 (1954)

LASALLE, J. P. (1960a): 'The time-optimal control problem' *in* Lefschetz, S. (Ed.): 'Contributions to the theory of nonlinear oscillations', **5,** p. 1 (Princeton University Press)

LASALLE, J. P. (1960b): 'The bang-bang principle'. *Proc. 1st IFAC Congress, Moscow,* p. 493, (London, Butterworths, 1961)

LAYTON, J. M. (1976): 'Multivariable control theory', IEE Control Engineering Series 1, (Peter Peregrinus)

LÉAUTÉ, H. (1885): Mémoire sur les oscillations à longue périodes', *J. Ec. Polytech.,* **55,** p. 1

LÉAUTÉ, H. (1891): 'Du mouvement troublé des moteurs', *J. Ec. Polytech.,* **61,** p. 1

LEE, E. B., and MARKUS, L. (1961): 'Optimal control for nonlinear processes', *Arch. Rat. Mech. Anal.,* **8,** p. 36

LEE, E. B., and MARKUS, L. (1967): 'Foundations of optimal control theory', (New York, Wiley)

LEE, W. K., and LUECKE, R. H. (1974): 'A direct search for time-optimal control in stochastic systems', *Int. J. Control,* **19,** p. 129

LEE, Y. S. (1967): 'A time-optimal adaptive control system via adaptive switching hypersurface', *IEEE Trans. Autom. Control.* **AC-12,** p. 367

LETOV, A. M. (1960a): 'Analytic controller design I', *Autom. & Remote Control,* **21,** p. 303

LETOV, A. M. (1960b): 'Analytic controller design II', *ibid.,* **21,** p. 389

LETOV, A. M. (1961): 'The analytical design of control systems', *ibid.,* **22,** p. 363

LEWIS, R. M. (1980): 'Definition of order and junction conditions in singular optimal control problems', *SIAM J. Control & Optim,* **18,** p. 21

LIONS, J. L. (1971): 'Optimal control of systems governed by partial differential equations' (Springer-Verlag)

MACCOLL, L. A. (1945): 'Fundamental theory of servomechanisms', (Van Nostrand)

MACDONALD, D. C. (1950): 'Nonlinear techniques for improving servo performance', *Proc. Natl. Electron. Conf.,* **6,** p. 400

MALEK-ZAVAREI, M. (1980): 'Time-optimal control of a class of unstable third-order plants', *J. Franklin Inst.,* **309,** p. 125

MARCHAL, C. (1973): 'Chattering arcs and chattering controls', *J. Optim. Theory & Appl.,* **11,** p. 441

MARKUS, L. (1976): 'Basic concepts of control theory' *in* 'Control theory and topics in functional analysis', **1,** p. 1 (International Atomic Energy Agency, Vienna)

MARKUS, L., and LEE, E. B. (1962): 'On the existence of optimal controls', *Trans. ASME, J. bas. Engng.* Ser. D, **84,** p. 13

McCAUSLAND, I. (1965): 'Time-optimal control of a linear diffusion equation', *Proc. IEE,* **112,** p. 543

McDANELL, J. P., and POWERS, W. F. (1971): 'Necessary conditions for joining optimal singular and nonsingular subarcs', *SIAM J. Control,* **9,** p. 161

MEDITCH, J. S. (1964a): 'On minimum fuel satellite attitude control systems', *IEEE Trans. Appl. Ind.,* **83,** p. 120

MEDITCH, J. S. (1964b): 'On the problem of optimal thrust programming for a lunar soft landing', *IEEE Trans. Autom. Control,* **AC-9,** p. 477

MEEKER, L. D. (1978): 'Time-optimal feedback control for small disturbances', *ibid.,* **AC-23,** p. 1099

MEEKER, L. D. (1980): 'Measurement stability of third-order time-optimal control systems', *J. Diff. Eqns.,* **36,** p. 54

MEEKER, L. D., and PURI, N. (1971): 'Closed-loop computerized time-optimal control of multivariable systems', Proc. JACC, St. Louis, Mo

MELSA, J. L., and SCHULTZ, D. G. (1965): 'Closed loop, approximately time-optimal control of linear systems', Proc. JACC, p. 220

MELSA, J. L., and SCHULTZ, D. G. (1967): 'A closed-loop, approximately time-optimal control method for linear systems', *IEEE Trans. Autom. Control,* **AC-12,** p. 94

MEREAU, P. M., and POWERS, W. F. (1976): 'A direct sufficient condition for free time optimal control problems', *SIAM J. Control & Optim.,* **14,** p. 613

MERKLINGER, K. J. (1963): 'Numerical analysis of non-linear control systems using the Fokker-Planck-Kolmogorov equation' Proc. 2nd. IFAC Congress, Basle, Theory, p. 81 (London, Butterworths/Munich, Oldenbourg, 1964)

MICHAELS, L. H., and FREDERICK, D. K. (1968): 'The relationship between periodic solutions and stability in a class of third-order relay control systems', *IEEE Trans. Autom. Control,* **AC-13,** p. 400

MIELE, A. (1962): 'The calculus of variations in applied aerodynamics and flight mechanics' *in* Leitman, G. (Ed.): 'Optimization techniques with applications to aerospace systems' (New York, Academic Press)

MIH YIN and GRIMMELL, W. C. (1968): 'Optimal and near-optimal regulation of spacecraft spin axes', *IEEE Trans. Autom. Control,* **AC-13,** p. 57

MIKHAILOV, N. N., and NOVOSELTSEVA, Z. A. (1964): 'Prediction for an optimal steering of an object containing an oscillating member by means of analogue computers'. 4th. Inter. Analogue Comp. Meeting, Brighton

MIRICA, S. (1969): 'On the admissible synthesis in optimal control theory and differential games', *SIAM J. Control*, **7**, p. 292

MIRICA, S. (1976): 'Time-optimal feedback control for linear systems' *in* 'Control Theory and Topics in Functional Analysis, vol. II', p. 1 (International Atomic Energy Agency, Vienna)

MITSUMAKI, T. (1960): 'Modified optimum non-linear control'. Proc. 1st. IFAC Congress, Moscow. p. 520, (London, Butterworths 1961)

MOHLER, R. R. (1973): 'Bilinear control processes' (Academic Press)

MOHLER, R. R., and RINK, R. E. (1969): 'Control with a multiplicative mode', *Trans. ASME, J. bas. Engng.*, **91D**, p. 201

MOHLER, R. R., and RUBERTI, A. (eds.) (1972): 'Theory and applications of variable structure systems' (Academic Press)

MOROZ, A. (1969a): 'Synthesis of time-optimal control for linear third-order systems: part I', *Autom. & Remote Control*, **5**, p. 657

MOROZ, A. (1969b): 'Synthesis of time-optimal control for linear third-order systems: part II', *ibid.*, **7**, p. 1032

MOROZ, A. (1970): 'Time-optimal control synthesis problem', *ibid.*, **1**, p. 18

NAHI, N. E. (1964): 'On design of time-optimal systems via the second method of Lyapunov', *IEEE Trans. Autom. Control*, **AC-9**, p. 274

NAYLOR, A. W., and SELL, G. R. (1971): 'Linear operator theory in engineering and science' (Holt, Rinehart & Winston)

NEISWANDER, R. S., and MACNEAL, R. H. (1953): 'Optimization of non-linear control systems by means of non-linear feedbacks', *Trans. AIEE*, **72**, p. 262

NEUSTADT, L. W. (1960): 'Synthesizing time-optimal control systems', *J. Math. Anal. Appl.*, **1**, p. 484

NEUSTADT, L. W., and PAIEWONSKY, B. H. (1963): 'On synthesizing optimal controls'. Proc. 2nd IFAC Congress, Basle, Theory, p. 283 (London, Butterworths/Munich, Oldenbourg, 1964)

NEWMANN, M. M., and ZACHARY, D. H. (1965): 'Analogue computation of the switching surfaces in three-dimensional phase space for the optimization of non-linear control systems', *Int. J. Control*, **2**, p. 149

NISHIMURA, H. (1980): 'Fuel minimal takeoff path of jet lift VTOL aircraft', *J. Aircraft*, **17**, p. 290

NOTLEY, M. G. (1971): 'A heuristic approach to optimal control', *Int. J. Control*, **13**, p. 429

O'DONNELL, J. J. (1964): 'Bounds on limit cycles in two-dimensional bang-bang control systems with an almost time-optimal switching curve', *IEEE Trans. Autom. Control*, **AC-9**, p. 448

OLDENBURGER, R. (1966a): 'Optimal control' (Holt, Rinehart & Winston)

OLDENBURGER, R. (1966b): 'Optimal and self-optimizing control' (MIT Press)

OLDENBURGER, R., and THOMPSON, G. (1963): 'Introduction to time-optimal control of stationary linear systems', *Automatica*, **1**, p. 177

OLSDER, G. J. (1975): 'Time-optimal control of multivariable systems near the origin', *J. Opt. Th. Appl.*, **16**, p. 497

OLSDER, G. J. (1980): 'Comments on "Time-optimal feedback control for small disturbances"', *IEEE Trans. autom. Control*, **AC-25**, p. 136

OWENS, D. H. (1978): 'Feedback and multivariable systems'. IEE Control Engineering Series 7 (Peter Peregrinus)

PAIEWONSKY, B. (1963): 'Time-optimal control of linear systems with bounded control' *in* 'International symposium on nonlinear differential equations and nonlinear mechanics' (Academic Press)

PARKS, P. C. (1976): 'Applications of the theory of moments in automatic control', *Int. J. Systems Science*, **7**, p. 177

PAVLOV, A. A. (1966): 'Synthesis of time-optimal relay systems', (in Russian) (Moscow, Izdatelstvo Nauka.)

PERSSON, E. V. (1963): 'Synthesis of control systems operating linearly for small signals and approximately 'bang-bang' for large signals'. Proc. 2nd IFAC Congress, Basle, Theory, p. 210 (London, Butterworths/Munich, Oldenbourg, 1964)

PETERSON, D. W., and ZALKIND, J. H. (1978): 'A review of direct sufficient conditions in optimal control theory', *Int. J. Control*, **28,** p. 589

PINCH, E. R. (1979): 'Time-optimal control to target sets', *IMA Bulletin*, **15,** p. 197

PONTRYAGIN, L. S., BOLTYANSKII, V. G., GAMKRELIDZE, R. V., and MISCHENKO, E. F. (1962): 'The mathematical theory of optimal processes' (New York, Wiley)

POSTLETHWAITE, I., and MACFARLANE, A. G. J. (1979): 'A complex variable approach to the analysis of linear multivariable feedback systems'. Lecture notes in Control & Info. Sciences, **12** (Springer Verlag)

PRABHU, S. S., and McCAUSLAND, I. (1970): 'Time-optimal control of linear diffusion processes using Galerkin's method', *Proc. IEE*, **117,** p. 1398

QUINN, J. P. (1971): 'Two examples in the time-optimal control theory of distributed parameter systems'. Proc. 1st IFAC Symposium on Control of Distributed Parameter Systems, Banff (Pergamon Press)

RAGADE, R. K., and SARMA, I. G. (1967): 'A game theoretic approach to optimal control in the presence of uncertainty', *IEEE Trans. Autom. Control*, **AC-12,** p. 395

RANG, E. R. (1963): 'Isochrone families for second-order systems', *ibid.*, **AC-8,** p. 64

REEVE, P. J. (1970): 'Optimal control for systems which include pure delays', *Int. J. Control*, **11,** p. 659

ROBERTS, J. A. (1970): 'Linear control of saturating control systems', *ibid.*, **12,** p. 239

ROBERTS, J. A. (1973): 'Optimal and linear sub-optimal control of second-order saturating control systems', *Int. J. Control*, **17,** p. 897

ROBINSON, P. N., and ÖNER YURTSEVEN, H. (1969): 'A Monte Carlo method for stochastic time-optimal control', *IEEE Trans. Autom. Control*, **AC-14,** p. 574

ROGAK, E. D., KAZARINOFF, N. D., and SCOTT-THOMAS, J. F. (1970): 'Sufficient conditions for bang-bang control in Hilbert space', *J. Opt. Th. Appl.*, **5,** p. 1

ROHRER, R. A., and SOBRAL, M. (1966): 'Optimal singular solutions for linear multi-input systems', *Trans. ASME, J. bas. Engng.*, **88D,** p. 323

ROOTENBERG, J. (1974): 'The sensitivity of optimally designed control systems with minimum fuel performance index', *Int. J. Control*, **20,** p. 101

ROSE, N. J. (1953): 'Theoretical aspects of limit control'. Rep. No. 459, Experimental towing tank, Stevens Institute of Technology, Hoboken, NJ

ROSENBROCK, H. H. (1970): 'State space and multivariable theory' (Nelson)

ROZONOER, L. I. (1959a): 'L. S. Pontryagin's maximum principle in the theory of optimum systems—I', *Autom. & Remote Control*, **20,** p. 1288

ROZONOER, L. I. (1959b): 'L. S. Pontryagin's maximum principle in the theory of optimum systems—II', *ibid.*, **20,** p. 1405

ROZONOER, L. I. (1959c): 'L. S. Pontryagin's maximum principle in the theory of optimum systems—III', *ibid.*, **20,** p. 1517

RYAN, E. P. (1974): 'Time-optimal feedback control laws for certain third-order relay control systems', *Int. J. Control*, **20,** p. 881

RYAN, E. P. (1975): 'Optimal and sub-optimal relay control by state variable transformation', *ibid.*, **22,** p. 329

RYAN, E. P. (1976a): 'A near-time-optimal control approach for third- and fourth-order relay control systems', *ibid.*, **23,** p. 741

RYAN, E. P. (1976b): 'Time-optimal control of certain plants with positive real eigenvalues', *ibid.*, **23,** p. 775

RYAN, E. P. (1977a): 'Minimum-time isochronal surfaces for certain third-order systems', *ibid.*, **26,** p. 421

RYAN, E. P. (1977b): 'Time-optimal feedback control of certain fourth-order systems', *ibid.*, **26,** p. 675

RYAN, E. P. (1977c): 'Synthesis of a third-order time-fuel-optimal control system'. Control Theory Centre Report No. 66, University of Warwick, UK

RYAN, E. P. (1978a): 'Quasi-time-optimal feedback control of some fifth-order systems', *Trans. ASME, J. Dyn. Sys. Meas. & Control*, **100,** p. 201

RYAN, E. P. (1978b): 'On the synthesis of a third-order time-fuel-optimal control system', *IEEE Trans. autom. Control*, **AC-23,** p. 952

RYAN, E. P. (1979): 'Singular optimal controls for second-order saturating systems', *Int. J. Control*, **30,** p. 549

RYAN, E. P. (1980a): 'Synthesis of time-fuel-optimal control: a second-order example', *Int. J. Control*, **31,** p. 379

RYAN, E. P. (1980b): 'On the sensitivity of a time-optimal switching function', *IEEE Trans. autom. Control*, **AC-25,** p. 275

RYAN, E. P., and DORLING, C. M. (1981): 'Minimization of non-quadratic cost functionals for third order saturating control systems', *Int. J. Control*, **34,** p. 231

SAKAWA, Y., and HAYASHI, C. (1963): 'Solution of optimal control problems by using Pontryagin's maximum principle'. Proc. 2nd IFAC Congress, Basle, Theory, p. 339 (London, Butterworths/Munich, Oldenbourg, 1964)

SARMA, I. G., and PRASAD, U. R. (1972): 'Switching surfaces in N-person differential games', *J. Opt. Th. Appl.*, **10,** p. 160; see also 'Multicriteria decision making and differential games', Leitmann, G. (Ed.) (Plenum Pub. Corp., 1976)

SARMA, I. G., and RAGADE, R. K. (1966): 'Some considerations in formulating optimal control problems as differential games', *Int. J. Control*, **4,** p. 265

SARMA, I. G., RAGADE, R. K., and PRASAD, U. R. (1969): 'Necessary conditions for optimal strategies in a class of non co-operative N-person differential games', *SIAM J. Control*, **7,** p. 637

SCHMITENDORF, W. E., and BARMISH, B. R. (1980): 'Null controllability of linear systems with constrained controls', *SIAM J. Control & Optim*, **18,** p. 327

SILJAK, D. D. (1969): 'Nonlinear systems' (New York, Wiley)

SINGH, S. N. (1977): 'Single axis gyroscopic motion with uncertain angular velocity about spin axis', *Trans. ASME, J. Dyn. Systems Meas. & Control*, **99,** p. 259

SIRISENA, H. R. (1968): 'Optimal control of saturating linear plants for quadratic performance indices', *Int. J. Control*, **8,** p. 65

SIRISENA, H. R. (1974): 'A gradient method for computing optimal bang-bang controls', *Int. J. Control*, **19,** p. 257

SMITH, F. B. (1961): 'Time-optimal control of higher-order systems', *IRE Trans. autom. Control.*, **6,** p. 16

SMITH, F. W. (1966): 'Design of quasi-optimal minimum-time controllers', *IEEE Trans. autom. Control*, **AC-11,** p. 71

SUGIURA, I. (1966): 'Adaptive optimizing control utilizing open-loop type optimum-controller and sensitivity coefficients'. Memoires of the Fac. of Engng., Nagoya University, **18,** p. 182

SUSSMANN, H. J. (1972): 'The bang-bang problem for certain control systems in $GL(n, R)$', *SIAM J. Control*, **10,** p. 470

TCHAMRAN, A. (1966): 'On a class of constrained control, linear regulator problems', *Trans. ASME, J. bas. Engng.*, **88D,** p. 385

TRIEU, K. L., and PIERRE, D. A. (1970): 'Multi-mode digital controller for insensitive near-time-optimal control'. Proc. Nat. Elec. Conference, Chicago, p. 391

USPENSKY, J. V. (1948): 'Theory of equations', Chap. 5 (New York, McGraw-Hill)

UTKIN, V. I. (1977): 'Variable structure systems with sliding modes', *IEEE Trans. Autom. Control*, **AC-22,** p. 212

UTKIN, V. I. (1978): 'Sliding modes and their application in variable structure systems' (Moscow, MIR Publishers)

UTTLEY, A. M., and HAMMOND, P. H. (1952): 'The stabilization of on-off controlled servomechanisms', *in* Tustin (Ed.): 'Automatic and manual control', p. 285 (London, Butterworths)

VAKILZADEH, I. (1974): 'Bang-bang control of a second-order unstable system', *Int. J. Control*, **20,** p. 49

VAKILZADEH, I. (1978): 'Bang-bang control of a plant with one positive real pole and one negative real pole', *J. Opt. Th. Appl.*, **24,** p. 315

VENA, P. A., and BERSHAD, N. J. (1971): 'Time-optimal control switching times using entire function theory', *Int. J. Control*, **14,** p. 529

VINTER, R. B., and LEWIS, R. M. (1978): 'A necessary and sufficient condition of optimality of dynamic programming type', *SIAM J. Control & Optim*, **16,** p. 571

VINTER, R. B., and LEWIS, R. M. (1980): 'A verification theorem which provés a necessary and sufficient condition for optimality', *IEEE Trans. Autom. Control*, **AC-25,** p. 84

WANG, P. K. C. (1975): 'Time-optimal control of time lag systems with time lag controls', *J. Math. Anal. Appl.*, **52,** p. 366

WEISS, H. K. (1946): 'Analysis of relay servomechanisms', *J. Aeronautical Sciences*, **13,** p. 364

WEISSENBERGER, S. (1966): 'Stability-boundary approximations for relay control systems via a steepest ascent construction of Lyapunov functions', *Trans. ASME, J. Bas. Engng.*, **83,** p. 419

WELLS, W. R., and KASHIWAGI, Y. (1969): 'Synthesis of a time-optimal control problem with delay', *IEEE Trans. Autom. Control*, **AC-14,** p. 99

WHEEDEN, R. L., and ZYGMUND, A. (1977): 'Measure and integral' (Marcel Dekker Inc.)

WINDALL, W. S. (1970): 'The minimum-time thrust vector control law in the Apollo lunar module autopilot', *Automatica*, **6,** p. 661

WOLEK, S. (1971): 'Determination of switching instants in minimum-time control'. *Bulletin de l'Académie Polonaise des Sciences, Série des Sciences Techniques*, **19,** p. 57

WOLOVICH, W. A. (1974): 'Linear multivariable systems', Applied Mathematical Sciences 11 (Springer-Verlag)

WONHAM, W. M. (1963): 'Note on a problem in optimal non-linear control', *J. Electron. Control*, **15,** p. 59

WONHAM, W. M. (1974): 'Linear multivariable control: a geometric approach'. Lecture Notes in Economics and Mathematical Systems, 101 (Springer-Verlag)

WONHAM, W. M., and JOHNSON, C. D. (1964): 'Optimal bang-bang control with quadratic performance index', *Trans. ASME, J. Bas. Engng.*, **86D,** p. 107

YASTREBOFF, M. (1969): 'Synthesis of time-optimal control by time interval adjustment', *IEEE Trans. Autom. Control*, **AC-14,** p. 707

YEUNG, D. S. (1977): 'Time-optimal feedback control', *J. Opt. Th. Appl.*, **21,** p. 71

YOUNG, K.-K. D., KOKOTOVIC, P. V., and UTKIN, V. I. (1977): 'A singular perturbation analysis of high-gain feedback systems', *IEEE Trans. Autom. Control*, **AC-22,** p. 931

YUVARAJAN, S., and RAMASWAMI, B. (1969): 'Sub-optimal approach to the time-optimal control of second-order systems with complex poles', *ibid.*, **AC-14,** p. 763

ZACH, F. C. (1972): 'Time/fuel optimal and adaptive control for gravity gradient spacecraft', *in* Balakrishnan, A. V. (Ed.): 'Techniques of optimization' (New York, Academic Press)

ZACHARY, D. H. (1966a): 'Further consideration of an optimal control problem', *Int. J. Control*, **4,** p. 251

ZACHARY, D. H. (1966b): 'Finite time optimization of bang-bang control systems', *ibid.*, **4,** p. 357

ZACHARY, D. H., ROBERTS, A. P., and NEWMANN, M. M. (1965): 'Optimum transient response of a saturating system which has a linear region of control', *Int. J. Control*, **2,** p. 353

ZINOBER, A. S. I. (1975): 'Adaptive relay control of second-order systems', *Int. J. Control*, **21,** p. 81

ZINOBER, A. S. I. (1977): 'Analysis of an adaptive third-order relay control system using non-linear switching surface theory'. *Proc. Roy. Soc. Edinburgh*, **76A,** p. 239

ZINOBER, A. S. I. (1979): 'The self-adaptive control of overhead crane operations'. Proc. IFAC Symposium on Identification and System Parameter Estimation, Darmstadt (Pergamon Press)
ZINOBER, A. S. I. (1980): 'Self-adaptive near-optimal control of diffusion equations', *Proc. IEE*, **127,** p. 290
ZINOBER, A. S. I., and FULLER, A. T. (1973): 'The sensitivity of nominally time-optimal control systems to parameter variation', *Int. J. Control*, **17,** p. 673

Index

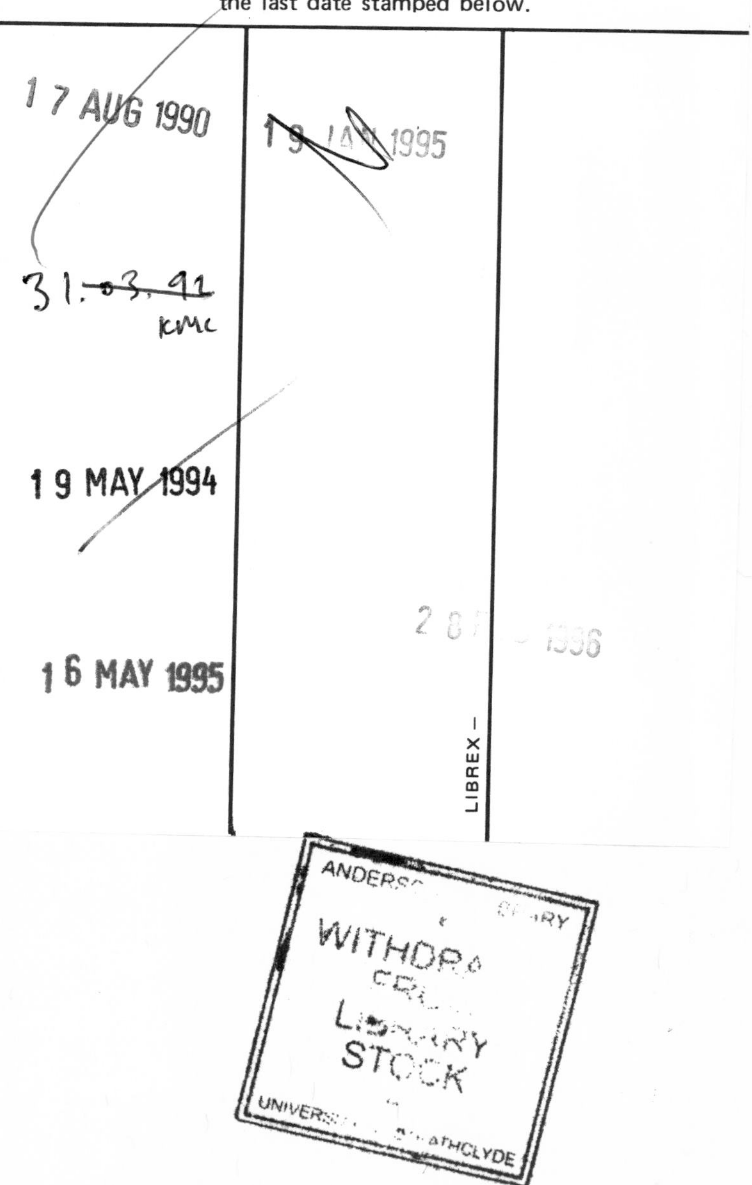